Satellite Altimetry and Earth Sciences

Satellite Altimetry and Earth Sciences

Editor: Israel Morris

Callisto Reference
www.callistoreference.com

Callisto Reference,
118-35 Queens Blvd., Suite 400,
Forest Hills, NY 11375, USA

Visit us on the World Wide Web at:
www.callistoreference.com

ISBN: 978-1-64116-809-0 (Hardback)

Cataloging-in-Publication Data

Satellite altimetry and earth sciences / edited by Israel Morris.
p. cm.
Includes bibliographical references and index.
ISBN 978-1-64116-809-0
1. Artificial satellites in earth sciences. 2. Satellite geodesy. 3. Earth sciences--Remote sensing.
4. Oceanography--Remote sensing. I. Morris, Israel.

QE33.2.A7 S28 2023
550--dc23

Table of Contents

Preface

The purpose of the book is to provide a glimpse into the dynamics and to present opinions and studies of some of the scientists engaged in the development of new ideas in the field from very different standpoints. This book will prove useful to students and researchers owing to its high content quality.

A satellite altimeter calculates the time taken by a short electromagnetic pulse for completing a round trip. This electromagnetic pulse is emitted from an orbiting spacecraft towards the planetary body and is reflected-off its surface. This instrument also measures the height of the satellite from the surface of the planet. Satellite altimetry can be used to obtain useful data related to ocean science. It also provides novel insights into plate tectonics, marine geodesy, ice sheet movements, high-resolution bathymetry for unmapped areas of the Earth's surface, and essential data on large-scale features of oceans. The data from satellite altimetry can be utilized to calculate the gravitational field of the Earth over oceans. This book unfolds the innovative aspects of satellite altimetry and its applications in Earth sciences, which will be crucial for the progress of this field in the future. It presents researches that have transformed this discipline and have aided its advancement. Researchers and students in the field of Earth sciences will be assisted by this book.

At the end, I would like to appreciate all the efforts made by the authors in completing their chapters professionally. I express my deepest gratitude to all of them for contributing to this book by sharing their valuable works. A special thanks to my family and friends for their constant support in this journey.

Editor

1

Elevation Changes of the Antarctic Ice Sheet from Joint Envisat and CryoSat-2 Radar Altimetry

Baojun Zhang [1,2], Zemin Wang [2], Quanming Yang [2,3], Jingbin Liu [1], Jiachun An [2,*], Fei Li [2], Tingting Liu [2] and Hong Geng [4]

1 State Key Laboratory of Information Engineering in Surveying, Mapping and Remote Sensing, Wuhan University, Wuhan 430079, China; bjzhang@whu.edu.cn (B.Z.); jingbin.liu@whu.edu.cn (J.L.)
2 Chinese Antarctic Center of Surveying and Mapping, Wuhan University, 129 Luoyu Road, Wuhan 430079, China; zmwang@whu.edu.cn (Z.W.); yangquanming@whu.edu.cn (Q.Y.); fli@whu.edu.cn (F.L.); ttliu23@whu.edu.cn (T.L.)
3 PowerChina Zhongnan Engineering Corporation Limited, Changsha 410014, China
4 School of Resource and Environment Science, Wuhan University, Wuhan 430079, China; genghong@whu.edu.cn
* Correspondence: jcan@whu.edu.cn

Abstract: The elevation changes of ice sheets have been recognized as an essential climate variable. Long-term time series of these changes are an important parameter to understand climate change, and the longest time-series of ice sheet elevation changes can be derived from combining multiple Ku-band satellite altimetry missions. However, unresolved intermission biases obscure the record. Here, we revise the mathematical model commonly used in the literature to simultaneously correct for intermission bias and ascending–descending bias to ensure the self-consistency and cohesion of the elevation time series across missions. This updated approach is applied to combine Envisat and CryoSat-2 radar altimetry in the period of 2002–2019. We tested this approach by validating it against airborne and satellite laser altimetry. Combining the detailed temporal and spatial evolution of elevation changes with firn densification-modeled volume changes due to surface processes, we found that the Amundsen Sea sector accounts for most of the total volume loss of the Antarctic Ice Sheet (AIS), mainly from ice dynamics. However, surface processes dominate the volume changes in the key regions, such as the Totten Glacier sector, Dronning Maud Land, Princess Elizabeth Land, and the Bellingshausen Sea sector. Overall, accelerated volume loss in the West Antarctic continues to outpace the gains observed in the East Antarctic. The total volume change during 2002–2019 for the AIS was -68.7 ± 8.1 km^3/y, with an acceleration of -5.5 ± 0.9 km^3/y^2.

Keywords: Antarctic ice sheet; satellite radar altimetry; long-term time series; elevation changes

1. Introduction

The elevation changes of ice sheets are the result of changes in ice sheet ice dynamics and weather-driven changes on the surface. The long-term evolution of ice sheet volume changes is an essential climate variable defined by the Global Climate Observing System (GCOS) to assess the impact of climate change and its contribution to sea level rise. Satellite altimetry can accurately measure the elevation of ice sheets and thereby provide valuable data for monitoring ice sheet changes at both local and continental scales. Since Zwally et al. (1983) [1] first used satellite altimetry measurements in ice sheet surveys, altimetry observations have been essential for studying ice sheet evolution [2–8]. However, limited by the effective lifetime of the satellite, a single mission cannot characterize the long-term evolution of the ice sheet [9,10]. Consequently, combining observations from multiple altimetric missions is needed to reconstruct the long-term evolution to improve our knowledge of the ice sheets' responses to present climate change [10,11].

Accounting for the intermission bias caused by differences in the position of the centers of gravity of the two satellites and the phase of the antennae and bias caused by orbital errors, is primary important problem for cross-calibrating observations from multiple altimeter missions [12]. This intermission bias was regarded as a constant and was corrected by calculating the global average in the open sea through the intersection [13]. However, Zwally et al. (2005) [14] found that there was significant intersatellite elevation bias with a distinct spatial pattern over the ice sheets that did not occur over the oceans. This finding was also confirmed by maps of the along-track centered mean bias between the ERS-2 and Envisat altimeter height over the ice sheets and oceans given in Frappart et al. (2016) [15]. Therefore, Schröder et al. (2019) [10], Paolo et al. (2016) [11], Adusumilli et al. (2018) [16], and Sørensen et al. (2018) [17] accounted for the intermission bias at each grid cell using the observations over overlapping epochs.

Additionally, the correction of systematic elevation crossover bias due to the angle between the radar polarization and wind-induced features of the firn is also not trivial for processing radar altimetry data over ice sheets [18,19]. In cross-calibrating the observations from multiple radar altimeter missions, the bias can be divided into an intramission component and an intermission component [15]. The former is known as the ascending–descending bias (A–D bias), which refers to the time-independent part of the difference in the elevation measured on the ascending and descending tracks for the same altimetry mission. Previous studies have shown that the A–D bias is as large as a few meters with a distinct spatial pattern over ice sheets in the elevation measurements [18–21] and can lead to large errors in the estimation of elevation changes from altimetric observations [22]. Fortunately, Armitage et al. (2014) [18] found that systematic bias varies as a function of the angle between the radar polarization direction and the wind-induced firn azimuthal anisotropy direction. The A–D biases of CryoSat-2 [3,5] or Envisat [23] were successfully corrected by introducing a term of satellite flight direction into the least-squares model for elevation changes. Moreover, Helm et al. (2014) [2] and Schröder et al. (2019) [10] reduced the A–D bias by retracking the return signal waveform of the altimetry observations with a threshold retracker focused on the lower part of the leading edge and thus the surface scattering part. It is because Helm et al. (2014) [2] found the A–D bias is strongly dependent on the choice of the retracker. As for the latter, the intermission component, we call the intermission A–D bias in this study. Khvorostovsky (2012) [12] found that it significantly affected the intermission bias of elevation and backscatter coefficient.

In summary, in the previous studies, the intermission bias and the A–D bias were always corrected separately. In this study, we present an updated plane-fitting least-squares regression model for obtaining a long-term elevation time series of the Antarctic Ice Sheet (AIS) by combining radar altimetry missions. A larger amount of observations are used in the integrated adjustment model to correct the intermission bias and the A–D bias simultaneously. The updated approach is then utilized to develop a monthly elevation time series for a 5 km grid from June 2002 to April 2019 over the AIS and compare the results with those derived by airborne and ICESat laser altimetry observations. Finally, we study and discuss the spatial and temporal variations in the derived volume change of the AIS.

2. Data

2.1. Envisat and CryoSat-2 Altimetry Data

Envisat was launched in March 2002 and operated in a 35-day repeat orbit with a latitude limit of 81.5° S [24]. Here, we applied the recently reprocessed Geophysical Data Record Full Mission Reprocessing V3.0, which only provides the retracked elevations using the ICE-1 retracker from June 2002 to April 2012, to provide better consistency in the data records. The ICE-1 retracker uses standard Offset Centre Of Gravity (OCOG) retracking using a threshold of 0.25 for the OCOG amplitude [25]. Moreover, Davis (1997) [26], Nilsson et al. (2016) [27], and Schröder et al. (2017) [28]

argued that such a threshold retracker is more precise in terms of ice sheet elevation measurements. The erroneous height records are culled using standard quality flags.

CryoSat-2 was launched in April 2010 using a 369-day full repeat orbit with a subcycle of 30 days and is still in operation. It operates at an altitude of 717 km with an inclination of 92° and provides coverage up to 88° S. CryoSat-2 measures the interior parts of the AIS in the Low-Resolution Mode (LRM), while using the novel Synthetic Aperture Radar Interferometric (SARIn) mode around the AIS margins. The LRM is equivalent to conventional Ku-band radar altimeters. SARIn is very suitable for measuring steep and complex topography across ice sheet margins because of the use of dual antennae operating in an interferometric mode [29]. The retracked LRM observations with the OCOG retracker from July 2010 to April 2019 were used in the following study to match the Envisat ICE-1 retracker elevations. As for the SARIn observations, the only retracked observations by the Wingham/Wallis model fit retracker provided by the product were used. The erroneous height records and SARIn records where the interferometric across-track location failed were removed.

Therefore, the period from July 2010 to April 2012 could be used to calibrate the intermission biases.

2.2. Laser Altimeter Elevation Datasets from the Operation IceBridge (OIB) Project and ICESat

In this study, based on previous studies, we presented an updated regression model to correct the intermission bias and the A–D bias simultaneously. To evaluate the performance of the model, we needed to validate its results. Here, we used laser altimeter elevation datasets obtained from NASA's OIB project and ICESat mission.

The OIB project operated since 2009 to bridge the gap between the ICESat and ICESat-2 missions and maintain a continuous observational record over the Arctic and the Antarctic. The Airborne Topographic Mapper (ATM) is an important airborne laser altimeter payload for OIB projects that can provide surface elevation measurements with an accuracy of 10 cm or better [30]. In addition, some pre-IceBridge ATM elevation observations were also collected during 1993–2008.

From 2003 to 2009, ICESat was the first satellite laser altimeter mission, and, by operating in the near inferred spectrum, it provided precise observations of the surface elevation. Here, the latest release of ICESat data (GLA12, Version 34) was used to derive surface elevation change estimates in a 20 km grid via a repeat-track analysis [31]. Intercampaign bias correction and saturation correction were also applied. The final elevation change rate for each grid was generated from the median of surface height change rates to minimize the impact of the remaining outliers, thereby providing a good dataset for benchmarking our elevation change estimates from Envisat.

3. Algorithm for Elevation Time Series

3.1. The Updated Least-Squares Regression Model

Previous studies have demonstrated that elevation time series' can be obtained by performing a least-squares regression on all elevation measurements within a grid cell, also known as "plane-fitting" [3,10,17,23]. Following these studies, the regression parameters can be expressed as

$$\begin{aligned} h(lon_i, lat_i, t_i) = \; & h_0 + a_0(lon_i - lon_0) + a_1(lat_i - lat_0) + a_2(lon_i - lon_0)^2 \\ & + a_3(lat_i - lat_0)^2 + a_4(lon_i - lon_0)(lat_i - lat_0) + a_5(t_i - t_0) \\ & + a_6\cos(2\pi(t_i - t_0)) + a_7\sin(2\pi(t_i - t_0)) + a_8\left(bs_i - \overline{bs}\right) + b_{AD}(-1)^{AD} \\ & + b_m(-1)^m + res(lon_i, lat_i, t_i), \end{aligned} \tag{1}$$

where $h(lon_i, lat_i, t_i)$ is the elevation measured by the satellite altimetry missions at longitude (lon_i), latitude (lat_i), and time (t_i); h_0 is the mean altitude of the bin with the center coordinates (lon_0, lat_0) in which the least-squares regression is performed; t_0 is the reference epoch set to May 2010; a_0~a_4 are the coefficients of a biquadratic surface model accounting for topography; a_5 is the elevation change rate constant; a_6~a_7 describe the seasonal variations in elevation; a_8 is for the anomaly of backscattered power $(bs_i - \overline{bs})$ to account for variations in the penetration depth of the radar signal; b_{AD} is for the intramission A–D bias, and AD is assigned a value of 1 for ascending tracks or 0 for descending tracks; b_m is only used in CryoSat-2 data processing to account for the possible bias caused by the different retracking algorithms used for LRM and SARIn, and m is 1 for LRM or 0 for SARIn; and $res(lon_i, lat_i, t_i)$ denotes the residuals.

However, the regression model above is designed for CryoSat-2. It is not suitable for multimission processing, unless observations from multiple radar altimeter missions are also cross-calibrated. Cross-calibration is used to correct the bias between different satellite radar altimeter missions. As mentioned in Section 1, the bias between two radar altimeter missions, including the intermission bias and the intermission A–D bias, can be regarded as invariants in the local area [10,14,15,18] and can be modeled like A–D in Equation (1). Hence, the intermission extension of Equation (1) is given by

$$\begin{cases} h(lon_i, lat_i, t_i) = h_0 + a_0(lon_i - lon_0) + a_1(lat_i - lat_0) + a_2(lon_i - lon_0)^2 \\ \quad + a_3(lat_i - lat_0)^2 + a_4(lon_i - lon_0)(lat_i - lat_0) + a_5(t_i - t_0) \\ \quad + a_6\cos(2\pi(t_i - t_0)) + a_7\sin(2\pi(t_i - t_0)) + a_8\left(bs_i - \overline{bs}\right) \\ \quad + b_{C_aC_d}(-1)^{C_aC_d} + b_{C_aE_a}(-1)^{C_aE_a} + b_{C_aE_d}(-1)^{C_aE_d} + b_{C_dE_a}(-1)^{C_dE_a} \\ \quad + b_{C_dE_d}(-1)^{C_dE_d} + b_{E_aE_d}(-1)^{E_aE_d} + res(lon_i, lat_i, t_i), \\ b_{C_aE_a} - b_{C_aC_d} - b_{C_dE_a} = 0, \\ b_{C_dE_d} - b_{C_dE_a} - b_{E_aE_d} = 0, \\ b_{C_aE_d} - b_{C_aC_d} - b_{C_dE_a} - b_{E_aE_d} = 0, \end{cases} \tag{2}$$

where $b_{E_aE_d}$ and $b_{C_aC_d}$ are for the intramission A–D bias of Envisat or CryoSat-2, and $b_{C_aE_a}$, $b_{C_dE_a}$, $b_{C_aE_d}$ and $b_{C_dE_d}$ are for the total static systematic biases of the intermission bias and the intermission A–D bias between Envisat and CryoSat-2. In this study, we referred to these total static systematic biases as the total intermission biases. The internal consistency between these biases (assuming they are constant over time) would give additional constraints, which were handled by the last three equations in Equation (2). Notably, here we discarded $b_m(-1)^m$, as this bias could be modeled as a part of the intramission A–D bias.

3.2. *Calculation of the Combined Elevation Time Series*

We first performed the least-squares regression, as shown in Equation (2), at the intersections of the ascending and descending satellite ground tracks of Envisat to estimate all the parameters. For each of the intersections, an independent least-squares regression was estimated based on all Envisat and Cryosat-2 observations within 2.5 km of the center of an intersection. The first-round estimated parameters require that the number of observations of each satellite in an ascending or descending orbit shall not be less than 50. Then, an iterative procedure continues until parameters are estimated that pass the 3σ outlier rejection criteria, as in [6], or until the remaining observations in the bin cannot meet the thresholds. The thresholds require that the total observations be more than 100 and that the number of ascending or descending observations of each satellite be no less than 20. The estimations of the systematic biases $b_{E_aE_d}$, $b_{C_aC_d}$, $b_{C_aE_a}$, $b_{C_dE_a}$, $b_{C_aE_d}$, and $b_{C_dE_d}$ over the AIS are determined at all intersections. To ensure a robust solution, we arbitrarily removed the intersections where one or more systematic biases exceeded 50 m. This editing step removed about 1.2% of the intersections over the AIS.

Then, we computed the least-squares regression shown in Equation (2) at a 2.5 km × 2.5 km grid to ensure the resolution of the results. For each of the grid nodes, all Envisat and Cryosat-2 observations within 2.5 km of the center of a grid node were used in the independent least-squares regression estimator. The same 3σ outlier rejection criteria and thresholds used in the least squares regression at the intersections above were also applied to ensure robust least square estimations. Due to a lack of observations, some of the 6 systematic bias parameters were not able to be estimated at some grid points. Here, interpolation of the corresponding parameters calculated at the intersections mentioned above were used. Then, the corrected elevation was calculated as

$$\begin{aligned}\Delta h(lon_0, lat_0, t_i) &= h(lon_i, lat_i, t_i) \\ &-\Big(h_0 + a_0(lon_i - lon_0) + a_1(lat_i - lat_0) + a_2(lon_i - lon_0)^2 \\ &+a_3(lat_i - lat_0)^2 + a_4(lon_i - lon_0)(lat_i - lat_0) + a_8\big(bs_i - \overline{bs}\big) \\ &+b_{C_aC_d}(-1)^{C_aC_d} + b_{C_aE_a}(-1)^{C_aE_a} + b_{C_aE_d}(-1)^{C_aE_d} + b_{C_dE_a}(-1)^{C_dE_a} \\ &+b_{C_dE_d}(-1)^{C_dE_d} + b_{E_aE_d}(-1)^{E_aE_d}\Big).\end{aligned} \tag{3}$$

Here, the mean altitude of the bin h_0 is removed to facilitate the study of elevation changes; then, the elevation for a given month t^j is derived from

$$\overline{\Delta h}\big(lon_0, lat_0, t^j\big) = \frac{1}{n}\sum_{i=1}^{n} \Delta h(lon_0, lat_0, t_i). \tag{4}$$

where n is the number of corrected elevations in month t^j. With $\overline{\Delta h}\big(lon_0, lat_0, t^j\big)$ computed for all months in a given bin, a time series of elevation can be formed.

A floating median is a robust low-pass filter and was used in this study to identify and remove remaining outliers, similar to that in Schröder et al. (2019) [10]. Then, the monthly elevation estimations were smoothed using a spatiotemporal median filter to generate the final 5 km grid of the elevation time series. In the temporal domain, median filtering was performed on the monthly elevation estimations over 3 months (i.e., the given month, the previous month, and the next month) to eliminate high frequency errors. In the spatial domain, median filtering was operated over the monthly elevation estimations within a radius of 10 km from a grid node to minimize the impact of possible outliers. If the number of monthly elevation estimations within the 10 km radius was less than 9, the filtering radius would be expanded in steps of 10km until the radius was 30km, and the scaled median absolute deviation was used as the realistic error uncertainty of the elevation of the grid node as in Ewert et al. (2012) [32]. Then, the trend, acceleration, and other spatial-temporal characteristics of elevation and volume over the AIS can be derived from the elevation time series [4].

3.3. *Calculation of Volume Changes for the AIS from the Combined Elevation Time Series*

Long-term time series of volume changes are of great scientific value for the assessment of changes in the AIS and its response to climate change. After correcting for Glacial Isostatic Adjustment (GIA)-induced vertical crustal deformation and elastic solid earth rebound effects, the altimetric volume time series can be calculated from the elevation time series. In this study, the ICE-6G_C model [33] was used to correct for the uplift due to GIA. Elastic contributions were corrected by multiplying the elevation change by a scale factor $\alpha = 1.0205$, which was inferred in Groh et al. (2012) [34]. For gridded values of a volume time series, the uncertainties were estimated using formal error propagation techniques. The GIA model does not have estimated uncertainties, so a standard deviation of 30% of the value for each grid point was used as a conservative estimate. For elastic deformation, the uncertainty of the scale factor α was considered negligible, similar to Schröder et al. (2019) [10]. Moreover, when summing up the volume time series for individual drainage basins and major sectors, the correlated uncertainties were accounted for according to the method in Schröder et al. (2019) [10], i.e., by applying a scaling factor derived from a 50 km autocorrelation radius to the squared uncertainties. To analyze and interpret volume changes of the AIS, we also calculated the surface volume changes due to surface processes (surface mass balance variations, liquid water processes, and firn compaction) using the Institute for Marine and Atmospheric Research Firn Densification Model (IMAU-FDM) [35].

4. Results

4.1. The Total Intermission Biases and the Intramission A–D Bias

A larger amount of observations from different missions are used simultaneously in the least-squares model to account for the total intermission biases and the intramission A–D bias is the highlight of our proposed algorithm. Figure 1 shows estimates of these biases derived at the intersections of the ascending and descending satellite ground tracks of Envisat. In this study, the mode mask boundary between CryoSat-2's LRM and SARIn modes is used to simply divide flat areas and steep areas: CryoSat-2 uses the SARIn mode to measure the steep areas at the margins of the ice sheets, and the LRM mode over the flat interior of the ice sheets [2,29]. Corresponding statistics can be found in Table 1.

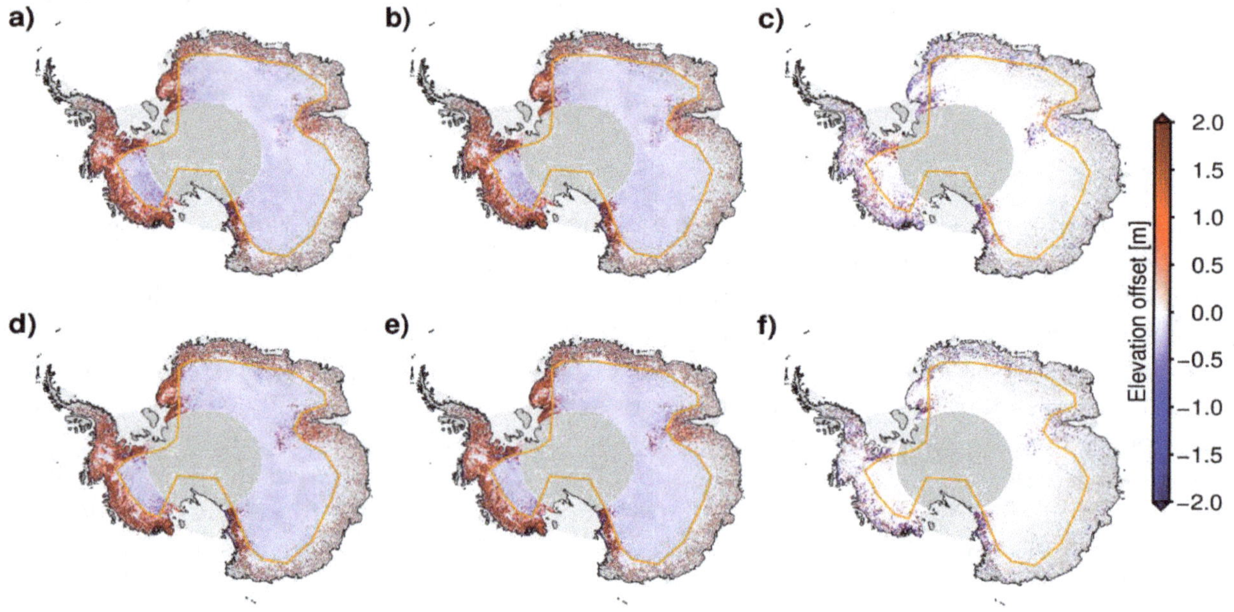

Figure 1. Maps of the total intermission biases between CryoSat-2 ascending tracks and Envisat ascending tracks ((**a**), $b_{C_aE_a}$), CryoSat-2 ascending tracks and Envisat descending tracks ((**b**), $b_{C_aE_d}$), CryoSat-2 descending tracks and Envisat ascending tracks ((**d**), $b_{C_dE_a}$), CryoSat-2 descending tracks and Envisat descending tracks ((**e**), $b_{C_dE_d}$) and the intramission ascending–descending (A–D) bias of Envisat ((**c**), $b_{C_aC_d}$) and CryoSat-2 ((**f**), $b_{E_dE_a}$) at the intersections of Envisat's ascending tracks and descending tracks. The mode mask boundary between CryoSat-2's LRM and SARIn modes [29] is shown in orange.

In terms of the intramission A–D bias, the spatial distributions of CryoSat-2 and Envisat exhibit similar patterns. Over the interior of the AIS, the bias is close to zero, while in the steeper SARIn area, the amplitude of bias exceeds 1 m (Figure 1c,f). The median, mean, RMS, and the 10th and 90th percentile range (P_{90}-P_{10}) over the SARIn mode area are 0.00 m, −0.14 m, 4.71 m, and 1.81 m for CryoSat-2 and 0.00 m, −1.38 m, 13.86 m, and 0.94 m for Envisat, while those over the LRM mode area are relatively close (Table 1). Whether using Cryosat-2 or Envisat, the mean and RMS of the intramission A–D bias over the SARIn area are much greater than those over the LRM area. This is mainly because the azimuthally anisotropic effects of polarization on altimetry measurements can be negligible over flat and homogeneous areas due to the similar dielectric conditions and low mispointing, but the opposite is true in steep and very active areas [27]. In general, the spatial distribution patterns and magnitude of the intramission A–D bias calculated in this study are consistent with those in Khvorostovsky (2012) [12], Davis (1997) [27], and McMillan et al. (2014) [5].

Given the total intermission biases between CryoSat-2 and Envisat, it can be seen from Figure 1 that all of them share the similar spatial distribution patterns. These biases are highly correlated to the topography slope of the ice sheets. In steep and complex areas, these biases can reach up to several meters, while in the flat interior of the AIS, these biases appear to be relatively uniform at a magnitude of about 0.2 m. According to Frappart et al. (2016) [15], the intermission bias in the flat interior of the AIS is mainly caused by orbital errors and differences in the position of the centers of gravity of the two satellites and the phase of the antennae, while the bias over the steep area originates from the geographical small-scale bias. Thus, the total intermission bias in the flat interior of the AIS shows a uniform distribution, but a higher amplitude of bias is found where the topography is steeper. Overall, the total intermission biases estimated in this study agree with those found in Frappart et al. (2016) [15]. This agreement gives us confidence in our updated least-squares model.

Table 1. Statistics of the total intermission biases and the intramission A–D bias of Envisat and CryoSat-2 at the intersections of Envisat's ascending tracks and descending tracks over the CryoSat-2 LRM and SARIn mask areas. The number of intersections is 51,221 in the Low-Resolution Mode (LRM) mode area and 15,476 in the SARIn mode area. The median, mean, RMS, and the 10th and 90th percentile range (P_{90}–P_{10}) of the biases are given.

Parameter	SARIn				LRM			
	Median (m)	Mean (m)	RMS (m)	P_{90}–P_{10} (m)	Median (m)	Mean (m)	RMS (m)	P_{90}–P_{10} (m)
$b_{C_aE_a}$	1.20	2.35	21.00	12.40	−0.29	−0.14	1.40	0.19
$b_{C_aE_d}$	1.13	0.96	25.36	11.44	−0.29	−0.14	1.45	0.21
$b_{C_aC_d}$	0.00	−0.14	4.71	1.81	0.00	−0.01	1.59	0.12
$b_{C_dE_a}$	1.16	2.49	20.47	12.58	−0.29	−0.13	1.48	0.16
$b_{C_dE_d}$	1.09	1.11	24.73	11.64	−0.29	−0.13	1.46	0.16
$b_{E_dE_a}$	0.00	−1.38	13.86	0.94	0.00	0.00	0.46	0.10

4.2. Comparison with OIB and ICESat

The results of our least-squares model are validated against the laser altimeter elevation datasets obtained from the OIB project and ICESat. Both datasets are taken from near-inferred laser altimeters, and their accuracy has been validated [36,37]. The results of our validation are shown in Figure 2.

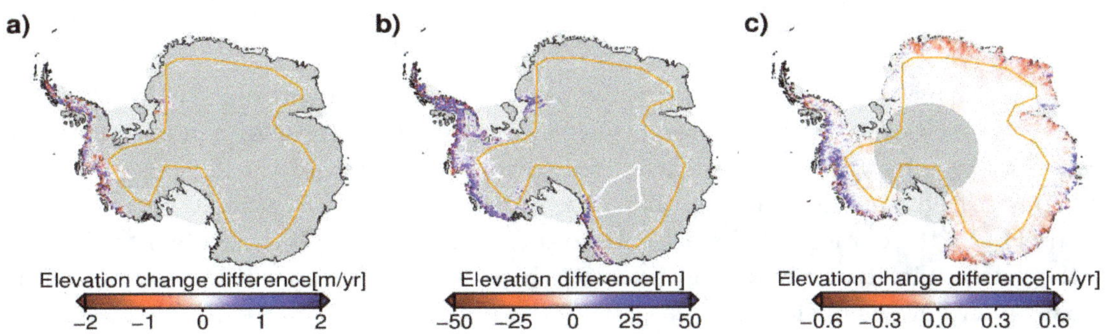

Figure 2. Validation with laser altimeter elevation data obtained from the OIB and ICESat. Differences between the elevation changes derived from OIB observations and the combined elevation time series between 2002 and 2016 (**a**). Elevation differences between OIB observations and the combined elevation time series over the period of 2009–2018 (**b**). Difference between the average elevation change rates from the combined elevation time series and the average ICESat surface elevation change rates between 2003 and 2009 (**c**). The mode mask boundary between the CryoSat-2 LRM and SARIn modes is shown in orange.

Surface elevation changes can be obtained wherever there are coincident OIB ATM elevation measurements. To validate our results, we used the surface elevation changes from OIB ATM L4 [38]. These changes were computed from the OIB ATM elevation measurements observed during 2002–2016. The differences between elevation changes derived from OIB and the combined elevation time series ($\Delta h = \Delta h_{OIB} - \Delta h_{ALT}$) are shown in Figure 2a. To gain insight into the results of the intercomparison, we separated the evaluation results according to where or when the OIB ATM elevation measurements were observed to derived the OIB surface elevation changes for a grid cell, hereby referred to as the steeper SARIn area and the flat LRM area, and the periods before 2010, after 2010, and from 2002 to 2016. The OIB surface elevation change from 2002 to 2016 here refers to that is computed from a pair of coincident measurements, with one obtained before 2010 and the other after 2010. This allowed us to evaluate our time series in terms of their temporal and spatial distribution. The mean and RMS of these differences are given in Table 2. Meanwhile, statistics of the comparison between the OIB product and time series of Schröder et al.'s (2019) [10] elevation time series are also given in Table 2 to demonstrate the effects of different approaches to account for the intermission biases on the combined elevation time series. However, as shown in Figure 2a, the validation results are mainly concentrated along the outlet glaciers of the West Antarctic Ice Sheet (WAIS) and the Antarctic Peninsula Ice Sheet (APIS). Thus, intercomparison over the LRM area may be less persuasive. Over the SARIn area, the RMS of our results is nearly 30% smaller than that in Schröder et al. (2019) [10]. This suggests that our method is effective in correcting the systematic biases in the SARIn area. The decrease of the RMS by nearly 40% of the comparison of from 2002 to 2016 between ours results and that in Schröder et al. (2019) [10] further demonstrates the effectiveness of our method in intermission bias correction.

Table 2. Statistics of the comparison between elevation changes derived from the combined elevation time series and those from Operation IceBridge (OIB) observations. The mean and RMS of the differences between elevation changes over the AIS and CryoSat-2 LRM and SARIn mask areas, as well as those acquired before 2010, after 2010, and from 2002 to 2016, are given. Note that the intercomparison results are mainly concentrated along the outlet glaciers of West Antarctica and the Antarctic Peninsula.

		Different Regions			**Difference Periods**		
		All	**SARIn**	**LRM**	**Before 2010**	**After 2010**	**From 2002–2016**
This study	Mean (m)	−0.83	−0.94	0.20	−0.59	−0.27	−1.72
	RMS (m)	4.41	4.62	1.58	5.18	2.00	5.83
Schröder et al. (2019)	Mean (m)	−0.81	−1.00	0.11	−0.85	−0.14	−2.11
	RMS (m)	5.88	6.43	1.17	5.79	2.40	9.60

To make a consistent intercomparison, we used all surface elevations during 2009–2018 from OIB ATM L2 [39] within a 2.5-kilometer radius and a 4-day interval of our radar altimetry grid nodes. When comparing with the OIB surface elevations, the mean altitude of the bin (i.e., the h_0 in Equation (2)) is added back into the combined elevation time series. Figure 2b shows the map of the differences between the combined elevation time series and the OIB measurements. The differences are again the greatest in steep regions of the AIS, which is likely because that the radar altimeter has larger measurement errors in steeper areas due to the higher surface slope and roughness. Over the AIS, the median and RMS of the differences are 5.7 m and 28.7 m, respectively. They are better than the median of 25.4 m and RMS of 138.6 m of the differences between the interpolated grid cells of DEM in Slater et al. (2018) [40] and the OIB measurements, which are comparable with our combined time series in their processing strategy and resolution.

However, OIB ATM elevation observations are strongly concentrated along the outlet glaciers of West Antarctica and the Antarctic Peninsula. As a supplement, we also used ICESat data to validate our merged time series. For ICESat, the average elevation change rates for the period coinciding with ICESat over the AIS are first derived from our combined elevation time series by fitting a first-order polynomial and a sine with a 1-year period to account for the average elevation change rate and seasonal variations. Subsequently, the average elevation change rates from our combined results within a 10 km radius of an ICESat grid node are compared with the corresponding average ICESat rate. A map of the differences is shown in Figure 2c. Again, the largest differences are observed at steep areas. In the flat interior of the AIS, the differences are small. The higher surface slope and roughness are still the main sources of these differences. Changes in the penetration depth of the microwave signal into the snow due to changes in the snow surface characteristics is another important source of difference [41]. The median and RMS of the differences over the AIS are −1.1 cm/y and 10 cm/y, respectively, while the median and RMS are −1 cm/y and 11 cm/y for the differences between the average ICESat elevation change rates and those in Schröder et al. (2019) [10]. Thus, even if Schröder et al. (2019) [10] had used ICESat observations to construct their elevation time series, our results would still be comparable to theirs.

4.3. Elevation Time Series

As mentioned above, the average rates of elevation change over any subinterval can be obtained from the elevation time series using a fitting method with a first-order polynomial and a sine with a 1-year period. Unlike the information obtained directly from elevation observations, this method provides more accurate average rates for specific periods. The average elevation change rates over the periods of 2002–2010, 2010–2019, and 2002–2019 for our elevation time series are shown in Figure 3. Well-known long-term thinning and thickening spatial distribution patterns can be seen in these maps [2,4,5,10,31]. Overall, significant signs of thinning mainly occur across the Amundsen Sea sector, the Getz sector, and the Totten Glacier. A continuation of the signals of elevation uplift are observable in the Kamb Ice Stream and across Dronning Maud Land, while there are no significant changes in elevation across the interior of the East Antarctic Ice Sheet (EAIS). Whether over the Envisat period (2002–2010) with sparse observations or over the CryoSat-2 period (2010–2019) with dense observations, we obtained elevation change results that are in general agreement with other published results (Figure 3a,b). Comparing the results of the two periods, we find that thinning across the Amundsen Sea sector, the Getz sector, and Totten Glacier has significantly accelerated and expanded. Signs of thickening across Dronning Maud Land also show modest acceleration and expansion. In the coastal regions of Princess Elizabeth Land and the Bellingshausen Sea sector, elevation changes over the two periods show an opposite trend. Such significant interannual variations in elevation changes were previously reported in Schröder et al. (2019) [10].

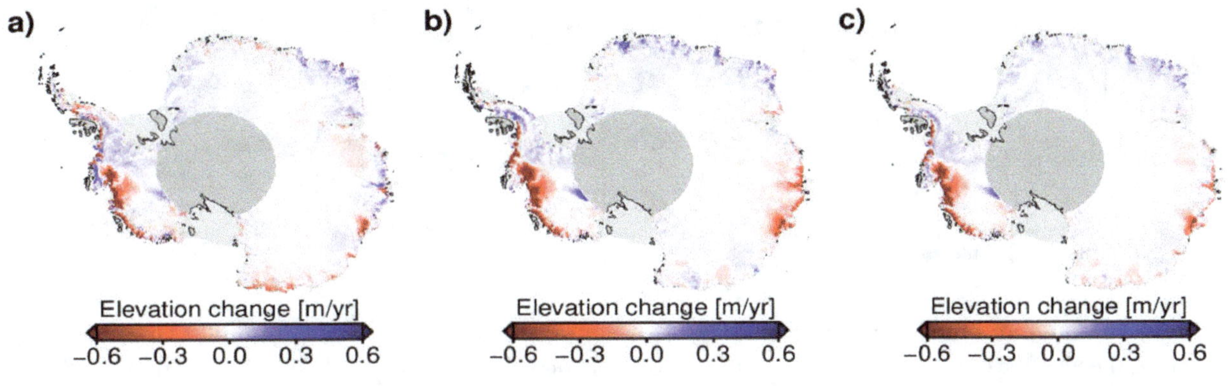

Figure 3. Maps of the long-term elevation changes from the combined elevation time series over the periods of 2002–2010 (**a**), 2010–2019 (**b**) and 2002–2019 (**c**).

Moreover, to be comparable to Schröder et al. (2019) [10], we provide some points' elevation time series in four representative regions in Figure 4. As shown in Figure 4a, Pine Island Glacier experienced a significant continuous thinning during 2002–2019. The rate of thinning decreases as its distance from the grounding line increases. The average rate of the elevation close to the grounding line (Point B) is −2.36 ± 0.02 m/y over 2002–2019, which far exceeds the thinning rate of −0.96 ± 0.01 m/y at Point C. The rate of thinning is not constant throughout the period but has slowed since 2010. This deceleration is most evident in the time series of Point B near grounding line but remains almost invisible in the time series of Point C far from grounding line. The spatial-temporal variations in the elevation change rates over Pine Island Glacier agree with those previously reported in Schröder et al. (2019) [10] and Flament and Remy [4].

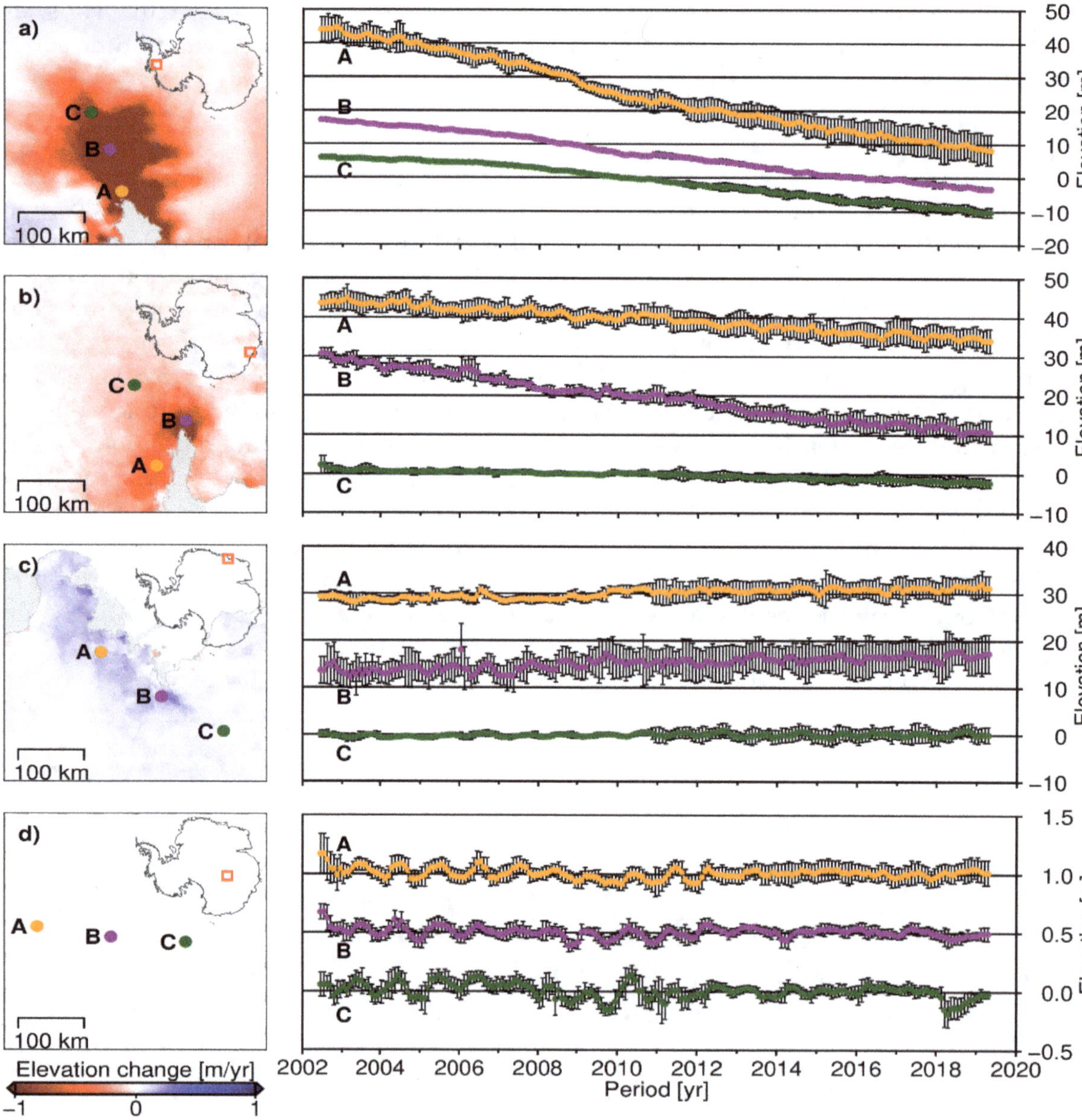

Figure 4. Elevation time series for Pine Island Glacier (**a**), Totten Glacier (**b**), Dronning Maud Land (**c**), and Lake Vostok (**d**). The geographical locations of the selected points (A, B, and C) are marked in green, purple, and orange on the left maps of elevation changes over 2002–2017. The time series and their 1σ uncertainty ranges for each point are given in the right maps. The time series of the points A and B are shifted along the Elevation axis for better visibility.

In the EAIS, Totten Glacier has the most significant thinning. Examples of this glacier's elevation time series are shown in Figure 4b. Similar to Pine Island Glacier, the largest thinning rates at Totten Glacier are observable near the grounding line. The mean thinning rate at Point B on the main tributary of Totten Glacier is −1.19 ± 0.01 m/y over 2002–2019. The thinning rate of Point A at the other tributary is −0.58 ± 0.01 m/y. Between the two tributaries, there is a ridge of thinning losing height much less rapidly than the surrounding fast-flowing areas. Flament and Remy [4] treat the 'ridge' as important evidence of the dynamic thinning of Totten Glacier. The time series of the examples indicates that there was no obvious acceleration or deceleration of the elevation change rates over 2002–2019. This confirms the previous results in Flament and Remy [4] and coincides with the flow velocity of Totten Glacier, which has remained constant since 2007 (as reported in Li et al. (2016) [42]).

Dronning Maud Land is located in the EAIS, which was chosen as an example of elevation uplift. In Figure 4c, we can observe a modest thickening over 2002–2019 in Dronning Maud Land. The largest thickening rate in the region is approximately 0.25 ± 0.01 m/y (Point B). The thickening rates at the other two points are 0.19 ± 0.01 m/y (Point A) and 0.05 ± 0.01 m/y (Point C). These thickening rates are in good agreement with the corresponding estimates in Schröder et al. (2019) [10]. Many previous studies have attributed the thickening of the ice sheet over the period in Dronning Maud Land to increased snowfall [5,10,15,43,44].

Lake Vostok is the largest of the subglacial lakes located far inland of the EAIS and is used to represent the slow-flowing interior. The corresponding elevation time series are shown in Figure 4d. The elevation change rates at the selected points of 3 ± 0.7 mm/y (Point A), 1.4 ± 0.5 mm/y (Point B), and 0.7 ± 0.5 mm/y (Point C) are in good agreement with the estimates of −3 ± 9 mm/y ~ 5 ± 9 mm/y from altimetry data in Schröder et al. (2019) [10] and −1 ± 5 mm/y from GNSS profiles in Schröder et al. (2017) [28].

4.4. Volume Time Series for the AIS

To quantitatively examine the changes in volume at a regional scale, the volume time series for six typical regions selected based on elevation change features are shown in Figure 5. It is very challenging to use radar altimetry data to estimate the surface elevation changes of ice sheets due to signal penetration into the snow. Fortunately, Legrésy and Rémy (1998) [45], Davis and Ferguson (2004) [46], and Slater et al. (2018) [40] have shown that the annual and subannual changes in the snowpack volume/surface scattering ratio can produce spurious changes in altimetric elevation. Therefore, in Figure 5, we also show the time series without seasonal oscillations. To clearly illustrate the relationship between surface volume changes and altimetric volume changes, time series with linear trends and seasonal oscillations removed are shown in Figure 6.

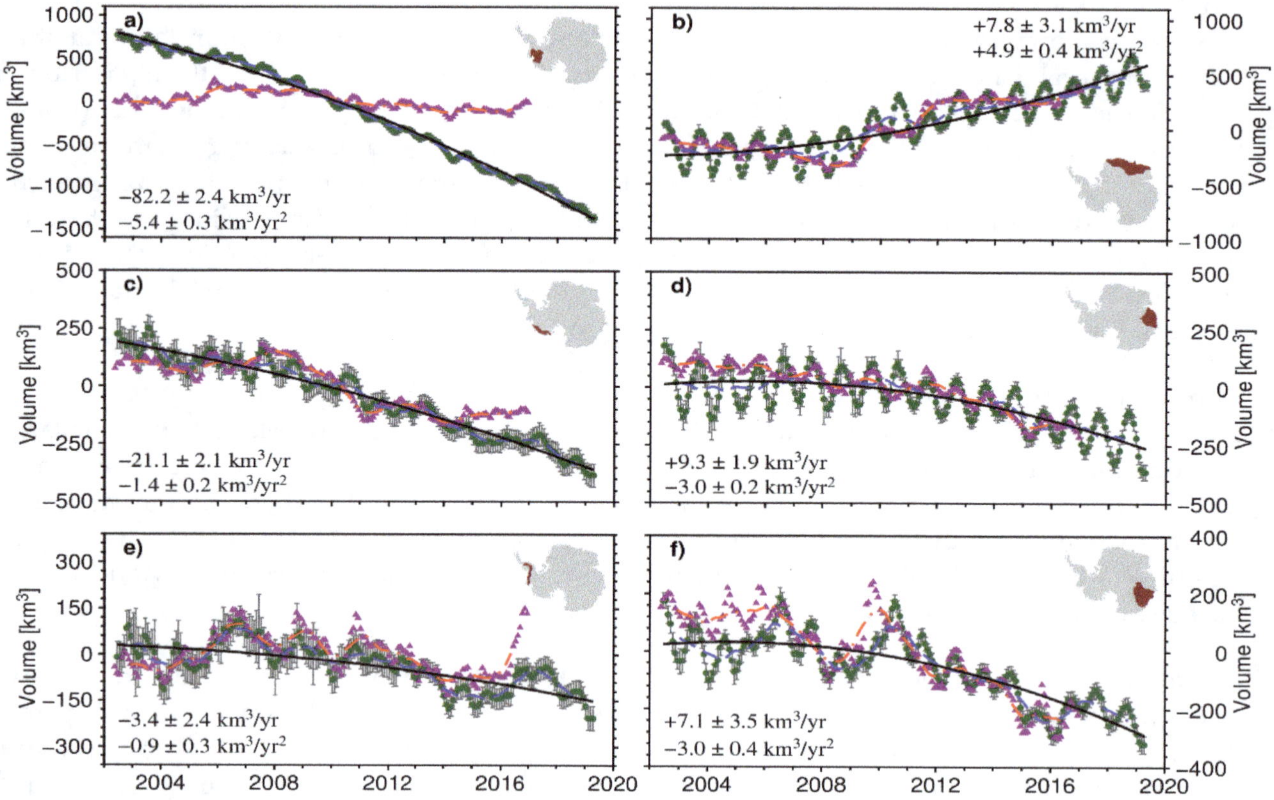

Figure 5. Volume change of some subregions north of 81.5° S (Amundsen Sea sector (**a**), Dronning Maud Land (**b**), Getz sector (**c**), Princess Elizabeth Land (**d**), Bellingshausen Sea sector (**e**), and Totten Glacier sector (**f**)) from our combined altimetric time series (green dots) and the Institute for Marine and Atmospheric Research Firn Densification Model (IMAU-FDM) (purple triangles). The blue dashed curves and red dash-dotted curves are the time series after removing the seasonal oscillations using a 13-month moving average. Due to the defects of the moving average algorithm, the 6-month values at the beginning and end of the time series are not displayed. The solid black lines are the best-fit quadratic curves for the altimetric volume time series. Additionally, the altimetric volume time series' linear and quadratic trend estimates are also included. The gray error bars show the 1σ uncertainty range of the altimetry data. The regions covered by each localization are shaded red in the inset. The boundaries refer to the Antarctic drainage systems in [47].

It can be clearly seen that there are strong interregional variations in the altimetric volume curves for smaller regions within Antarctica during 2002–2019. The Amundsen Sea sector (Figure 5a) displays a large coherent trend featuring a volume loss of -82.2 ± 2.4 km^3/y with a significant acceleration of -5.4 ± 0.3 km^3/y2. The time series reveals that this region experienced significant acceleration around 2006. Surface processes cannot cover the large coherent trends of accelerating volume loss. This means that sustained increases in ice discharge are the main reason for accelerated thinning, which agrees with many studies using observations from the Gravity Recovery and Climate Experiment (GRACE) [41–44]. There is a good consistency among the interannual changes in both volume time series (Figure 6a), with altimetric volume changes lagging behind those of the surface volume by several months.

The Getz sector and the Bellingshausen Sea sector are the other two regions of volume loss in the WAIS. The Getz sector has experienced a rapid volume loss of 21.1 ± 2.1 km^3/y and acceleration of -1.4 ± 0.2 km^3/y2 (Figure 5c). Figure 5e suggests that the Bellingshausen Sea sector has progressed to a less negative volume change at a rate of -3.4 ± 2.4 km^3/y and an acceleration of -0.9 ± 0.3 km^3/y2. The time series reveals that these two regions both experienced small losses in 2002–2006, followed by an increase in volume loss in 2007–2015 and regained volume in 2016. However, in contrast with the

Bellingshausen Sea sector, discrepancies between altimetric and surface volume changes are very large in some subintervals in the Getz sector. This can be clearly seen from the time series of interannual changes (Figure 6c,e). Moreover, the interannual changes in altimetric volume in the Bellingshausen Sea sector are in good agreement with those of the surface volume change, except for a slight lag. This means that surface processes control the volume change in the Bellingshausen Sea sector, while ice discharge sometimes dominates the volume changes in the Getz sector.

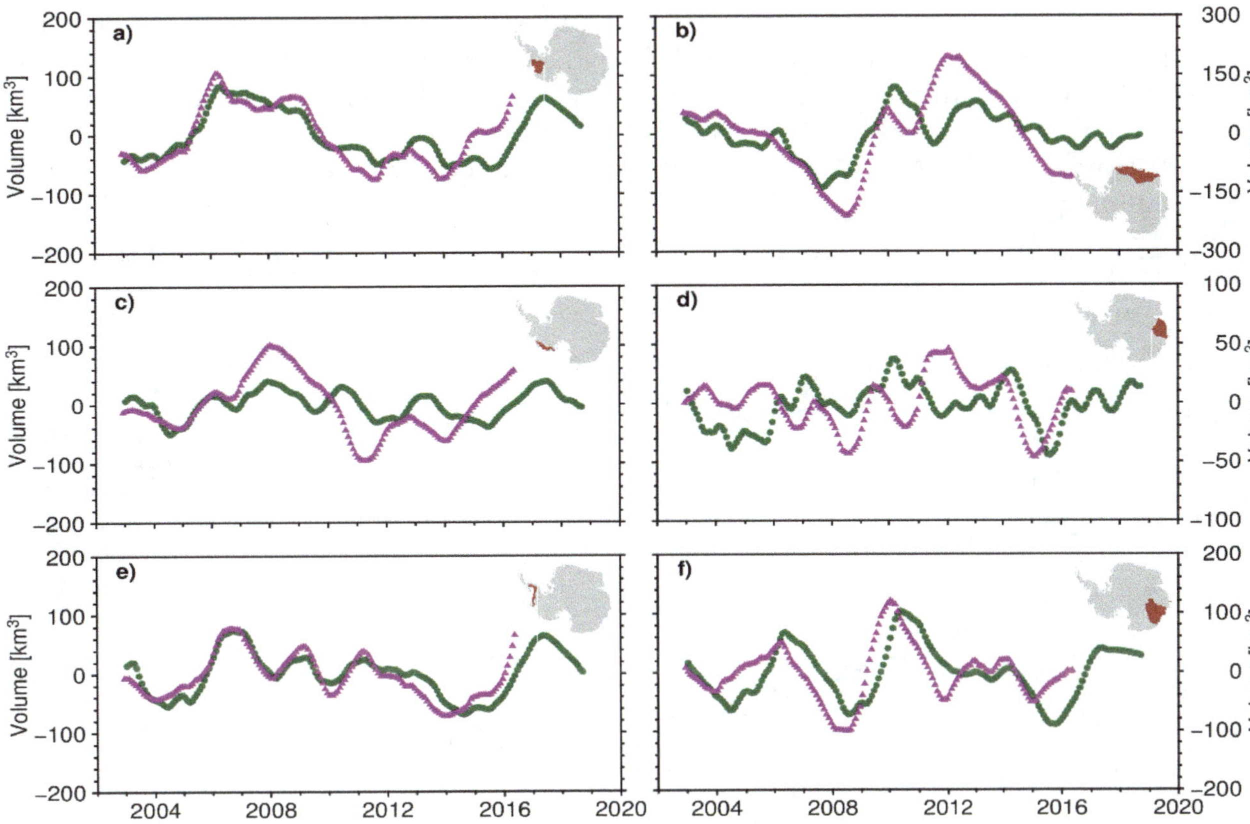

Figure 6. Volume changes of some subregions north of 81.5° S (Amundsen Sea sector (**a**), Dronning Maud Land (**b**), Getz sector (**c**), Princess Elizabeth Land (**d**), Bellingshausen Sea sector (**e**), and Totten Glacier sector (**f**)) from our combined altimetric time series (green dots) and IMAU-FDM (purple triangles). After removing the long-term trends, seasonal oscillations were removed using a 13-month moving average. Due to the defects of the moving average algorithm, the 6-month values at the beginning and end of the time series are not displayed. The regions covered by each localization are shaded red in the inset. The boundaries refer to the Antarctic drainage systems in [47].

In Dronning Maud Land, the volume change and acceleration for the entire period are 7.8 ± 3.1 km^3/y and -4.9 ± 0.4 km^3/y^2, respectively. However, the relevant time series reveals that there is a significant change in its volume trend partway through the period of 2002–2019 (Figure 5b). Prior to the end of 2008, this area showed a slow increase in volume. Subsequently, the Dronning Maud Land experienced a continuous increase in volume gain. It can be seen from Figure 5b that surface processes dominate the volume change trends in Dronning Maud Land. The impact of two extreme accumulation events in 2009 and 2012 [48,49] can also be clearly seen in the volume time series, especially in Figure 6b. Several months after the surface volume changes, the altimetric volume responds.

In contrast with Dronning Maud Land, Princess Elizabeth Land, and the Totten Glacier sector show a gradual change from initial volume accumulation to volume loss (Figure 5d,f). The accelerations in volume change of these sectors are -3.0 ± 0.2 km^3/y^2 and -3.0 ± 0.4 km^3/y^2, respectively. The fluctuations of volume changes in 2008–2009 caused by surface process anomalies are most obvious in the Totten

Glacier sector, as previously reported in Li et al. (2016) [42]. The trends of volume changes, especially accelerations, are dominated by surface processes in the two regions. The rapid volume loss of the Totten Glacier sector began in 2010. The difference is that the interannual oscillations of the Totten Glacier sector are consistent with its surface volume changes for the whole period, while those in Princess Elizabeth Land are not.

Figure 7 shows the derived volume time series for the entire AIS and the subregions of EAIS, WAIS, and APIS north of 81.5° S over 2002–2019. An overall volume loss of -68.7 ± 8.1 km^3/y and an acceleration in loss of -5.5 ± 0.9 km^3/y^2 during 2002–2019 were detected. Overall, increasing volume loss is mainly a result of volume changes of the WAIS. The volume loss of the WAIS during 2002–2019 was -88.4 ± 3.9 km^3/y with an acceleration of -8.1 ± 0.4 km^3/y^2. The APIS contributes to a small part of volume loss at a lower rate of -18.8 ± 2.2 km^3/y, and its loss gradually slows, with an acceleration of $+1.9 \pm 0.2$ km^3/y^2. The EAIS has experienced a steady volume gain at a rate of 38.5 ± 6.7 km^3/y. The acceleration of the EAIS is 0.8 ± 0.8 km^3/y^2.

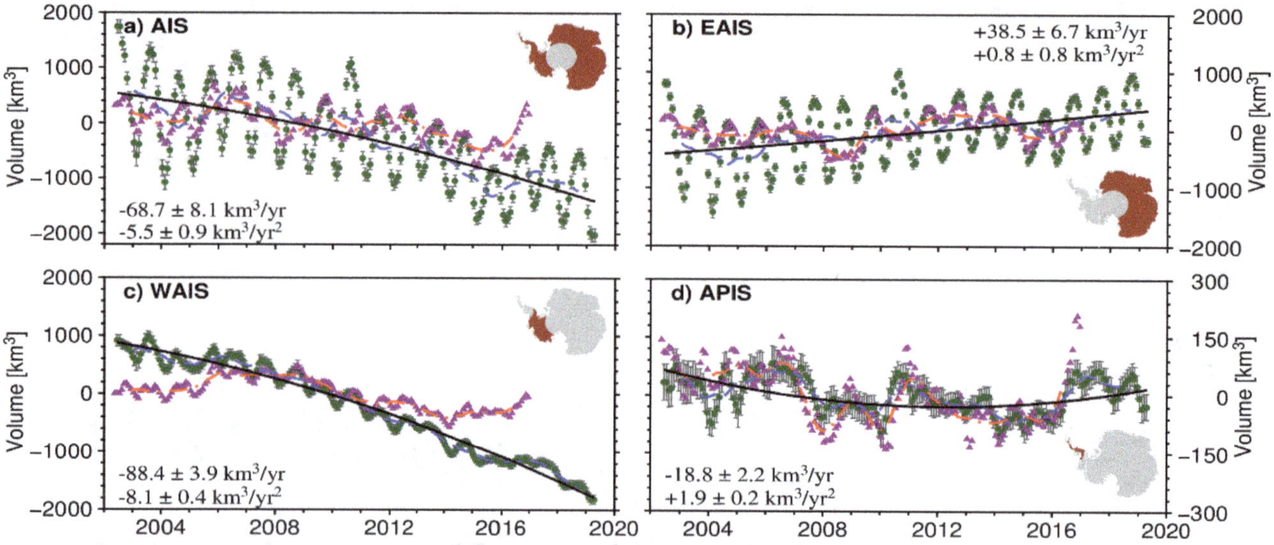

Figure 7. Volume changes in the AIS north of 81.5° S (**a**) and the three subregions ((**b**): EAIS, (**c**): WAIS, and (**d**): APIS) from our combined altimetric time series (green dots) and IMAU-FDM (purple triangles). The blue dashed curves and red dash-dotted curves are the time series after removing seasonal oscillations using a 13-month moving average. Due to the defects of the moving average algorithm, the 6-month values at the beginning and end of the time series are not displayed. The solid black lines are the best-fit quadratic curves for the altimetric volume time series. Included are the altimetric volume time series linear and quadratic trend estimates. The gray error bars show the 1σ uncertainty range of the altimetry data. The regions covered by each localization are shaded red in the inset. The boundaries refer to the Antarctic drainage systems in [47].

Figure 8 shows the time series for the entire AIS and the subregions of EAIS, WAIS, and APIS with linear trends and seasonal oscillations removed. Figures 7 and 8 show that surface volume changes can only explain a small proportion of the overall volume changes in the AIS. The biggest difference in long-term trends comes from the WAIS, although its interannual variations in surface and altimetric volume changes show good consistency. According to previous studies, the differences in long-term trends in the WAIS are possibly due to ice dynamics [4,5,44,50]. Similarly, because of dynamic thinning, surface processes cannot fully explain the altimetric volume changes in the APIS, especially during 2002–2006. Following the disintegration of the Larson B Ice Shelf in 2002, increased ice discharge occurred [51,52]. Although surface volume changes in some subregions are in good agreement with their altimetric volume changes, there are differences in both the long-term trends and the interannual variations between the two integrated volume time series for the EAIS. These differences can also be seen in Schröder et al. (2019) [10].

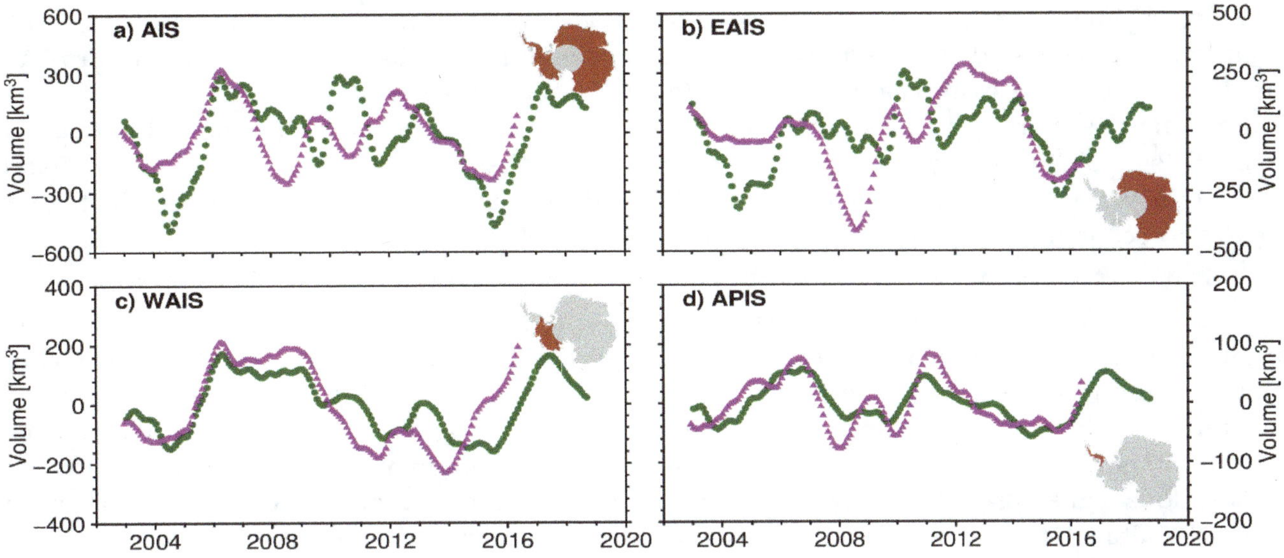

Figure 8. Volume change of the AIS north of 81.5° S (**a**) and the three subregions ((**b**): EAIS, (**c**): WAIS, and (**d**): APIS) from our combined altimetric time series (green dots) and IMAU-FDM (purple triangles). After removing the long-term trends, seasonal oscillations were removed using a 13-month moving average. Due to the defects of the moving average algorithm, the 6-month values at the beginning and end of the time series are not displayed. The regions covered by each localization are shaded red in the inset. The boundaries refer to the Antarctic drainage systems in [47].

5. Discussion

Accurately correcting intermission biases is crucial for combining observations from different radar altimetry missions to study the long-term changes of ice sheets. Imperfect methods may lead to non-self-consistent combined results. This is why there are leap or step signals at the junctions between ERS-2 and Envisat and between Envisat and CryoSat-2 in the elevation time series shown in Figure 10 of Schröder et al. (2019) [10]. This can be attributed to the uncertainties in the intermission cross-calibration due to the short overlapping period. In this study, we constructed an updated plane-fitting least-squares regression model by introducing the comprehensive parameters of the intermission bias and the A–D bias and their internal consistency into the existing least-squares regression model. In contrast with previous methods, it allowed us to use a larger amount of observations in the same regression model for the integrated least-squares adjustment to improve the accuracy of cross-calibration of the data from different missions, and thereby to ensure a self-consistent and accurate elevation time series. We used Envisat and Cryosat-2 observations to test our model and successfully established a 5 km grid of monthly elevation time series over the AIS. The estimations of the total intermission biases and the intramission A–D bias are consistent with those in previous studies [5,15,18]. Moreover, the self-consistent elevation time series at the joint of the two satellites (Figure 4) show that our algorithm is effective in cross-calibration. The intercomparison between the surface elevation changes from OIB ATM L4 and from the combined elevation time series also demonstrate the effectiveness of our method in creating time series using data from different missions. In addition, comparisons with OIB-derived surface elevations and ICESat surface elevation changes also increased confidence in our merged time series. It should be noted that the difference in footprint size between Envisat data and CryoSat-2 SARIn-mode data may be a source of uncertainty in the estimates of intermission bias and in creating of the merged time series. Following the study by Adusumilli et al. (2018) [16], we believe that the point of closest approach (POCA) applied to generate Envisat and CryoSat-2 Level 2 products reduces the difference between topography sampled with the different satellite radar altimeter techniques. Certainly, it still needs further study to accurately calibrate the different techniques against each other in the future.

Our gridded monthly elevation time series allow for the analysis of temporal and spatial evolutions over 2002–2019. Well-known patterns of thinning and thickening were also confirmed in this study. Volume losses are mainly concentrated in the coastal sectors of the WAIS, especially in the Amundsen Sea sector, while the volume gains are mainly observable in the Dronning Maud Land of the EAIS [2,4,5,10,14]. A simple comparison of the elevation change rates of the two periods (2002–2010 vs. 2010–2019) reveals an expansion of thinning in the Pine Island Glacier and interannual fluctuations of elevation changes in Princess Elizabeth Land and the Bellingshausen Sea sector. Although these findings were previously reported [10,53,54], they indicate that our time series are reliable.

Due to signal penetration into the snow, the elevation changes derived from radar altimeters might not reflect the actual surface elevation changes [17,55]. However, including the backscatter coefficient in the least-squares regression model can only account for a part of the effect of surface penetration, as shown in Simonsen and Sorensen (2017) [3] and Sørensen et al. (2018) [17]. Indeed, there were spurious seasonal fluctuations caused by surface penetration in our combined results. However, this spurious signal also exists in Schröder et al. (2019) [10], even though the authors reprocessed their radar altimetry observations with a low threshold retracker able to produce spurious changes in elevation due to changes in surface penetration [27,28]. Therefore, radar altimetry data is generally considered to be very difficult to use for studying the changes caused by surface processes directly, especially for surface accumulation events. Nevertheless, the Greenland 2012 melt event provides us with an extreme example for which radar altimeters can detect the changes caused by surface processes. The formation of a refrozen melt layer can increase the reflective surface for radar altimeters instantaneously [55]. Comparably, other surface processes, such as firn compaction, can also form ice lenses to change the reflecting surface, allowing them to be observed by radar altimeters. That is to say, after a time-consuming process of firn compaction, accumulation events can be observed with time lags by radar altimeters. After removing seasonal variations to minimize the effects of surface penetration, more reliable estimates of the interannual and long-term variations caused by surface processes can be clearly seen from our combined altimetric time series, such as snowfall in Dronning Maud Land in 2009 and 2012 and snow accumulation in 2008–2009 in the Totten Glacier sector.

Therefore, together with surface volume changes, the integrated volume time series over larger regions enabled us to further improve our understanding of elevation change processes. The agreement between altimetric and surface volume changes in Dronning Maud Land indicates that surface processes dominate the variability of its volume changes. In other words, changes in ice dynamics do not significantly affect the time variability of its volume signal. This confirms that they are distal to areas with fast ice flow [56]. Modest thickening in Dronning Maud Land was associated with increases in snowfall due to its extreme accumulation events in 2019 and 2012 [54,57]. The long-term trend of altimetric volume changes of Princess Elizabeth Land are consistent with those of the surface volume. This result agrees with that in Schröder et al. (2019) [10], showing that thinning here is related to less snowfall. In the Totten Glacier sector and the Bellingshausen Sea sector, the altimetric and surface volume changes also agree well. The difference is that excessive ice discharge and grounding line retreats have made Totten Glacier and several Bellingshausen Sea glaciers thin [42,56]. However, the surface volume changes are so large that they still dominate the volume changes in these regions, such as the accelerated thinning of Totten Glacier over 2010–2015 [42].

As reported in previous studies [4,5,10,44,54], changes in the Amundsen Sea sector ice sheet are mainly driven by rapid ice discharge. Mouginot et al. (2014) [53] and Harig et al. (2015) [57] reported that flow acceleration and widening of Pine Island Glacier and Thwaites Glacier led to rapid acceleration loss in the Amundsen Sea sector. These factors agree with the pronounced differences that we observed between altimetric and surface volume changes here. In the Getz sector, we found that neither ice dynamics nor surface processes dominate volume changes over the entire period of 2002–2019. As reported in Shepherd et al. (2019) [54], the patterns of elevation changes in this sector reflect a complex mixture of surface processes and dynamic ice imbalances.

Overall, the accelerated ice volume loss in the coastal regions of WAIS continues to outpace the gains made in the EAIS. The volume loss of the AIS is mainly caused by ice discharge in the Amundsen Sea sector. Meanwhile, in Princess Elizabeth Land, the Totten Glacier sector of the EAIS, and the Bellingshausen Sea sector and Getz sector of the WAIS, surface processes also make a considerable contribution to volume loss. The contribution of surface processes in Princess Elizabeth Land and the Totten Glacier sector of the EAIS to the acceleration of volume loss over the AIS also cannot be ignored. Volume gains due to surface processes were mainly observed in the EAIS, especially in Dronning Maud Land.

6. Conclusions

In this study, we presented an updated plane-fitting least-squares regression model that combines different radar altimetry missions. Using this approach, we constructed a monthly elevation time series using 5 km grid cells over the AIS. Further analysis of the output estimates for the systematic biases and elevation time series shows that the intermission bias and the A–D bias were successfully corrected. Validation with airborne and ICESat laser altimetry observations indicates that our combined gridded monthly elevation time series are reliable.

Further analysis shows that our resulting elevation time series can provide detailed insight into temporal and spatial evolution of elevation for the AIS over 2002–2019. In addition to the well-known patterns of thinning and thickening, acceleration and increases in thinning in Pine Island Glacier and Thwaites Glacier, as well as interannual fluctuations in Princess Elizabeth Land and the Bellingshausen Sea sector, were shown very clearly. Overall, the volume loss in the WAIS continues to outpace the gains made in the EAIS. From 2002 to 2019, the total volume change estimates for the AIS were -68.7 ± 8.1 km^3/y, with an acceleration of -5.5 ± 0.9 km^3/y^2. Most of the volume loss was from the Amundsen Sea sector and the Getz sector, and most of the acceleration was from the Amundsen Sea sector, Princess Elizabeth Land, and the Totten Glacier sector.

After removing the seasonal variations, we obtained the interannual and long-term variations caused by surface processes in our combined time series. With the help of additional surface volume changes derived from IMAU-FDM, we determined that ice discharge in the Amundsen Sea sector provides the largest accelerated volume loss of the AIS. However, surface processes still play an important role in the volume changes of the AIS. In addition to thickening in Dronning Maud Land, surface processes in regions such as the Totten Glacier sector, Princess Elizabeth Land, and the Bellingshausen Sea sector are the main sources of acceleration in volume loss for the AIS.

Author Contributions: B.Z., Q.Y., Z.W., and F.L. conceived the study, supervised the experiments, and improved the manuscript. J.L., J.A., T.L., and H.G. contributed to the analysis and discussion. All authors have read and agreed to the published version of the manuscript.

Acknowledgments: We would like to thank organizations that shared their datasets and software for use in this study. The Envisat and CryoSat-2 observations were obtained from the European Space Agency (ESA) Earth Online, and the airborne elevation data was provided by the National Snow and Ice Data Center (NSIDC). All geographical plots were produced using the Generic Mapping Tools (GMT). We also specifically thank Sebastian B. Simonsen, René Forsberg and Louise S. Sørensen for their comments and suggestions.

References

1. Zwally, H.J.; Bindschadler, R.A.; Brenner, A.C.; Martin, T.V.; Thomas, R.H. Surface elevation contours of greenland and antarctic ice sheets. *J. Geophys Res. Ocean.* **1983**, *88*, 1589–1596. [CrossRef]
2. Helm, V.; Humbert, A.; Miller, H. Elevation and elevation change of Greenland and Antarctica derived from CryoSat-2. *Cryosphere* **2014**, *8*, 1539–1559. [CrossRef]

3. Simonsen, S.B.; Sorensen, L.S. Implications of changing scattering properties on Greenland ice sheet volume change from Cryosat-2 altimetry. *Remote Sens. Environ.* **2017**, *190*, 207–216. [CrossRef]
4. Flament, T.; Remy, F. Dynamic thinning of Antarctic glaciers from along-track repeat radar altimetry. *J. Glaciol.* **2012**, *58*, 830–840. [CrossRef]
5. McMillan, M.; Shepherd, A.; Sundal, A.; Briggs, K.; Muir, A.; Ridout, A.; Hogg, A.; Wingham, D. Increased ice losses from Antarctica detected by CryoSat-2. *Geophys. Res. Lett.* **2014**, *41*, 3899–3905. [CrossRef]
6. Sørensen, L.S.; Simonsen, S.B.; Nielsen, K.; Lucas-Picher, P.; Spada, G.; Adalgeirsdottir, G.; Forsberg, R.; Hvidberg, C.S. Mass balance of the Greenland ice sheet (2003–2008) from ICESat data–The impact of interpolation, sampling and firn density. *Cryosphere* **2011**, *5*, 173–186. [CrossRef]
7. Zwally, H.J.; Li, J.; Brenner, A.C.; Beckley, M.A.; Cornejo, H.G.; DiMarzio, J.P.; Giovinetto, M.B.; Neumann, T.A.; Robbins, J.; Saba, J.L.; et al. Greenland ice sheet mass balance: Distribution of increased mass loss with climate warming; 2003–07 versus 1992–2002. *J. Glaciol.* **2011**, *57*, 88–102. [CrossRef]
8. Pritchard, H.D.; Arthern, R.J.; Vaughan, D.G.; Edwards, L.A. Extensive dynamic thinning on the margins of the Greenland and Antarctic ice sheets. *Nature* **2009**, *461*, 971–975. [CrossRef]
9. Wouters, B.; Bamber, J.L.; van den Broeke, M.R.; Lenaerts, J.T.M.; Sasgen, I. Limits in detecting acceleration of ice sheet mass loss due to climate variability. *Nat. Geosci.* **2013**, *6*, 613–616. [CrossRef]
10. Schröder, L.; Horwath, M.; Dietrich, R.; Helm, V.; van den Broeke, M.R.; Ligtenberg, S.R.M. Four decades of Antarctic surface elevation changes from multi-mission satellite altimetry. *Cryosphere* **2019**, *13*, 427–449. [CrossRef]
11. Paolo, F.S.; Fricker, H.A.; Padman, L. Constructing improved decadal records of Antarctic ice shelf height change from multiple satellite radar altimeters. *Remote Sens. Environ.* **2016**, *177*, 192–205. [CrossRef]
12. Khvorostovsky, K.S. Merging and analysis of elevation time series over Greenland ice sheet from satellite radar altimetry. *IEEE Trans. Geosci. Remote Sens.* **2012**, *50*, 23–36. [CrossRef]
13. Fricker, H.A.; Padman, L. Thirty years of elevation change on Antarctic Peninsula ice shelves from multimission satellite radar altimetry. *J. Geophys. Res. Ocean.* **2012**, *117*, C02026. [CrossRef]
14. Zwally, H.J.; Giovinetto, M.B.; Li, J.; Cornejo, H.G.; Beckley, M.A.; Brenner, A.C.; Saba, J.L.; Yi, D.H. Mass changes of the Greenland and Antarctic ice sheets and shelves and contributions to sea-level rise: 1992–2002. *J. Glaciol.* **2005**, *51*, 509–527. [CrossRef]
15. Frappart, F.; Legrésy, B.; Niño, F.; Blarel, F.; Fuller, N.; Fleury, S.; Birol, F.; Calmant, S. An ERS-2 altimetry reprocessing compatible with ENVISAT for long-term land and ice sheets studies. *Remote Sens. Environ.* **2016**, *184*, 558–581. [CrossRef]
16. Adusumilli, S.; Fricker, H.A.; Siegfried, M.R.; Padman, L.; Paolo, F.S.; Ligtenberg, S.R.M. Variable basal melt rates of Antarctic Peninsula ice shelves, 1994–2016. *Geophys. Res. Lett.* **2018**, *45*, 4086–4095. [CrossRef]
17. Sørensen, L.S.; Simonsen, S.B.; Forsberg, R.; Khvorostovsky, K.; Meister, R.; Engdahl, M.E. 25 years of elevation changes of the Greenland Ice Sheet from ERS, Envisat, and CryoSat-2 radar altimetry. *Earth Planet. Sci. Lett.* **2018**, *495*, 234–241. [CrossRef]
18. Armitage, T.W.K.; Wingham, D.J.; Ridout, A.L. Meteorological origin of the static crossover pattern present in low-resolution-mode CryoSat-2 data over Central Antarctica. *IEEE Geosci. Remote Sens. Lett.* **2014**, *11*, 1295–1299. [CrossRef]
19. Remy, F.; Legresy, B.; Benveniste, J. On the azimuthally anisotropy effects of polarization for altimetric measurements. *IEEE Trans. Geosci. Remote Sens.* **2006**, *44*, 3289–3296. [CrossRef]
20. Arthern, R.J.; Wingham, D.J.; Ridout, A.L. Controls on ERS altimeter measurements over ice sheets: Footprint-scale topography, backscatter fluctuations, and the dependence of microwave penetration depth on satellite orientation. *J. Geophys. Res. Atmos.* **2001**, *106*, 33471–33484. [CrossRef]
21. Legresy, B.; Remy, F.; Schaeffer, P. Different ERS altimeter measurements between ascending and descending tracks caused by wind induced features over ice sheets. *Geophys. Res. Lett.* **1999**, *26*, 2231–2234. [CrossRef]
22. Yang, Q.M.; Yang, Y.D.; Wang, Z.M.; Zhang, B.J.; Jiang, H. Elevation change derived from SARAL/ALtiKa altimetric mission: Quality assessment and performance of the Ka-Band. *Remote Sens.* **2018**, 10. [CrossRef]
23. Yang, Y.D.; Li, F.; Hwang, C.; Ding, M.H.; Ran, J.J. Space-Time evolution of Greenland Ice Sheet elevation and mass from Envisat and GRACE data. *J. Geophys. Res. Earth Surf.* **2019**, *124*, 2079–2100. [CrossRef]
24. Batoula, S.; Urien, S.; Picard, B.; Muir, A.; Roca, M.; Garcia, P. Envisat Altimetry Level 2 Product Handbook. *CLS-ESLF-18-0003 2018, Issue 2.0*. Available online: https://earth.esa.int/documents/700255/3528455/PH_light_3528452rev3528450_ESA.pdf/ (accessed on 22 July 2020).

25. Wingham, D.; Rapley, C.; Griffiths, H. New techniques in satellite altimeter tracking systems. In *Proceedings of the IGARSS Symposium, Zurich, Switzerland, 8–11 September 1986*; Guyenne, T.D., Hunt, J.J., Eds.; European Space Agency Publications Division: Noordwijk, The Netherlands, 1986; Volume SP-254, pp. 1339–1344.
26. Davis, C.H. A robust threshold retracking algorithm for measuring ice-sheet surface elevation change from satellite radar altimeters. *IEEE Trans. Geosci. Remote Sens.* **1997**, *35*, 974–979. [CrossRef]
27. Nilsson, J.; Gardner, A.; Sørensen, L.S.; Forsberg, R. Improved retrieval of land ice topography from CryoSat-2 data and its impact for volume-change estimation of the Greenland Ice Sheet. *Cryosphere* **2016**, *10*, 2953–2969. [CrossRef]
28. Schröder, L.; Richter, A.; Fedorov, D.V.; Eberlein, L.; Brovkov, E.V.; Popov, S.V.; Knöfel, C.; Horwath, M.; Dietrich, R.; Matveev, A.Y.; et al. Validation of satellite altimetry by kinematic GNSS in central East Antarctica. *Cryosphere* **2017**, *11*, 1111–1130. [CrossRef]
29. Bouzinac, C. CryoSat Product Handbook. 2014. Available online: https://earth.esa.int/documents/10174/125272/CryoSat_Product_Handbook (accessed on 30 June 2020).
30. Krabill, W.; Hanna, E.; Huybrechts, P.; Abdalati, W.; Cappelen, J.; Csatho, B.; Frederick, E.; Manizade, S.; Martin, C.; Sonntag, J.; et al. Greenland ice sheet: Increased coastal thinning. *Geophys. Res. Lett.* **2004**, *31*, L24402. [CrossRef]
31. Zhang, B.; Wang, Z.; Li, F.; An, J.; Yang, Y.; Liu, J. Estimation of present-day glacial isostatic adjustment, ice mass change and elastic vertical crustal deformation over the Antarctic ice sheet. *J. Glaciol.* **2017**, *63*, 703–715. [CrossRef]
32. Ewert, H.; Groh, A.; Dietrich, R. Volume and mass changes of the Greenland ice sheet inferred from ICESat and GRACE. *J. Geodyn.* **2012**, *59–60*, 111–123. [CrossRef]
33. Argus, D.F.; Peltier, W.R.; Drummond, R.; Moore, A.W. The Antarctica component of postglacial rebound model ICE-6G_C (VM5a) based on GPS positioning, exposure age dating of ice thicknesses, and relative sea level histories. *Geophys. J. Int.* **2014**, *198*, 537–563. [CrossRef]
34. Groh, A.; Ewert, H.; Scheinert, M.; Fritsche, M.; Rülke, A.; Richter, A.; Rosenau, R.; Dietrich, R. An investigation of glacial isostatic adjustment over the Amundsen Sea sector, West Antarctica. *Glob. Planet. Chang.* **2012**, *98–99*, 45–53. [CrossRef]
35. Ligtenberg, S.R.M.; Helsen, M.M.; van den Broeke, M.R. An improved semi-empirical model for the densification of Antarctic firn. *Cryosphere* **2011**, *5*, 809–819. [CrossRef]
36. Martin, C.F.; Krabill, W.B.; Manizade, S.S.; Russell, R.L.; Sonntag, J.G.; Swift, R.N.; Yungel, J.K. Airborne Topographic Mapper Calibration Procedures and Accuracy Assessment. Nasa Technical Memorandum TM2012-21(February) 2012, 20120008479. Available online: https://ntrs.nasa.gov/archive/nasa/casi.ntrs.nasa.gov/20120008479.pdf (accessed on 15 July 2020).
37. Magruder, L.A.; Webb, C.E.; Urban, T.J.; Silverberg, E.C.; Schutz, B.E. ICESat altimetry data product verification at white sands space harbor. *IEEE Trans. Geosci. Remote Sens.* **2007**, *45*, 147–155. [CrossRef]
38. Studinger, M. *National Snow and Ice Data Center. IceBridge ATM L4 Surface Elevation Rate of Change*; Version 1; updated 2018; USA NASA National Snow and Ice Data Center Distributed Active Archive Center: Boulder, CO, USA, 2014; Available online: https://nsidc.org/data/IDHDT4/versions/1 (accessed on 15 July 2020). [CrossRef]
39. Studinger, M. *IceBridge ATM L2 Icessn Elevation, Slope, and Roughness*; Version 2; updated 2020; ASA National Snow and Ice Data Center Distributed Active Archive Center: Boulder, CO, USA, 2014; Available online: https://nsidc.org/data/ILATM2/versions/2 (accessed on 15 July 2020). [CrossRef]
40. Slater, T.; Shepherd, A.; McMillan, M.; Muir, A.; Gilbert, L.; Hogg, A.E.; Konrad, H.; Parrinello, T. A new digital elevation model of Antarctica derived from CryoSat-2 altimetry. *Cryosphere* **2018**, *12*, 1551–1562. [CrossRef]
41. Sørensen, L.S.; Simonsen, S.B.; Meister, R.; Forsberg, R.; Levinsen, J.F.; Flament, T. Envisat-derived elevation changes of the Greenland ice sheet, and a comparison with ICESat results in the accumulation area. *Remote Sens. Environ.* **2015**, *160*, 56–62. [CrossRef]
42. Li, X.; Rignot, E.; Mouginot, J.; Scheuchl, B. Ice flow dynamics and mass loss of Totten Glacier, East Antarctica, from 1989 to 2015. *Geophys. Res. Lett.* **2016**, *43*, 6366–6373. [CrossRef]
43. Groh, A.; Ewert, H.; Rosenau, R.; Fagiolini, E.; Gruber, C.; Floricioiu, D.; Abdel Jaber, W.; Linow, S.; Flechtner, F.; Eineder, M.; et al. Mass, volume and velocity of the Antarctic ice sheet: Present-Day changes and error effects. *Surv. Geophys.* **2014**, *35*, 1481–1505. [CrossRef]

44. Velicogna, I.; Sutterley, T.C.; van den Broeke, M.R. Regional acceleration in ice mass loss from Greenland and Antarctica using GRACE time-variable gravity data. *Geophys. Res. Lett.* **2014**, *41*, 8130–8137. [CrossRef]
45. Legrésy, B.; Rémy, F. Using the temporal variability of satellite radar altimetric observations to map surface properties of the Antarctic ice sheet. *J. Glaciol.* **1998**, *44*, 197–206. [CrossRef]
46. Davis, C.H.; Ferguson, A.C. Elevation change of the Antarctic ice sheet, 1995–2000, from ERS-2 satellite radar altimetry. *IEEE Trans. Geosci. Remote Sens.* **2004**, *42*, 2437–2445. [CrossRef]
47. Zwally, H.J.; Giovinetto, M.B.; Beckley, M.A.; Saba, J.L. Antarctic and Greenland Drainage Systems. GSFC Cryospheric Sciences Laboratory. 2012. Available online: http://icesat4.gsfc.nasa.gov/cryo_data/ant_grn_drainage_systems.php (accessed on 15 July 2020).
48. Boening, C.; Lebsock, M.; Landerer, F.; Stephens, G. Snowfall-driven mass change on the East Antarctic ice sheet. *Geophys. Res. Lett.* **2012**, *39*, L21501. [CrossRef]
49. Lenaerts, J.T.M.; van Meijgaard, E.; van den Broeke, M.R.; Ligtenberg, S.R.M.; Horwath, M.; Isaksson, E. Recent snowfall anomalies in Dronning Maud Land, East Antarctica, in a historical and future climate perspective. *Geophys. Res. Lett.* **2013**, *40*, 2684–2688. [CrossRef]
50. Shepherd, A.; Ivins, E.; Rignot, E.; Smith, B.; van den Broeke, M.; Velicogna, I.; Whitehouse, P.; Briggs, K.; Joughin, I.; Krinner, G.; et al. Mass balance of the Antarctic Ice Sheet from 1992 to 2017. *Nature* **2018**, *558*, 219–222. [CrossRef]
51. Shuman, C.A.; Berthier, E.; Scambos, T.A. 2001–2009 elevation and mass losses in the Larsen A and B embayments, Antarctic Peninsula. *J. Glaciol.* **2011**, *57*, 737–754. [CrossRef]
52. Berthier, E.; Scambos, T.A.; Shuman, C.A. Mass loss of Larsen B tributary glaciers (Antarctic Peninsula) unabated since 2002. *Geophys. Res. Lett.* **2012**, *39*, L13501. [CrossRef]
53. Mouginot, J.; Rignot, E.; Scheuchl, B. Sustained increase in ice discharge from the Amundsen Sea Embayment, West Antarctica, from 1973 to 2013. *Geophys. Res. Lett.* **2014**, *41*, 1576–1584. [CrossRef]
54. Shepherd, A.; Gilbert, L.; Muir, A.S.; Konrad, H.; McMillan, M.; Slater, T.; Briggs, K.H.; Sundal, A.V.; Hogg, A.E.; Engdahl, M.E. Trends in Antarctic ice sheet elevation and mass. *Geophys. Res. Lett.* **2019**, *46*, 8174–8183. [CrossRef]
55. Nilsson, J.; Vallelonga, P.; Simonsen, S.B.; Sørensen, L.S.; Forsberg, R.; Dahl-Jensen, D.; Hirabayashi, M.; Goto-Azuma, K.; Hvidberg, C.S.; Kjær, H.A.; et al. Greenland 2012 melt event effects on CryoSat-2 radar altimetry. *Geophys. Res. Lett.* **2015**, *42*, 3919–3926. [CrossRef]
56. Gardner, A.S.; Moholdt, G.; Scambos, T.; Fahnstock, M.; Ligtenberg, S.; van den Broeke, M.; Nilsson, J. Increased West Antarctic and unchanged East Antarctic ice discharge over the last 7 years. *Cryosphere* **2018**, *12*, 521–547. [CrossRef]
57. Harig, C.; Simons, F.J. Accelerated West Antarctic ice mass loss continues to outpace East Antarctic gains. *Earth Planet. Sci. Lett.* **2015**, *415*, 134–141. [CrossRef]

2

Monitoring Large-Scale Inland Water Dynamics by Fusing Sentinel-1 SAR and Sentinel-3 Altimetry Data and by Analyzing Causal Effects of Snowmelt

Ya-Lun S. Tsai [1,*], Igor Klein [1], Andreas Dietz [1] and Natascha Oppelt [2]

[1] German Remote Sensing Data Center (DFD), German Aerospace Center (DLR), Muenchener Strasse 20, 82234 Wessling, Germany; Igor.Klein@dlr.de (I.K.); Andreas.Dietz@dlr.de (A.D.)

[2] Department of Geography, Earth Observation and Modelling, Kiel University, Ludewig-Meyn-Str. 14, 24118 Kiel, Germany; oppelt@geographie.uni-kiel.de

* Correspondence: Ya-Lun.Tsai@dlr.de

Abstract: The warming climate is threatening to alter inland water resources on a global scale. Within all waterbody types, lake and river systems are vital not only for natural ecosystems but, also, for human society. Snowmelt phenology is also altered by global warming, and snowmelt is the primary water supply source for many river and lake systems around the globe. Hence, (1) monitoring snowmelt conditions, (2) tracking the dynamics of snowmelt-influenced river and lake systems, and (3) quantifying the causal effect of snowmelt conditions on these waterbodies are critical to understand the cryo-hydrosphere interactions under climate change. Previous studies utilized in-situ or multispectral sensors to track either the surface areas or water levels of waterbodies, which are constrained to small-scale regions and limited by cloud cover, respectively. On the contrary, in the present study, we employed the latest Sentinel-1 synthetic aperture radar (SAR) and Sentinel-3 altimetry data to grant a high-resolution, cloud-free, and illumination-independent comprehensive inland water dynamics monitoring strategy. Moreover, in contrast to previous studies utilizing in-house algorithms, we employed freely available cloud-based services to ensure a broad applicability with high efficiency. Based on altimetry and SAR data, the water level and the water-covered extent (WCE) (surface area of lakes and the flooded area of rivers) can be successfully measured. Furthermore, by fusing the water level and surface area information, for Lake Urmia, we can estimate the hypsometry and derive the water volume change. Additionally, for the Brahmaputra River, the variations of both the water level and the flooded area can be tracked. Last, but not least, together with the wet snow cover extent (WSCE) mapped with SAR imagery, we can analyze the influence of snowmelt conditions on water resource variations. The distributed lag model (DLM) initially developed in the econometrics discipline was employed, and the lagged causal effect of snowmelt conditions on inland water resources was eventually assessed.

Keywords: snow cover; water level; surface area; flooded area; hypsometry; distributed lag model; water resource

1. Introduction

Inland water resources play an essential role in not only the prosperity and stability of human society but, also, the sustainability and balance of various ecosystems. According to the reports of the United Nations Environment Programme (UNEP) [1], UN Food and Agriculture Organization (FAO) [2], and the latest UN World Water Development Report (WWDR) 2019 [3], transboundary surface water resources largely influence the socioeconomic-ecological systems; therefore, it is enlisted as the sixth 2030 Sustainable Development Goal (SDG). Expressly, the importance of local rivers and lakes should be noted, as they are critical freshwater sources for many regions [3] and, also, influence the climate [4]. However, as studies suggested [5,6], these vague SDGs provide no dependable guideline on how to achieve them. As a result, to elaborate the practical research gap of SDG target 6.6, i.e., protect and restore water-related ecosystems, an investigation of the main drivers of change in the high-mountain cryosphere is recommended [6]. The importance of snowmelt water to downstream freshwater is widely recognized. Assessing this cryo-hydrosphere interaction is indispensable to understand the future trend of water resources we have in severer climate change scenarios. Due to global warming, the shrinking of snow cover extent (SCE), earlier snowmelt season, and shorter snow cover duration have been observed almost globally, as stated in the Synthesis Report of the Intergovernmental Panel on Climate Change (IPCC) Fifth Assessment Report (AR5) [7]. Moreover, in the same report, it was also indicated that the change of melting snow is altering hydrological systems and affecting water resources in both quantity and quality. Thus, monitoring snowmelt conditions and the dynamics of snowmelt-influenced river and lake systems and quantifying the causal effect of snowmelt water on these waterbodies are critical. It would allow us not only to understand the impact of global warming to inland water resources but, also, to assess the vulnerability and variability of regional freshwater supplies.

For the observation and investigation of the influence of snowmelt on global waterbodies, the employment of spaceborne remote-sensing data is preferred, owing to its high efficiency and broad applicability. As mentioned in the latest report published by UN-Water assessing the progress of SDG indicator 6.6.1 [8], i.e., tracking time series changes in the extent (including quality, quantity, and area) of water-related ecosystems (including rivers and lakes), the utilization of globally transferable satellite data is strongly advocated. For monitoring the snowmelt conditions, as summarized in Tsai et al. [9], it is hardly possible to discriminate wet and dry snow with conventional multi-spectral data sources solely based on the reflectance differences, as the spectral characteristics of both snow types share high similarity. SAR data, on the other hand, are excellent for wet SCE (WSCE) mapping: The snowmelt-caused wetness in the snowpack would change the dielectric constant of the snowpack and, thus, dramatically shorten the penetration depth of the SAR signal, which eventually lead to a significant decrease of the backscatter coefficient when the snowpack starts to melt [9–11].

For monitoring the dynamics of river and lake systems, conventional studies focus on either the water level or surface area. For tracking the surface area, multi-spectral sensor-based band ratio water indices, such as the Normalized Difference Water Index (NDWI) [12], Modified NDWI (MNDWI) [13], and Automated Water Extraction Index (AWEI) [14], are commonly used due to their simplicities. However, in addition to the difficulty of the selection among these various band ratios, these indices are based on a multi-spectral sensor. Therefore, they are inevitably affected by cloud cover and polar darkness, which frequently happen for alpine regions and high-latitude zones [9]. Moreover, the spectral feature of water in the multi-spectral sensor is also affected by the variation of atmospheric/illumination conditions and water dynamics, such as water depth, sediment load, eutrophication degree, turbidity, sun angle, and sensor view angle [15,16]. On the contrary, SAR data can be utilized for waterbody monitoring thanks to its cloud-penetrating applicability and illumination-independent characteristics. Yet, it might suffer from relatively coarse resolution compared to optical sensors, the higher cost of the imagery, and most importantly, the requirement of computation-intensive processing.

Regarding the monitoring of the water level of lakes or rivers, using in-situ gauging station measurements is the optimal approach. However, although it provides the chance of continuous observations, it is not widely achievable in all lake and river systems because of the cost of instrument setup, intensive maintenance requirements, or the difficult accessibility of remote waterbodies. In fact, the number of available gauging stations has been decreasing globally [17,18]. Additionally, the data might not be publicly available. Another alternative is combining the surface area of a waterbody with the bathymetry of the lake or river channel to derive the water surface height. Nevertheless, accurate information of the bathymetry is still needed, which is only available for very few waterbodies, which were surveyed thoroughly [4,18]. On the contrary, spaceborne altimetry provides a favorable alternative for waterbody surface height monitoring due to its global availability. Many altimetry databases have therefore been established, including the Database for Hydrological Time Series of Inland Waters (DAHITI) (https://dahiti.dgfi.tum.de/en/) [19], The Global Reservoir and Lake Global Reservoir and Lake Monitor (G-REALM) (https://ipad.fas.usda.gov/cropexplorer/global_reservoir/) [20], and Hydroweb (http://hydroweb.theia-land.fr/) [21]. Unfortunately, because the design of spaceborne altimetry sensors is generally targeted for ocean surface surveying, i.e., a flat, isotropic, and ample reflecting surface, the footprint sizes of their echoes are enormously coarse (>2 km), and the sampling distance between these echo points is vast [22]. These characteristics hinder the viability of inland waterbody monitoring, especially for rivers, which usually have a more inhomogeneous neighboring topography [23]. Thus, to overcome the waveform difference to the standard Brown model [24], a computation-intensive waveform reprocessing (retracking) analysis is usually required for inland waterbodies [4,25–27]. However, the selection of retrackers (ground processing techniques that estimate the range to the point of closest approach on the surface) varies in different studies, and most of them use in-house algorithms or refinements [4,22,28,29], which largely limits the transferability of the method to other regions.

With the aim to overcome these limitations for monitoring the surface area and water level, in the present study, we utilized the new generation of high-resolution, freely accessible spaceborne SAR and altimetry data of Sentinel-1 and Sentinel-3, together with freely available online cloud-based processing services, to achieve a comprehensive inland waterbody dynamics (time series surface area and water level) tracking strategy with broad applicability. Although there have been some studies employing either spaceborne SAR or altimetry sensors for waterbody monitoring, few studies have yet integrated them to achieve an in-situ measurement-free approach. Additionally, in contrast to studies relying on multi-spectral sensors, this altimetry-SAR sensor combination provides a cloud-free and illumination-independent monitoring method. Furthermore, as the Sentinel fleet is the newest generation of spaceborne satellites with a freely accessible data policy, data availability is ensured for the following decades. Last, but not least, based on the depicted comprehensive waterbody dynamics, together with the SAR-based WSCE maps, we can further analyze the relationship between snowmelt conditions and inland water resource variations. By employing a sophisticated regression model adopted from the econometrics domain, we guarantee a quantified and unbiased lagged influence of the snowmelt condition on the waterbody dynamics; hence, the cryo-hydrosphere interaction can be well-identified.

2. Study Areas and Data

2.1. Study Areas

Since the goal of the present study is to (1) monitor the dynamics (both water level and surface area) of inland waterbodies and (2) quantify the lagged influence of snowmelt on inland water dynamics, we selected the study sites based on the following criteria: (1) their spatial scales were wide enough to be detected by spaceborne SAR and altimetry sensors, (2) they demonstrated a considerable magnitude of variations in both water level and surface area, and (3) their watersheds included snow-covered regions. Therefore, in our study, Lake Urmia in Iran and the Brahmaputra River were chosen.

2.1.1. Lake Urmia

Lake Urmia (or Orumiyeh Lake) is located in the northwest of Iran. It is the largest lake in Iran and is the largest perennial hyper-saline lake worldwide [30]. This endorheic lake is situated 1273 m above the mean sea level and is situated in a semi-arid climate zone with average annual precipitation and evaporation of around 341 and 1200 mm/y, respectively [31]. The highest river inflows can be observed in the springtime due to snowmelt [32]. Based on its geological location and climatic characteristics, Lake Urmia and the basin encompass a high biodiversity and are one of the most critical ecological habitats in the Middle East. For instance, Lake Urmia is the largest habitat of brine shrimp *Artemia urmiana* [33], which is the primary food source for the migratory birds [34,35]. The islands in the lake are critical destinations for these migratory birds and, also, the shelter for rare species of mammals and reptiles [36,37]. The vibrant ecosystem led to Lake Urmia being entitled as a Wetland of International Importance by the Ramsar Convention in 1971 and a Biosphere Reserve by UNESCO in 1976 [38].

However, recent studies have shown that Lake Urmia has been experiencing dramatic water depletion in recent decades, mainly due to intensive anthropogenic activities [39–41]. The variation of its surface area is also significant, as shown in Figure 1a. The construction of dams and causeways, as well as the extraction of groundwater for agricultural irrigation, are responsible for a large part of water resource losses [39,42,43]. Together with the fact that Lake Urmia is a terminal lake with a maximum depth of only 16 m, it is even more vulnerable to evaporation [44]. Thus, the salinity has reached a dangerous level for species [42], and the water equilibrium has fallen to a new low standard [43]. According to Abbaspour et al. [45], the lake might eventually dry up within a decade. The lake desiccation endangers not only the natural environment and habitat but, also, the nearby population of seven million inhabitants in both economic, as well as health, aspects [46]. For instance, the depleted lake bed would reveal dissolved salts crusts [47], which would be exposed to the wind and cause salt storms to the surrounding residential areas, the shrinking size of the lake would reduce its function as a mediator in the extreme climate [37], and the dehydrated bed would also lead to species migration [36]. Moreover, the shortage of water resources might also result in a political crisis [48].

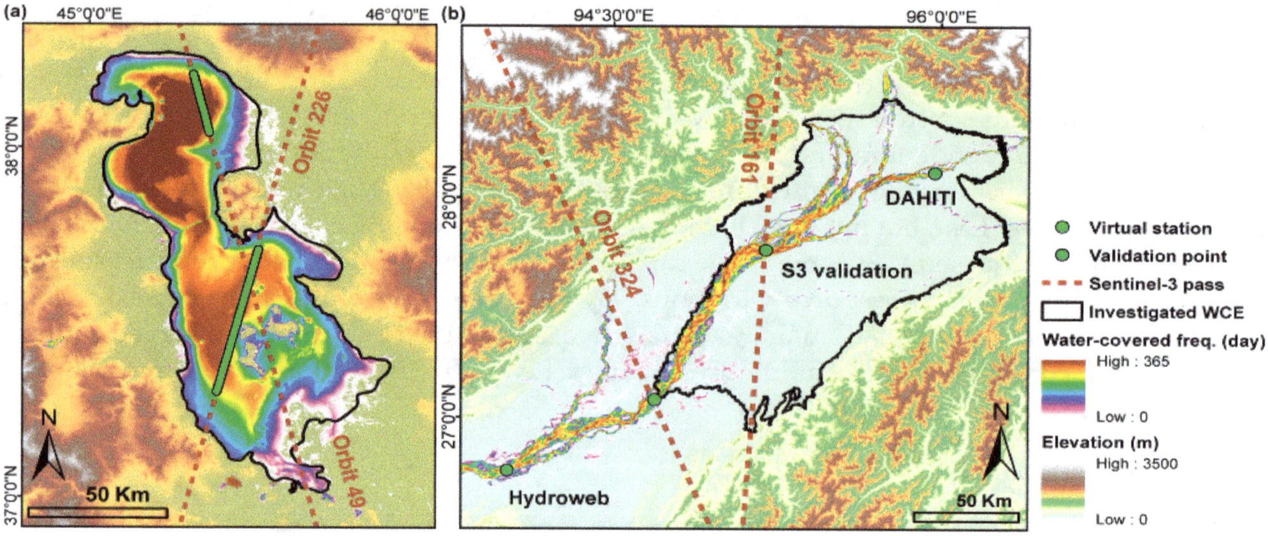

Figure 1. The locations of the study areas with the Sentinel-3 altimetry satellite pass ground tracks (dotted red lines) and the corresponding virtual stations (VSs) (green points). The investigated water-covered extent (WCE) (black polygon) and the average water-covered frequency of 2003–2018 derived from the Global WaterPack (GWP) are illustrated for (**a**) Lake Urmia and (**b**) Brahmaputra River.

2.1.2. Brahmaputra River

The Brahmaputra River originates in the Himalayan mountain range, which is mainly fed by snowmelt water and is the third-largest river in the world by discharge [49–51]. It is a trans-boundary river; by order, it passes by China, India, Bangladesh, and eventually merges with the Ganges river and flows into the Bay of Bengal. It is called by different names in each country, including Yarlung Zangbo, Brahmaputra, and Jamuna in China, India, and Bangladesh, respectively. As the youngest major river among the world [50], the Brahmaputra River has at least three characteristics: abundant tributaries, highly dynamic fluvial activities, and a varying river width, as shown in Figure 1b. Since the sediments of the Brahmaputra riverbed are composed of medium-to-fine sand and silt that is uniformly graded and has poor transport resistance [50], the riverbed and banks area are considerably mobile, which leads to frequent morphological changes affected by fluvial processes [50,52,53]. Therefore, river braiding, division, and shifting often happen [54]. However, although it is one of the most critical water resources in Asia, there is barely any in-situ river measurement available publicly. Due to the imminent danger of flooding, such in-situ data is regarded as classified information, and even the Global Runoff Data Center (GRDC) keeps no recent observations [55,56].

2.2. Data

2.2.1. Sentinel-1 SAR

The Sentinel-1 mission consists of two-satellites, with Sentinel-1A/B launched in April 2014 and April 2016, respectively. Its advanced terrain observation with progressive scans in the azimuth (TOPS) image acquisition technique enables a high signal-to-noise (SNR) ratio in the long-track direction without a scalloping effect while maintaining a wide coverage like conventional ScanSAR mode [57]. As in the present study, we only utilized the intensity information, the Level-1 Ground Range Detected (GRD) product, for which all the sub-swath have already been merged and de-bursting was selected. To increase the efficiency of the image analysis, we processed the images on Google Earth Engine (GEE). It provides the archive of Sentinel-1 GRD scenes and stores them in decibel (dB) units. Further calibrations, such as the orbit file application, noise removal, radiometric calibration, and terrain corrections, are processed as well. Furthermore, this cloud-based geospatial processing platform enables user-defined processing, which largely accelerates the time series SAR data analysis in a wide spatial scale [58].

2.2.2. Sentinel-3 Altimetry

Similar to Sentinel-1, Sentinel-3 is also composed of two satellites, with Sentinel-3A/B launched in February 2016 and April 2018, respectively. When it comes to inland water monitoring, compared to previous spaceborne altimetry sensors, the biggest advantage provided by the dual-frequency SAR radar altimeter (SRAL) instrument aboard Sentinel-3 is the much finer along-track resolution. Inherited from CryoSat, Sentinel-3 uses the high pulse repetition frequency (PRF) SAR mode (or delay Doppler) processing, which greatly improves the along-track resolution of the commonly used low-resolution mode (LRM) altimeters. It significantly increases the SNR ratio and, thus, enables the detection of much smaller targets [59]. Moreover, the echo reception window positioning method is also upgraded. Thanks to these improvements, Sentinel-3 is capable of monitoring finer inland waterbodies. To process the altimetry data via the European Space Agency (ESA) online service, Level-1 Non-Time Critical (NTC) products are selected.

2.2.3. Auxiliary Data

To include hydrological factors, including rainfall and evapotranspiration, the latest European Centre for Medium-Range Weather Forecasts (ECMWF) ERA5-Land data are used. This newly released reanalysis dataset provides global meteorological variables in a much higher spatial resolution (nine kilometers) compared to previous ERA-Interim (79 km) and ERA5 (31 km) datasets. Currently, it covers data from 1981 to the present, and eventually, it would cover the same period as ERA, i.e., 1950 to near real time.

For the examination of our water level results, the data provided by DAHITI, G-REALM, and Hydroweb are used. To compare our water-covered extent estimations, the Global WaterPack (GWP) was selected. It utilizes daily Moderate Resolution Imaging Spectroradiometer (MODIS) reflectance data at a 250-m spatial resolution and dynamically decides the threshold values, as well as the temporal interpolation techniques, to achieve the daily global cloud-free water/no water classification [15].

3. Methodology

3.1. Water Level Calculation with Sentinel-3 Altimetry

To derive the water level from Sentinel-3 altimetry data, we utilize the ESA's SAR Versatile Altimetric Toolkit for Ocean Research & Exploitation (SARvatore) service. It is based on the G-POD (Grid Processing on Demand) distributed platform, which guarantees high-speed processing and timely delivery. Another highlight of the SARvatore service is that it provides the inland water-customized processing configuration, which utilizes the higher posting rate (HPR) of 80-Hz data and processes with the new SAMOSA+ analytical retracker algorithm [60,61]. By utilizing this advanced setting mode, we can reduce the separation between two observations in along-track from the original 300 m to 80 m and, thus, provide more valid points over the targeted waterbody. The output files include not only L2 data in NetCDF format but a KML file containing the satellite pass ground track location. The KML file facilitates the identification of the interested regions, as shown in Figure 1, i.e., virtual station (VS) (the intersection of the altimetry's ground track and the waterbody), in the following analysis of the Multi-Mission Radar Altimetry Toolbox (BRAT). It must be noted that, because both of our study targets are characterized by significant seasonal WCE change, we first utilized the GWP to identify the subregion of a waterbody having (near) permanent water and, then, analyzed the altimetry data of that region to minimize the signal pollution. This step largely enhanced the accuracy by avoiding taking account river/lakebed-caused signals in the dry season. To derive the unbiased orthometric height over the inland waterbody H_{ORTHO}, i.e., the water level referred to the geoid, the following correction equation was used:

$$H_{ORTHO} = H_{SAT} - R_{OBS} - (\Delta R_{WTC} + \Delta R_{DTC} + \Delta R_{IONO} + \Delta R_{ET} + \Delta R_{PT}) - N_{GEOID} \quad (1)$$

where H_{SAT} is the satellite altitude above the reference geoid, R_{OBS} is the observed range, $\Delta R_{DTC}/\Delta R_{WTC}$ is the dry/wet tropospheric correction, ΔR_{IONO} is the ionospheric correction, $\Delta R_{ET}/\Delta R_{PT}$ is the earth tide and pole tide, respectively, and N_{GEOID} is the geoid height. For a full explanation of each correction, refer to [59]. Based on this formula, both propagation and geophysical biases can be compensated. Other geophysical terms, including lake tides, hydrostatic variations, thermal expansion, and wind piling-up effects are neglected, as suggested by a previous study [25]. For detailed information of each correction, refer to [59,62]. Finally, based on the corrected range, we calculated the average and standard deviation of the water level over the VS.

3.2. Water-Covered Extent (WCE) Calculation with Sentinel-1 SAR

To map the water-covered extent, i.e., the lake's surface area or river's channels and flooded area, the sensitivity of the SAR data to the surface wetness was utilized. As the backscatter coefficient would be much lower in wet and smooth surface compared to dry ground, it is viable to depict the water surfaces. Since our goal is to maximize the differentiation of water/nonwater regions, i.e., a binary classification, the task can be separated into two parts: (1) selection of the targeted image and (2) determination of the threshold. For the first point, based on the previous studies [63,64] and our testing, the value difference between waterbody and the land region is more significant in VV polarization (co-polarization) than VH (cross-polarization), so VV imagery was selected. Furthermore, we used Otsu's algorithm to decide the threshold [65]. It is favored for its simplicity and suitability for SAR image-based binary classification [63,66]. Practically, the critical point is to ensure that the manually selected samples include an equal number of water/nonwater pixels, as Otsu's algorithm can only handle bimodal distribution. However, sometimes, it is difficult to distinguish lake/river boundaries visually in some flood-caused vague scenes (as discussed in Discussions 5.1); we only used scenes having clear waterbody boundaries for training the Otsu's algorithm.

To define the targeted WCE for investigation, for Lake Urmia, we used the waterbody boundary provided by the Global Lakes and Wetlands Database (GLWD) (https://www.worldwildlife.org/pages/

global-lakes-and-wetlands-database) [67]. For the Brahmaputra River, we targeted the part of the watershed of the VS where the terrain was a flat plain (elevation lower than 200 m) and frequent river braiding and shifting events happened, as illustrated in Figure 1.

3.3. Lake Water Volume Estimation by Hypsometry Calculation and Detrended Volume Retrieval

To estimate the water volume variation of a waterbody, it is common to use either the simplest truncated pyramid model or the power-function model; however, they are only suitable for waterbodies characterized by a regular morphology (such as reservoirs) or a bowl-shaped morphology, respectively [36,68]. We utilized a universally applicable approach by modeling the water volume-water level relationship from the observed surface area-water level relationship [18]. The processing steps included: (1) calculation of the water level difference (or water depth) relative to the minimum water level in the sensing period ($\Delta L = L - L_{min}$); (2) establishing a scatter plot of surface area-water level difference and modeling it with a polynomial function (typically, the second, third, or quadratic order is chosen) ($A = f(\Delta L)$); (3) integration of the surface area-water level difference function to derive the water volume-water level difference function ($V = \int_{\Delta L_0}^{\Delta L} A d\Delta L$); and (4) the ingestion of water level observations to the resultant function to derive the corresponding water volume. Details about these steps can be found in [18,69]. In short, by using this approach, we can estimate the "dynamic" water volume above the "static" water volume, i.e., the water volume of the time when the water level is the lowest for any waterbody having a horizontal water surface [70].

Nevertheless, it must be noted that, compared to any single point of the river, the lake is a "container" that can store the water over time instead of merely letting the water flow by. Namely, the results of each previous hydrological year's water budget balance (increase with snowmelt and rainfall and decrease with evapotranspiration) would affect the current "stock" of the water volume. Hence, if we directly use the original time series water volume as the interannual variation, the intra-annual "container effect" could lead to a misinterpretation. Therefore, in the present study, we employed the Seasonal-trend decomposition procedures based on Local Regression (Loess) (STL) [71] to remove the intra-annual trend. This approach is suggested by previous studies [72–74]. STL is an iterative nonparametric filtering procedure that uses repeated Loess (Local Regression) smoothing. The strength of STL is its robustness and computational efficiency, as well as the capacity to depict a nonlinear pattern in time series data. For details about STL, refer to [71,75]. Practically based on the inner and outer loops with Loess processing, STL can decompose time series data into three components: a low-frequency long-term trend, a high-frequency seasonal variation, and residuals (or remainder). After processing STL, we subtracted the decomposed long-term trend from the original time series water volume to derive the detrended water volume, which represents the real inter-annual water volume variation.

3.4. WSCE Mapping with Sentinel-1 SAR and Hydrological Factors Areal Calculation

Based on the finding that the backscatter coefficient of the SAR signal decreases significantly when the snowpack starts to melt and, thus, increases the containing liquid water [9], Nagler and Rott [10] proposed a ratio-thresholding approach in 2000. It utilizes two SAR scenes (one is a wet snow-covered period image, and the other is a referenced snow-free image) and calculates their ratio image, which is then thresholded using a fixed value to derive the binary WSCE. For a detailed description of the processing steps and value setting, refer to [9,76,77].

Since the final regression analysis is watershed-based, all hydrological factors, including WSCE, rainfall, and evapotranspiration, need to be converted. Firstly, we delineated the watershed based on the Shuttle Radar Topography Mission Digital Elevation Model (SRTM DEM). For the Brahmaputra River, the VS where the Sentinel-3 track passes was used as the pour point, and the watershed of that point of that tributary was delineated. Based on the watershed, the WSCE was ratioed to derive the WSCE%; the rainfall and evapotranspiration were summed up to derive the mean amount.

3.5. Hydrological Econometrics Regression Analysis

To quantify the influence of the investigated hydrological factors (WSCE%, mean rainfall, and mean evapotranspiration) on the river/lake water amount, the utilization of a regression analysis is an optimized approach. However, some factors, especially the snowmelt, would pose a lagged effect on the downstream water level/volume as the snowmelt water would not only form a direct surface streamflow but, also, infiltrate into the ground and soil [78–80]. Thus, a conventional simple regression is not viable. Consequently, we employed the advanced distributed lag model (DLM) (or finite distributed model (FDL)) in the present study for handling multiple variables' dynamic influences with lags. DLM was firstly developed in the econometrics discipline [81] and was recently widely applied in biomedical or environmental exposure-caused mortality studies [82,83]. Yet, DLM is barely utilized in the remote sensing and cryo-hydrology domains. DLM is a dynamic model that assumes that the effect of each independent factor on the dependent variable spreads over some time instead of a single time point. To quantify it, it defines the temporal factor by assigning a lag dimension, which is an equally spaced and ordered time point series [84] (in the present study, we defined the minimum unit of the lag dimension as ten days). The lag dimension enables the effect of a single exposure event to be distributed over a specific period of time; therefore, the holistic understanding of the cross-temporal exposure-lag-response relationship can be revealed [84–86]. The multivariable DLM can be written in the formula:

$$y_t = \alpha + \sum_{s=0}^{q1} \beta 1_s x1_{t-s} + \sum_{s=0}^{q2} \beta 2_s x2_{t-s} + \cdots + \sum_{s=0}^{qN} \beta N_s xN_{t-s} + u_t \tag{2}$$

where α is the intercept, q and β are the lag length and lag coefficient (or lag weight or short-run multiplier) of each independent variable, respectively, and u_t is the error term. The lag coefficient is solved by the ordinary least squares (OLS) technique and represents the expected change in y_t stems from the change of xN_{t-s} by one unit, holding constant the other independent variables [87]. The sum of each independent variable's lag coefficient, i.e., $\sum_{s=0}^{qN} \beta N_s$, is called the long-run multiplier or long-run propensity (LRP), which is the cumulative effect of xN on y [88]. In the present study, we focused on the LRP value to quantify the causal effect of each factor on the water resource variations.

Practically, the most critical issue when implementing DLM is selecting a suitable lag length for each independent variable. As in the present study, we included three variables (WSCE%, mean rainfall, and mean evapotranspiration), and based on the background knowledge, it was assumed that their lag lengths should be different; therefore, it is impractical to use the conventional length decision method, i.e., successively adding/reducing lags. Previous studies utilized the prior knowledge to define the plausible lag lengths when multi-regressors were included [89,90]. However, in our study, the lagged effect of the same hydrological factor on the water amount varied from place to place as it was affected by the distance and the regional characteristics. Thus, there is no universal predefined lag length estimation available. Instead, we applied the following processing steps to decide the proper lag length of each hydrological factor: (1) building of the regression of the water amount (water level for the river and water volume for the lake) with each hydrological factor individually with different lag lengths and recording the resultant adjusted R^2 ($\overline{R}^2$) value and the direction of the coefficient (positive or negative) and (2) selection of the final lag length (starting and ending lags) of each hydrological factor based on two criteria: (a) having the coefficient direction fitting the knowledge (the coefficient of WSCE% and rainfall should be positive, while the coefficient of evapotranspiration should be negative, because more snowmelt and rainfall would increase the downstream water amount, while more evapotranspiration would reduce the water supply) and (b) having the ascending value of $\overline{R}^2$ (until the highest value). The theory behind our two-step approach is that the use of the $\overline{R}^2$ value helps to identify the appropriate set of regressors that can explain the variations of the dependent variable well, and only selecting the lags having ascending $\overline{R}^2$ values avoids selecting an unnecessarily long lag length.

Based on the techniques and processing steps mentioned above, by employing Sentinel-1 images, as well as Sentinel-3 data, together with a DEM and ERA5-Land records, we can quantify the causal effect of snowmelt conditions on any large-scale inland waterbody. The overall workflow of our study is illustrated in Figure 2.

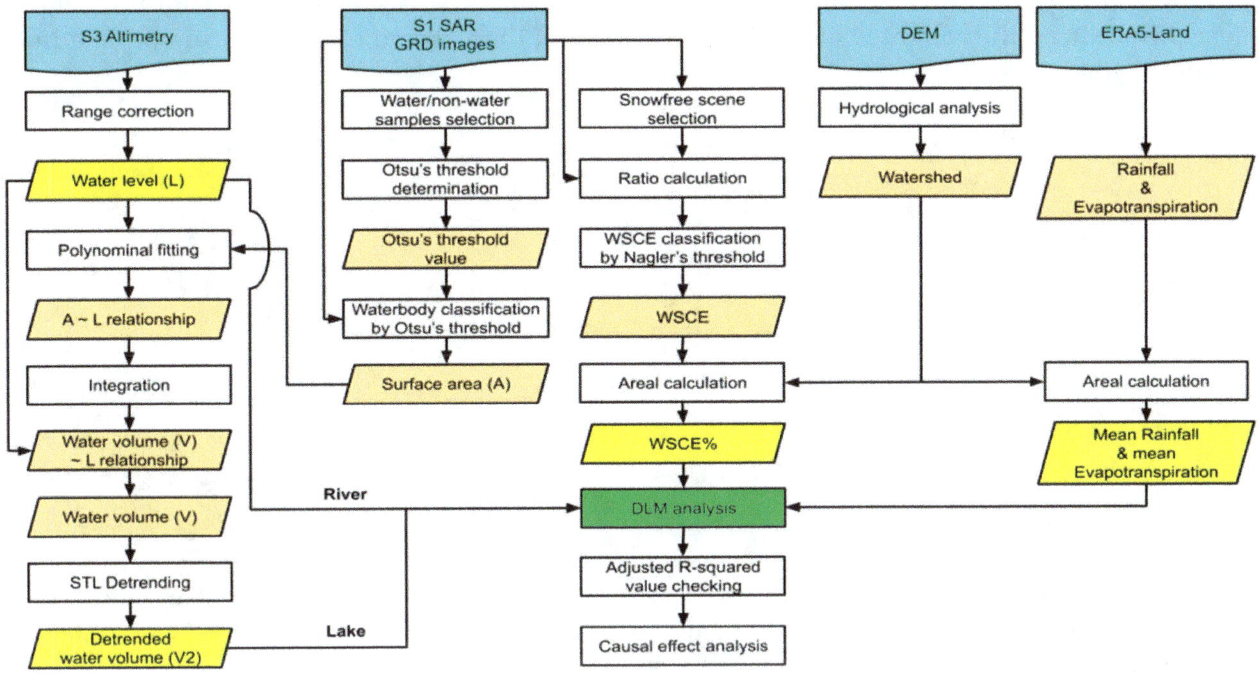

Figure 2. The overall workflow of the present study, including the processing of the Sentinel-1 synthetic aperture radar (SAR) and Sentinel-3 altimetry data for deriving the wet snow-covered extent (WSCE), surface area, and water level. The water volume estimated from polynomial fitting and integration is detrended by the Seasonal-trend decomposition procedures based on Loess (STL). Together with the hydrological factors derived from the ERA5-Land dataset, the distributed lag model (DLM) is eventually conducted. GRD: Ground Range Detected. The digital elevation model is abbreviated as DEM.

4. Results

4.1. Lake Urmia

4.1.1. Water Level Retrieval

Two passes of Sentinel-3 over Lake Urmia are processed via SARvatore to estimate the time series average water level of VSs. To examine the results, the water level records provided by altimetry databases are plotted together in Figure 3, with the error bar representing the standard deviation of the water level among the virtual stations (VSs). As our goal is to compare their long-term trends and short-term noisy levels, and because different altimeter sensors have different instrumental biases [91], in this figure, we show the water level variation relative to each record's mean water level during our sensing period. Firstly, we can observe that there is a high consistency between the two passes' results (the locations of two passes are illustrated in Figure 1). Most of the time, the differences of their values are within one standard deviation. It is reasonable that two passes' results still have some differences owing to the reasons including (1) different wind-caused lake surface wave conditions, as their sensing times are not identical, (2) varying water depths on the altimetry signal penetration depth, and (3) and the different surrounding topography-caused signal pollution conditions. Nevertheless, the internal consistency of each pass and their cross-consistency remain high when comparing to DAHITI and G-REALM's records. In Figure 3, it is evident that both of our two passes' raw results have a much smoother trend compared to the smoothed DAHITI result (no raw data is provided by the DAHITI database; thus, no uncertainty is plotted) and the raw G-REALM results, as these databases' results show more short-term high-frequency fluctuations. The reason is that the along-track resolution of the HPR Sentinel-3 (80 m) we employed is much higher than the Jason-3 altimetry sensor (>2000 m), which is used for DAHITI and G-REALM. Hence, based on these comparisons, it is confirmed that our Sentinel-3 results generally have a higher reliability. To fuse two passes' observations for the following analysis, we utilized the weighted average calculation and then used the Gaussian smooth method to filter out the noises.

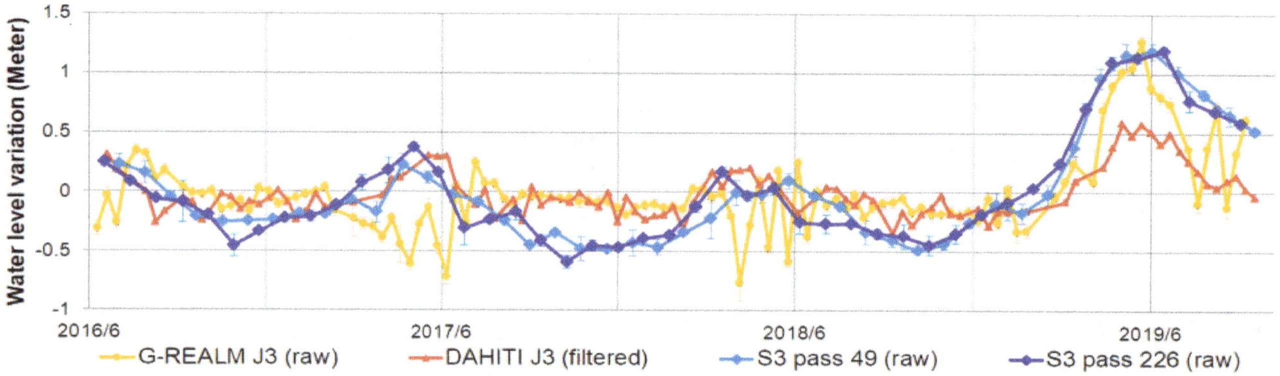

Figure 3. The time series water level variation of Lake Urmia estimated with Sentinel-3. The internal cross-comparison with different Sentinel-3 passes and the external comparison with altimetry databases, including DAHITI and Hydroweb (based on Jason-3 altimetry), are illustrated. The error bars represent the standard deviation of the water level among the virtual stations (VSs). Note: each of the variations is relative to each record's average height in the studying period.

4.1.2. Surface Area Estimation

To define the optimized Otsu's threshold value for mapping the surface area of Lake Urmia, we manually selected water/nonwater samples on each VV polarization scene, which showed a clear waterbody boundary. With the aim to densify the usable scenes, both ascending and descending Sentinel-1 image sets were used and processed individually. Their overall backscatter coefficient distributions and the resultant Otsu's threshold values are shown in Figure 4. Firstly, we can observe that both the flight direction image sets' sample intensity distributions are bimodal, which confirms the suitability of the Otsu's method. Secondly, it is found that both sets have nearly the same threshold values; therefore, we average two values to select a fixed value, i.e., −20.8 dB, to be the waterbody classification threshold for Lake Urmia.

Based on the defined threshold value, we can classify the water-covered area of Lake Urmia, i.e., surface area, from both stacks of ascending and descending Sentinel-1 images. For cross- and external comparisons, we plotted the same date's surface areas estimated with ascending/descending Sentinel-1 images and the GWP product in Figure 5. It is evident that both of our ascending and descending results match perfectly with GWP with an R-square value of around 0.98. These high values prove that the surface area classified with our SAR-Otsu's approach is highly reliable. Additionally, it is also interesting to realize that there is a systematic offset between our SAR-based estimation and the GWP product, which is based on a coarse optical sensor. Based on a further analysis, we can confirm that it is because of an underestimation of the GWP product due to a high reflectance of lake soil when the water level is low, as well as due to a mixed pixel effect of the coarser spatial resolution (see Discussion 5.1). Thus, the reliability of our SAR Otsu's approach for accurately delineating the surface area of Lake Urmia is ensured and the time series surface area dynamics of Lake Urmia can be mapped. It is obvious that the change of the surface area of Lake Urmia is dramatic. In the sensing period, the minimum and maximum surface areas are around 1126.86 (2017/10/31) and 3569.77 (2019/7/11) km^2. The areal difference is significant, as the maximum size is almost equal to 3.2 times the minimum surface area.

Flight direction	Ascending	Descending
Otsu's threshold value (dB)	−20.678	−20.962
Distribution of backscatter coefficient	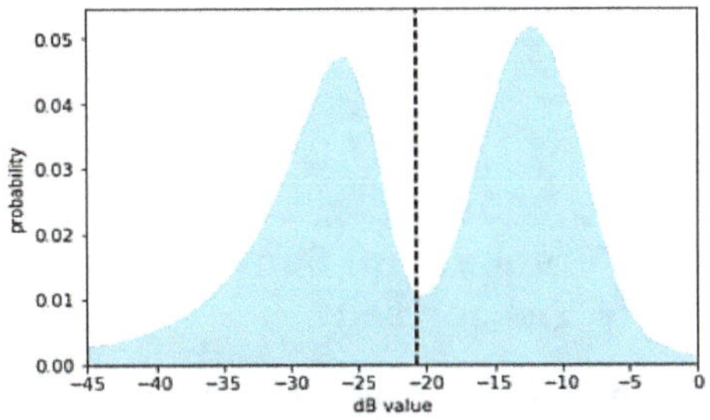	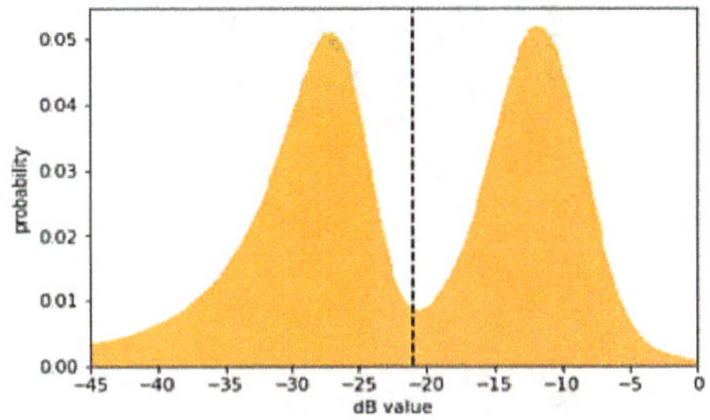

Figure 4. The distributions of the backscatter coefficient and the decided Otsu's threshold values of the ascending and descending Sentinel-1 SAR imagery for the Lake Urmia waterbody classification.

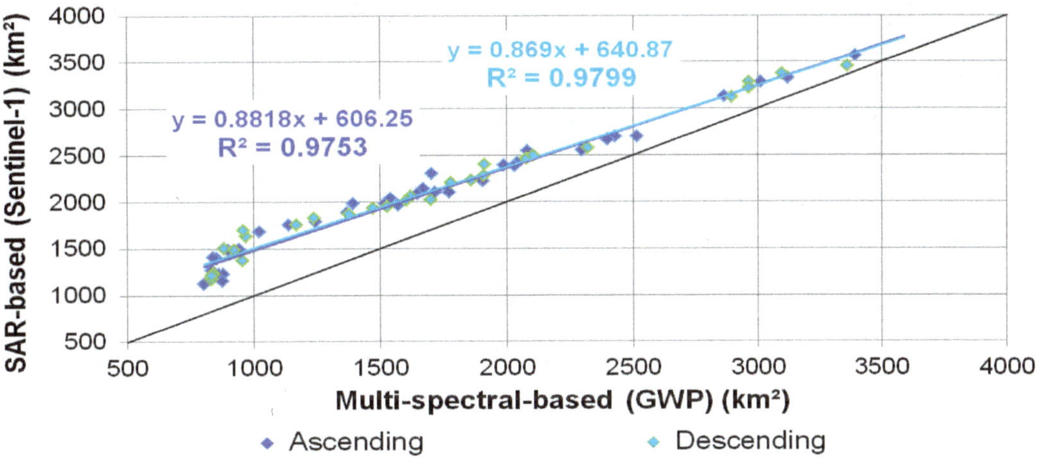

Figure 5. The internal cross-comparison of Lake Urmia's surface area detected by ascending and descending imagery of Sentinel-1, and the external comparison with the GWP product.

4.1.3. Hypsometry Estimation and Detrended Volume Retrieval

To derive the time series water volume, the hypsometry estimated from the water level-surface area relationship is necessary. Hence, we first plot the time series water level and surface area (derived from both ascending and descending Sentinel-1 SAR as well as the GWP) in Figure 6. A third-order polynomial function is used for fitting as it provides the best R-squared value. Moreover, previous in-situ surveying [47] and study [36] also suggest the utilization of a third-order function for Lake Urmia's hypsometry. It is found that both of our Sentinel-1 SAR ascending (0.98) and descending (0.99) show a perfect R-squared value, while the GWP (0.94) shows a slightly poorer matching. These values suggest that both our Sentinel-3 altimetry-based water level and Sentinel-1 SAR-Otsu' approach-based surface area are reliable as they share a high consistency. The slightly lower R-square value of GWP also agrees with the previous finding of the under-estimation of GWP. Eventually, to densify the available observations, we integrate both ascending and descending SAR stacks-derived surface area to build a surface area-water level relationship, which can be written as:

$$A = 1164.4 \times dL^3 - 3459.1 \times dL^2 + 4267.1 \times dL + 1102.5 \quad (3)$$

where A and dL represents the surface area and water level difference, respectively. It yields a high fitting R-squared value around 0.99.

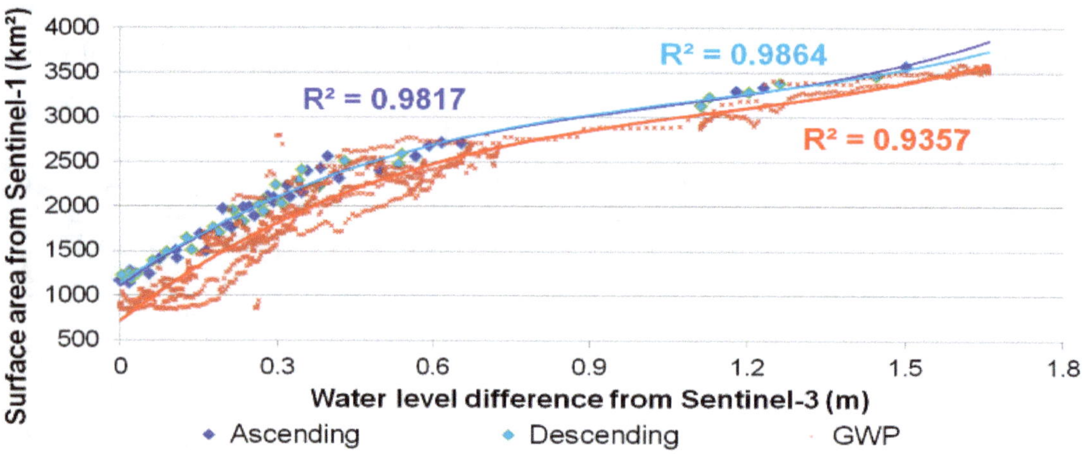

Figure 6. The hypsometry of Lake Urmia estimated by fitting the surface area detected by ascending and descending Sentinel-1 SAR imagery and the Global WaterPack (GWP) product with the water level estimated by Sentinel-3 altimetry using third-order polynomial functions.

Based on the fact that the dynamic water volume should be equal to zero when the water depth (water level difference relative to the lowest water level during the sensing period) is zero, we can derive the fourth-order water volume-water level function by integrating the surface area-water level formula. Thus, Lake Urmia's water volume-water level function can be written as:

$$V = 291.10 \times dL^4 - 1153.03 \times dL^3 + 2133.55 \times dL^2 + 1102.50 \times dL \quad (4)$$

where V represents the dynamic water volume. Based on this equation, the time series water volume of Lake Urmia can be derived. Nevertheless, to mitigate the bias of the container effect of the lake for the following analysis, as described in the Section 3.3, the STL is then applied to estimate the detrended water volume, as plotted in Figure 7. It is clear to observe that the trend of the water volume was almost stable in the first two years, while a peak can be identified in 2019. By subtracting the trend from the estimated water volume, we can obtain the detrended water volume, which represents the real interannual water volume variation.

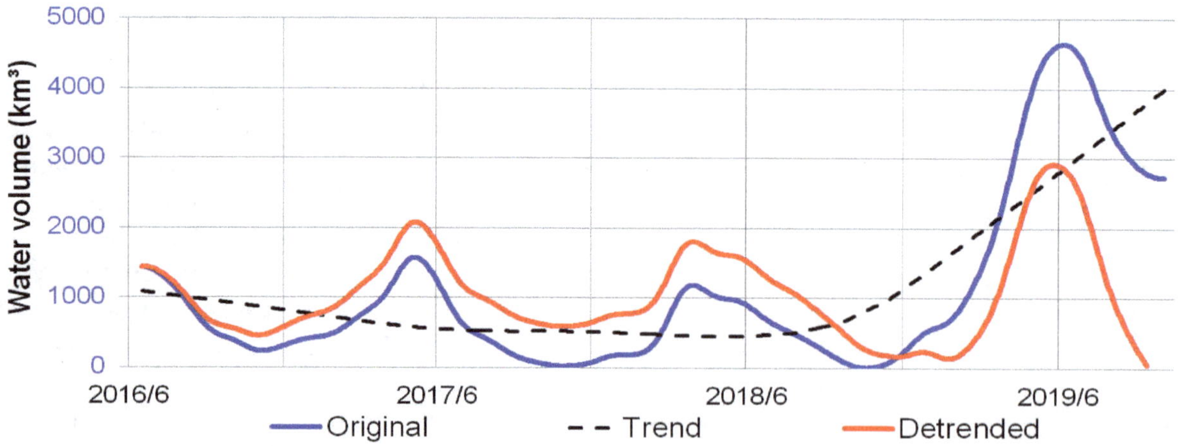

Figure 7. The detrended processing for Lake Urmia's time series water volume estimation using the Seasonal-trend decomposition procedures based on Loess (STL). The estimated trend value and the original/detrended water volume are illustrated with the black, blue, and red lines, respectively.

4.1.4. WSCE Mapping and Hydrological Factors Calculation

Both ascending and descending stacks of images of Sentinel-1 SAR are used to map the WSCE. The summertime imagery of 2018 is selected as the referenced snow-free image for calculating the ratio image. By applying the threshold value proposed by Nagler and Rott [9,10], the time series WSCE of each flight direction image stack can be estimated. To examine their consistency, we plotted their time series WSCE% relative to Lake Urmia's watershed in Figure 8. It is found that both flight directions' results show a high agreement, which proves the internal robustness of our approach. As there is no solid way to externally validate the WSCE in a comparable high resolution [9,76], this cross-track comparison is used for the examination.

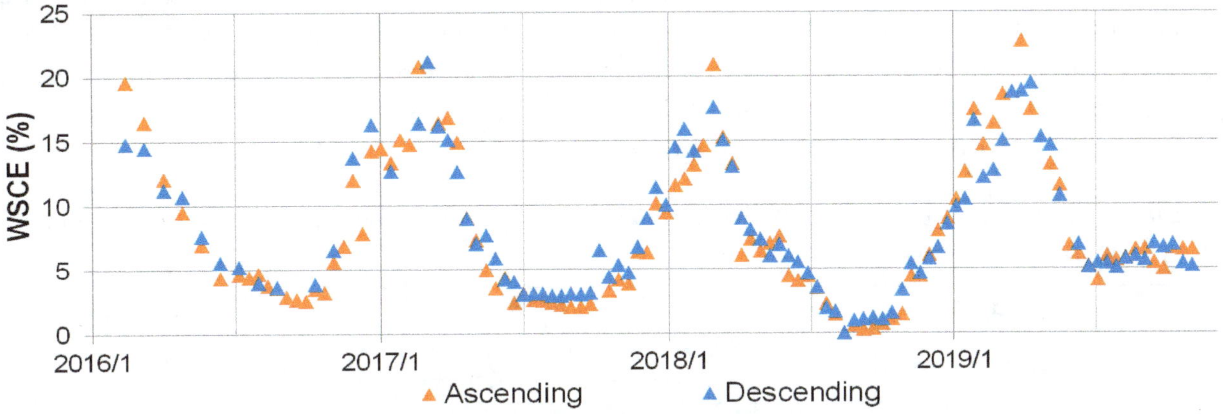

Figure 8. The estimation of the wet snow-covered extent (WSCE) percentage of the Lake Urmia watershed using ascending and descending Sentinel-1 SAR imagery.

Based on the previous data preparation, before processing the DLM analysis, we needed to convert all hydrological factors to the watershed-based unit, i.e., to derive the mean evapotranspiration/rainfall and WSCE%. Together with the estimated detrended water volume, we illustrated all the time series data in Figure 9 to observe their relationships. Firstly, it was found that excluding rainfall shows two peaks each year; other data, including WSCE%, evapotranspiration, and detrended water volume, have only one peak annually. Additionally, the pattern of the evapotranspiration is nearly identical each year. Thus, we can assume that the variation of water resources of Lake Urmia cannot be perfectly explained by rainfall and evapotranspiration data only. Secondly, by comparing the time of the peak of each data, it is found that, in the temporal aspect, the water volume always reaches each year's maximum value later than the WSCE%. To quantify these factors' lagged and overlapping influences on the water volume variation, a hydrological analysis based on DLM is required.

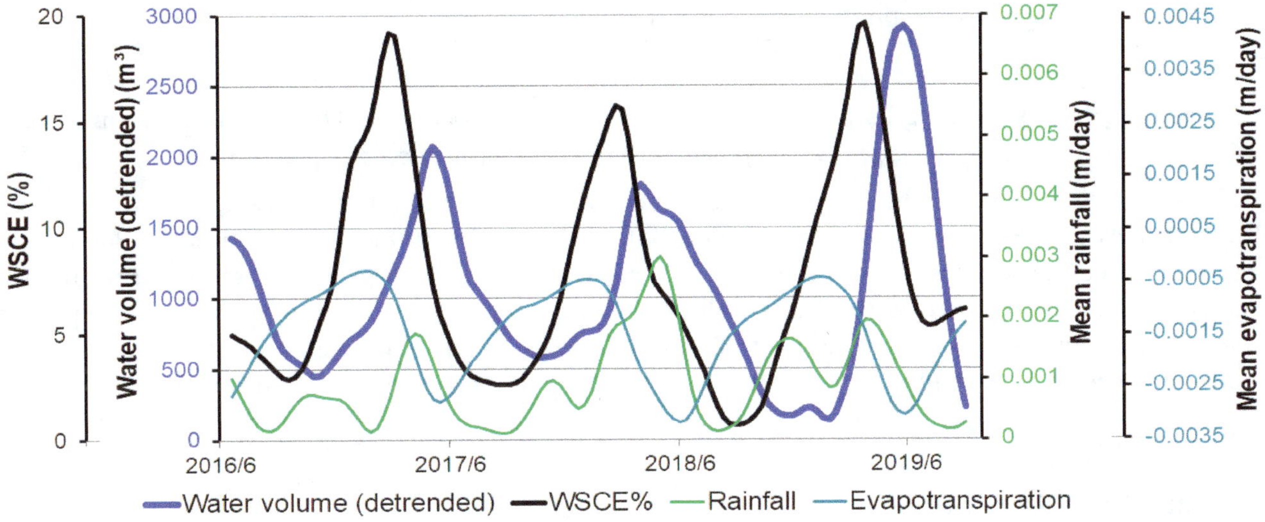

Figure 9. The time series detrended water volume of Lake Urmia and the WSCE%, as well as hydrological factors, including liquid rainfall and evapotranspiration.

4.1.5. DLM Hydrological Analysis

To process the DLM for quantifying the causal effects of hydrological factors on the detrended water volume variation, it is necessary to decide the proper lag length for each factor. Based on the steps and criteria mentioned in Section 3.5, we summarized the determined lag length in Table 1. Firstly, it was found that the WSCE% has the longest lag length of 80 days, rainfall has 30 days delay, and evapotranspiration shows no delay longer than ten days. As the directions of the coefficient of WSCE%, rainfall, and evapotranspiration are always positive, positive, and negative in different lag lengths testing, respectively, we selected the lag starting from day zero for all factors. Additionally, we can observe that the WSCE% has the highest $\overline{R}^2$ of 0.85, followed by 0.23 and 0.61 for rainfall and evapotranspiration, respectively. The high/low $\overline{R}^2$ values of the WSCE%/rainfall indicate that their influential magnitudes on the water volume differ considerably.

Table 1. The summaries of the distributed lag model (DLM) analysis of Lake Urmia and the Brahmaputra River. WSCE: wet snow cover extent.

Study Area	**Lake Urmia**		**Brahmaputra River**
Dependent Variable	**Water Volume (Original)**	**Water Volume (Detrended)**	**Water Level**
PRE-ANALYSIS (LAG LENGTH DECISION)			
Highest adjusted R^2 (individual)			
WSCE%	0.64	0.85	0.91
Mean rainfall	0.25	0.23	0.94
Mean evapotranspiration	0.22	0.61	0.90
Lag length (days)			
WSCE%	0–60	0–80	70–150
Mean rainfall	0–20	0–30	0–20
Mean evapotranspiration	0	0	0–10
DISTRIBUTED LAG MODEL (DLM) REGRESSION ANALYSIS			
Model adjusted R^2 (overall)	0.67	0.91	0.97
Long-run propensity (LRP)			
WSCE%	135.68	108.49	0.10
Mean rainfall	60,695.88	−88,737.6	149.186
Mean evapotranspiration	−76,581.40	−120,327.76	−446.02

Based on the decided lag lengths, the DLM can thus be processed. The results are summarized in Table 1. A significant regression equation is found ($F(14, 81) = 70.78$, $p < 2.2e^{-16}$) with an $\overline{R}^2$ of 0.91. The high $\overline{R}^2$ indicates that our model contains a clear explanation of the variations of the water volume. The overall F-test value of 70.78 proves that our hydrological factors with different lag lengths can reliably predict the water volume. Eventually, based on the LRP of WSCE% estimated from DLM, we can conclude the causal effect of snowmelt on the water resource as, when increasing one percent of the WSCE% during the period of zero to 80 days before, the detrended water volume of Lake Urmia would increase 108.5 cubic meters while holding constant other independent variables.

4.2. Brahmaputra River

4.2.1. Water Level Retrieval

The same Sentinel-3 altimetry data processing approach is employed to estimate the time series water level of the Brahmaputra River. Since the Brahmaputra River is a long river spreading across various geomorphologies, and the satellite pass ground track location of each spaceborne altimetry differs, there is no direct way to validate our results, as the water level differs in different river sections. Nevertheless, we considered that the water level of each location of the same river should share a similar pattern, with a higher correlation in shorter distance differences. Thus, in total, three VSs were utilized for comparison, including another Sentinel-3 pass (pass 161, processed by the same approach) and two Jason-3 passes (data provided by DAHITI and Hydroweb individually). Their locations and the time series water level results are illustrated in Figures 1 and 10, respectively. Firstly, it is clear that four time series data have highly similar trends and seasonality. Nevertheless, the magnitude of the water level variation of each location shows an ascending order from the upstream to downstream, i.e., around three (DAHITI), four (Sentinel-3 pass 161's results), seven (Sentinel-3 pass 324's result), and eight (Hydroweb) meters, respectively. The reason for the more significant water level change in the downstream may be due to the flatter terrain when compared to the upstream. In addition, as we observed in the Lake Urmia case, the DAHITI and Hydroweb databases contain more noises than our HPR Sentinel-3 results, while the comparable patterns remain.

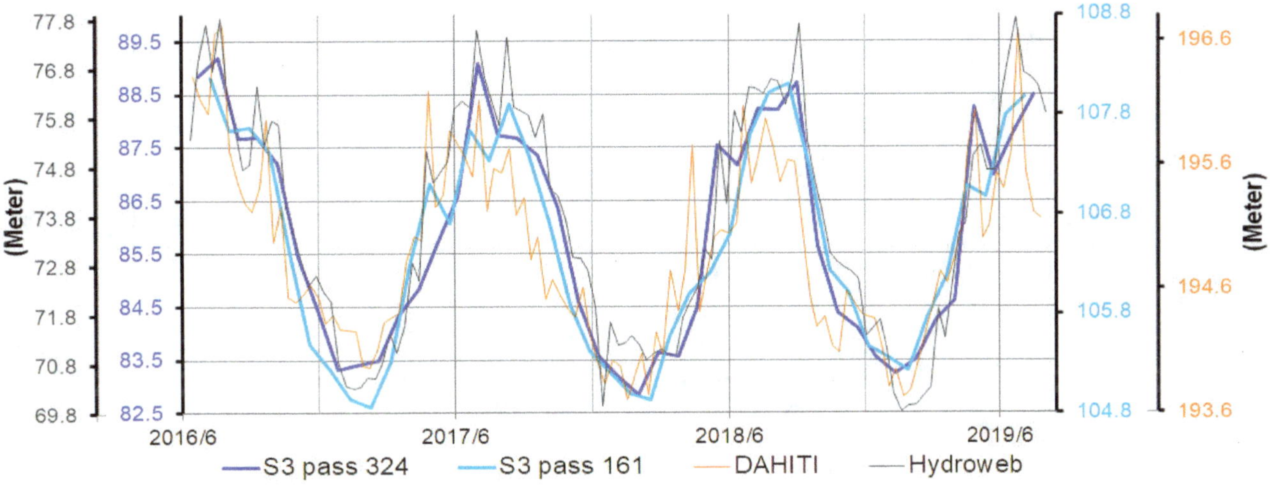

Figure 10. The time series water level of the Brahmaputra River estimated with Sentinel-3. The internal cross-comparison with different Sentinel-3 passes, and the external comparison with altimetry databases are illustrated. Note that the VS of each record is different, so the magnitude of water level variations differ, while the similar trends are still identifiable.

4.2.2. WSCE Mapping and Hydrological Factors Calculation

Both Sentinel-1 ascending and descending image stacks are used to calculate the WSCE of the Brahmaputra River. The summertime image of 2018 was selected as the referenced snow-free image.

To investigate the internal consistency, the WSCE% derived from each stack is plotted in Figure 11. Similar to Lake Urmia's case, a high agreement between both flight directions is observed. It suggests that the quality of our WSCE results should be trustworthy.

We also analyzed the relationship between the hydrological factors and water level before the DLM processing. Their patterns during the sensing period are illustrated in Figure 12. Compared with the Lake Urmia case, the Brahmaputra River shows some remarkable differences. Firstly, it is observed that all four parameters have only one peak per year. Secondly, the temporal delay between the peak of the water level and WSCE% is longer than the time delay of the water level and rainfall. Based on these two findings, we can assume the importance of the rainfall to water amount variation is higher in the Brahmaputra River than Lake Urmia. Nevertheless, since all hydrological factors and water levels show comparable seasonal patterns, the employment of DLM is indispensable to analyze their lagged causal effects on the water level.

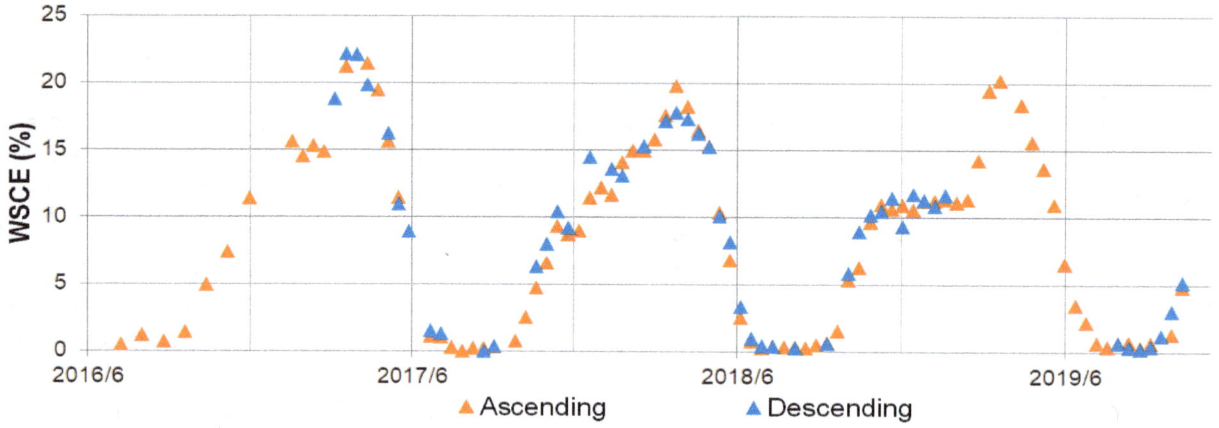

Figure 11. The estimation of the wet snow-covered extent (WSCE) percentage of the Brahmaputra River watershed using ascending and descending Sentinel-1 SAR imagery.

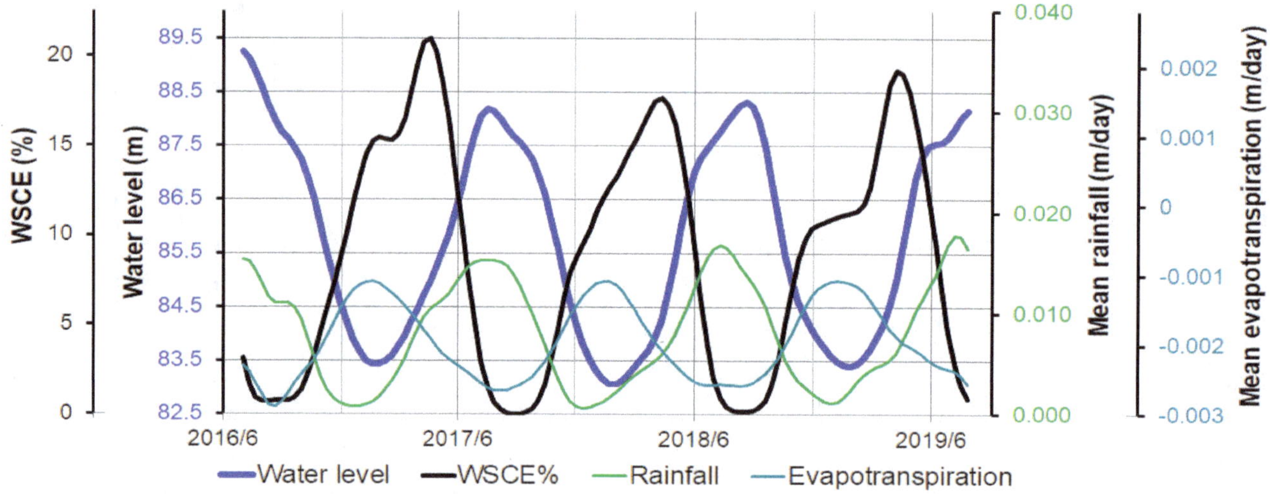

Figure 12. The time series water level of the Brahmaputra River and the WSCE%, as well as the hydrological factors, including liquid rainfall and evapotranspiration.

4.2.3. DLM Hydrological Analysis

To decide the suitable lag length for each hydrological factor, we built regressions with each factor individually, and the resultant lag length is summarized in Table 1. Firstly, like Lake Urmia, WSCE% has the longest lag length, and evapotranspiration has the shortest lag length. However, it is found that the lag length of WSCE% is nearly two times longer than for Lake Urmia. Another critical difference is that the starting lag length of WSCE% is not zero days. It is because, in the period of lag length between zero to 60 days, the direction of the WSCE%'s coefficient is negative, which violates the background knowledge. Thus, we only selected the lag length from 70 to 150 days (based on the criteria described in Methodology Section 3.5). Therefore, using the prior knowledge to aid the decision of the lag length is critical. Secondly, when comparing the $\overline{R}^2$ values, all three hydrological factors show comparable results, i.e., higher than 0.90.

The DLM can then be processed based on the decided lag lengths of each factor; the results are summarized in Table 1. A significant regression equation is found (F(14, 73) = 198.7, $p < 2.2e^{-16}$) with a $\overline{R}^2$ of 0.97. The high $\overline{R}^2$ indicates that the variation of the water level can be perfectly explained by our model, and the F-test value of 198.7 suggests that hydrological factors with different lag lengths can reliably predict the water level. Based on these examinations, we can conclude our DLM is robust. Hence, on the basis of the estimated LRP of WSCE%, the causal effect of snowmelt to the water resource can be concluded as, when increasing one percent of the WSCE% during the period of 70 to 150 days before, the water level of our VS of the Brahmaputra River would increase 0.1 m while holding constant other independent variables.

4.2.4. Flooded Area Estimation

The SAR-Otsu's approach is also utilized in the Brahmaputra River to delineate the flooding area (or the change of river channels). Both ascending and descending image stacks of Sentinel-1 are processed individually, with their backscatter coefficient distributions and the resultant Otsu's threshold values shown in Figure 13. Like in Lake Urmia's case, we can find comparable Otsu's threshold values. Hence, we average two values to select −16.5 dB as the fixed threshold for the waterbody classification.

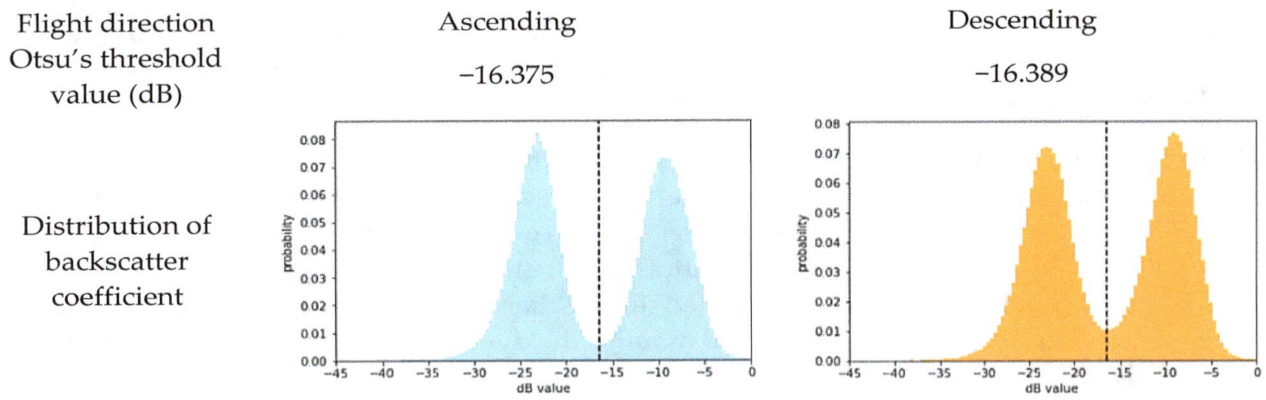

Figure 13. The distributions of the backscatter coefficient and the decided Otsu's threshold values of the ascending and descending of Sentinel-1 SAR imagery for mapping the flooded area of the Brahmaputra River.

For Lake Urmia, we combined the surface area with the water level to derive the hypsometry; on the contrary, for the Brahmaputra River, we can use the water level as the external data to examine our mapped flooded area. It is based on the fact that, when the flooded area is enlarging, the water level will rise simultaneously. Thus, we plotted the time series flooded area, water level, and WSCE% in Figure 14. Firstly, it is found that, generally, the trends of the flooded area derived from both ascending and descending Sentinel-1 SAR imagery stacks are agreeable, although internal noises and the differences between them are also clear. Moreover, their noisy levels are higher when compared to the same descending/ascending stack-based WSCE%, as illustrated in Figure 11. We consider it is because (1) the short-term variation of the flooded area is more abrupt and dynamic compared to the long-term gradient snowmelt process, and (2) the WSCE% mapping is by thresholding the ratio image relative to a referenced image (as described in Methodology Section 3.4), while the flooded area is mapped by directly applying the Otsu's threshold to each imagery, so the influence of the local incidence angle variation of each image is severer. Hence, to reduce the fluctuations, we average the ascending and descending stacks' results to derive a smoother trend. Secondly, by comparing the average flooded area with the water level, we can observe a temporally highly correlated pattern. It agrees with our prior knowledge and, also, indirectly proves the credibility of our Sentinel-1 SAR-based flooded area estimations, as well as Sentinel-3 altimetry-based water level records. Finally, the synchronized flooded area and water level trend temporally lag when compared to the WSCE% trend. Around 110 days of the temporal gap between the peak of WSCE% and water level/flooded area are noted, which fall within the estimated lag length of WSCE% (70 to 150 days) for the DLM analysis (Table 1). This agreement also strongly proves the reliability of our DLM lag length decision.

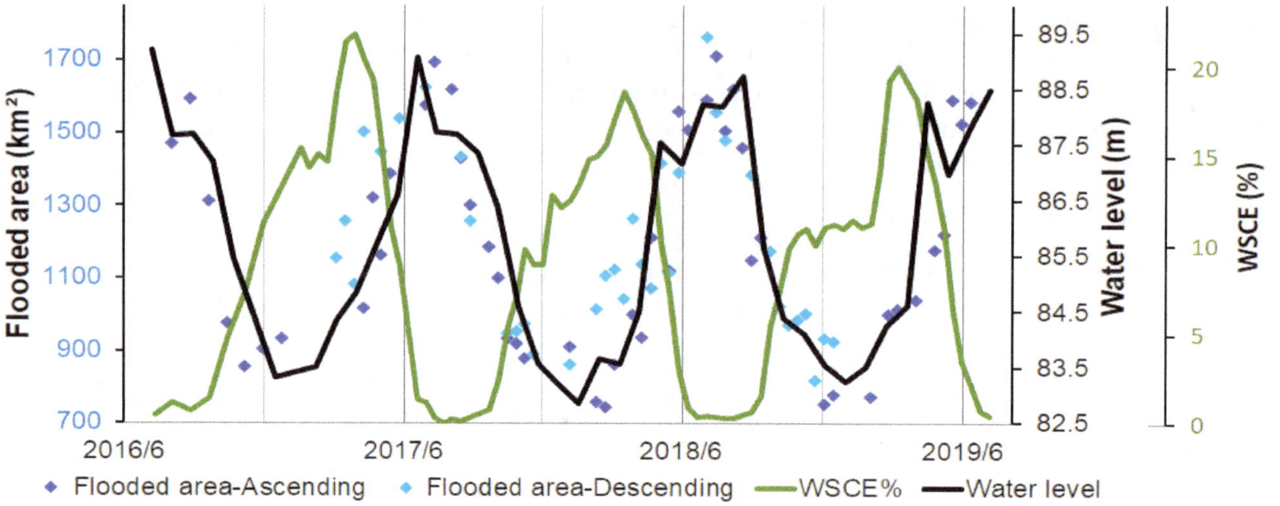

Figure 14. The internal cross-comparison of the Brahmaputra River's flooded area detected by the ascending and descending imagery of Sentinel-1 and the comparisons with the time series water level estimated by Sentinel-3 altimetry and the Sentinel-1 SAR-based WSCE%.

Based on the externally examined flooded area, the time series flooding dynamics of the Brahmaputra River can be depicted. For instance, we illustrate the maximum (2019/8/9) and minimum (2018/12/12) flooded area of the middle and lower sections of our study area in Figure 15. It is obvious that their flooded areas (river channels) significantly differ, as, in the dry season, the river shrinks back to the center of the wide channels mapped during the wet season. This example suggests that, by using SAR-Otsu's approach, it is viable to monitor the dynamics of the river channel migrations and seasonality.

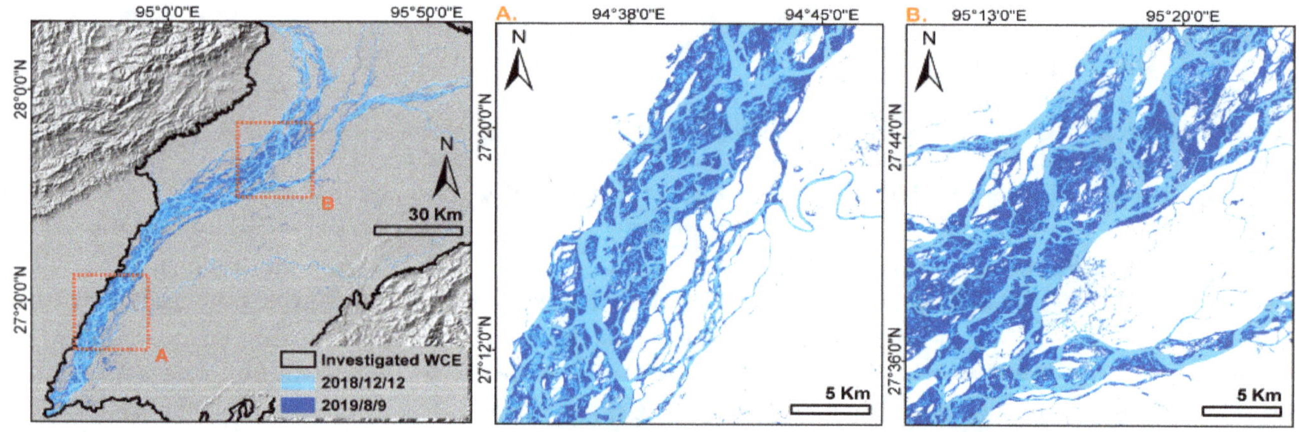

Figure 15. The maximum (2019/8/9) and minimum (2018/12/12) flooded area of the Brahmaputra River during the sensing period detected by the Sentinel-1 SAR imagery using the decided Otsu's threshold value.

5. Discussions

5.1. *The Cause of the Different Lake Surface Areas Detected by Sentinel-1 SAR and GWP Product*

To compare the surface area mapped by SAR (Sentinel-1) and the GWP, we plotted their estimations for the same dates in Figures 5 and 16. It can be observed that the GWP systematically underestimates the surface area compared to SAR-derived estimations. To examine that this effect is not caused by the overestimation of our SAR-based results, we plot the same date's high-resolution Sentinel-2 imagery in Figure 16 for comparison. It must be noted that, due to the reasons that (1) there is no absolute definition of the lake boundary as stated in [4], (2) the calibration procedures of optical imagery would also affect the waterbody delineation, and (3) our intention is to not include another threshold-based classification approach (such as NDWI), as it would cause another uncertainty; we compare them visually. It is clear that our SAR-based surface area agrees well with the blueish waterbody boundary observed in the Sentinel-2 scene. On the contrary, the GWP underestimates the surface area. Especially on 2017/10/31, the upper and lower parts of the lake obviously remained connected in the Sentinel-2 image, which was also delineated correctly in our Sentinel-1-based results. Nevertheless, the GWP suggests the lake is separated into two parts. This underestimation-caused separation frequently happens in the dry season (as marked in Figure 16b). The reason for this underestimation is due to the high reflectance signal in MODIS near-infrared (NIR) and red bands when water is very shallow. During the low water level period, Lake Urmia's bright lake soil with a high salt content dominates the spectral reflectance sensed by the optical MODIS sensor. Furthermore, the global operating GWP approach is treating mixed pixels that result from coarser spatial resolutions rather conservatively. Therefore, these shallow waters remain undetected, which results in systematic bias during low water levels, because the GWP is based on a universal approach operating on a global scale and cannot fit all waterbodies perfectly. On the contrary, as mentioned in Methodology, we manually selected water/nonwater samples to decide the customized Otsu's threshold value for Lake Urmia.

Another finding is that the SAR-based approach can only be implemented during the lake water depletion period, i.e., the dry season. As shown in Figure 16b, we found that all valid SAR-based surface area estimations occur between May and October. To investigate the reasons, firstly, we plotted the rainfall records provided by the ERA5-Land dataset. It is found that the water depletion period is highly correlated but slightly delayed due to the rainfall decreasing period. Hence, we assume the reason why the SAR imagery cannot map the lake boundary properly in the lake water accumulation period, i.e., rainfall-dominated wet season, is that a series of rainfall events would cause the surrounding regions of the lake to have a wet surface due to (1) sudden flooding events of the lake and/or (2) direct ground dampening caused by the rainfall. As we classify the waterbody based on the backscatter coefficient of the SAR signal, which is mainly influenced by the surface roughness and wetness, we can hardly distinguish whether the water is in the lake or on the surface of the surrounding damped/flooded areas. Thus, in the lake water accumulation period, i.e., wet season, the SAR-based observations would dramatically overestimate the surface area. For instance, given the example of 2019/4/7 shown in Figure 16a, in the Sentinel-2 scene, it is found that the surrounding area of the lake is much wetter (more blueish) and shows more small flood-caused tributaries and braids compared to the other three dates of the dry season. On the contrary, the GWP estimations during the wet season reveal the increasing surface water area more accurately and do not overestimate due to the wet soil.

Based on the above-mentioned discussions, we summarized the pros and cons of using SAR and multispectral sensors for waterbody mapping in Table 2.

Table 2. The comparison of the advantages and disadvantages of using SAR and the multispectral sensor for waterbody detection.

Sensor	SAR	Multi-Spectral
Classification basis	Backscatter coefficient	Illumination reflectance
Primary sensing factor	Surface roughness Surface dielectric property	Bands' reflectance differences Imagery calibration
Possible noise source	Winds, flooding, soil moisture, vegetation, ice, and snow cover	Cloud, mist, floating vegetation, turbidity, highly eutrophicated waters, high sediment load, ice, and snow cover
Advantage	All-weather and illumination-independent sensing ability Cloud penetration	Straightforward to interpret The abundance of openly accessible sensors and images less sensitive to soil moisture and usable during wet season
Disadvantage	Affected by wind, flooding, soil moisture	Affected by cloud and illumination conditions

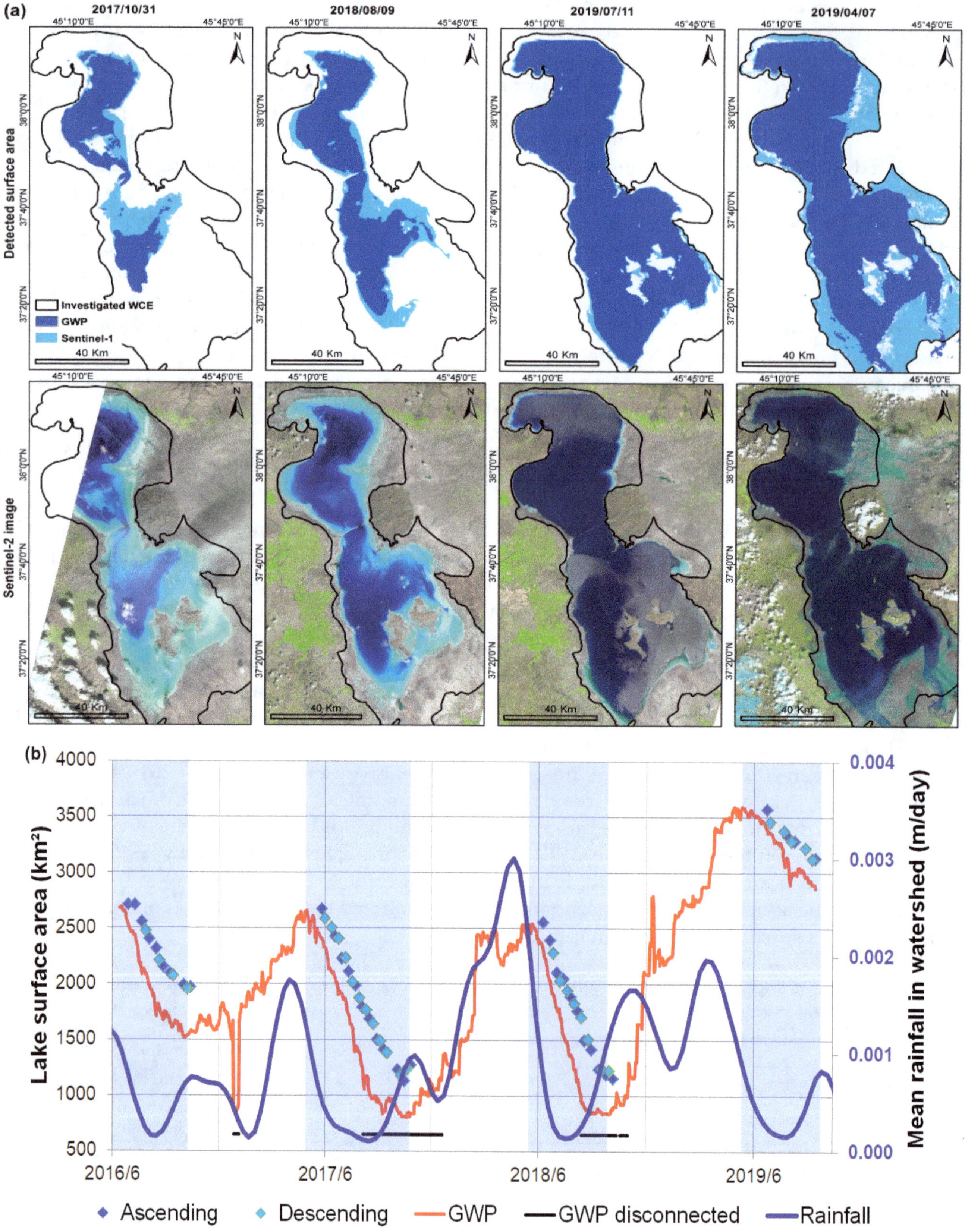

Figure 16. (**a**) The comparison of the surface area of Lake Urmia detected by Sentinel-1 SAR imagery and the GWP product and the same date high-resolution Sentinel-2 imagery. The mean rainfall volume is plotted in (**b**) to identify the dry season, with the lake water depletion periods marked in blue boxes. The date that shows lake disconnection in the GWP is marked with black points in (**b**).

5.2. The Necessity of the Detrending Process for the Lake Water Volume before a Hydrological Analysis

In the present study, we employed the STL seasonal decomposition technique for detrending, i.e., removing the annual aggregation effect. The necessity of detrending can be proven by the much-higher $\overline{R}^2$ value of detrended water volume (0.91) when compared to the original data (0.67), as summarized in Table 1. The significant enhancement of the water volume variation explanation agrees with our hypothesis: the "container effect" of the lake, i.e., the intra-annual trend of water volume, would hinder the accuracy of the DLM analysis between the water volume and hydrological factors.

5.3. The Lag Length Difference between Each Hydrological Variable and Study Area

As mentioned in Methodology Section 3.5, based on the $\overline{R}^2$ value, we can decide the suitable lag length of each hydrological factor for the lake and river DLM analyses. The final lag lengths of each case are summarized in Table 1. It is found that, for both cases, the lag length of the WSCE% is always the longest among the three factors, and evapotranspiration is the shortest (nearly instant). This finding agrees with the prior knowledge that the snowmelt water takes a much longer time to aggregate and eventually influences the downstream waterbody, because snow-covered regions are usually located at much higher elevation zones, which have longer distances to the downstream waterbody. Moreover, snowmelt water would not only form a direct surface streamflow but, also, infiltrate into the ground and soil [78–80]. In contrast, although the rainfall would also infiltrate into the ground, its distribution is usually broader but not limited in the high elevation zone. It might also spread over the whole watershed of the waterbody, including the waterbody itself. Hence, the lag length of the rainfall is much shorter than the snowmelt. For evapotranspiration, it is straightforward that its influence is rapid, as it would immediately reduce the amount of water of the waterbody.

Additionally, it is also found that the lag length of the WSCE% of the Brahmaputra River is nearly two-fold of Lake Urmia. We assume this is due to the Brahmaputra River being characterized by a much narrower and longer watershed along the tributary when compared to Lake Urmia. Consequently, the distance between the snow-covered area and the waterbody/VS is much more distant, and thus, the lagged effect of the snowmelt is more significant.

5.4. The Cause of the Long Lag Length of the SAR Sensor-Based WSCE%

Another reason for the long lag length of SAR sensor-based WSCE% is because of the unique way that SAR detects snowmelt processing and shows the temporal offset to the real condition. Based on the in-situ measurements and modeling simulation [11], it is known that the whole snow melting process can be divided into three phases, i.e., moistening, ripening, and runoff. In each phase, the contents of the snow water equivalent (SWE) (the total mass of water, including both liquid and solid water, stored in the form of snow) and the liquid water content (LWC) (the mass of liquid water inside the snowpack) differ, and the backscatter coefficient of SAR also changes accordingly and results in a U-shaped curve (Figure 17). In detail, firstly, in the moistening stage, the diurnal melting-freezing cycles gradually increase the LWC and lead to the gentle decreasing of the backscatter coefficient; in the ripening stage, the LWC significantly increases (while the SWE still remains the same), and the backscatter coefficient rapidly decreases and reaches the minimum value at the end of the ripening stage when the snowpack is saturated; and, finally, during the runoff stage, as the snowmelt water releases, both the LWC and SWE decrease, and the backscatter coefficient increases accordingly [11].

It must be noted that, as the backscatter coefficient of SAR would already slightly drop in the middle of the first moistening stage, the snow would already be detected and regarded as wet snow by the backscattering ratio threshold method [9–11]. However, the snowmelt water is actually released only until the final runoff stage starts. Namely, the SAR-detected wet snow-covered period would (1) happen much earlier and (2) last much longer than the real snowmelt water-releasing period, as illustrated in Figure 17. As a result, the lag length of the WSCE% would be longer than expected, as it acts more like a leading indicator of the real snowmelt water generation time.

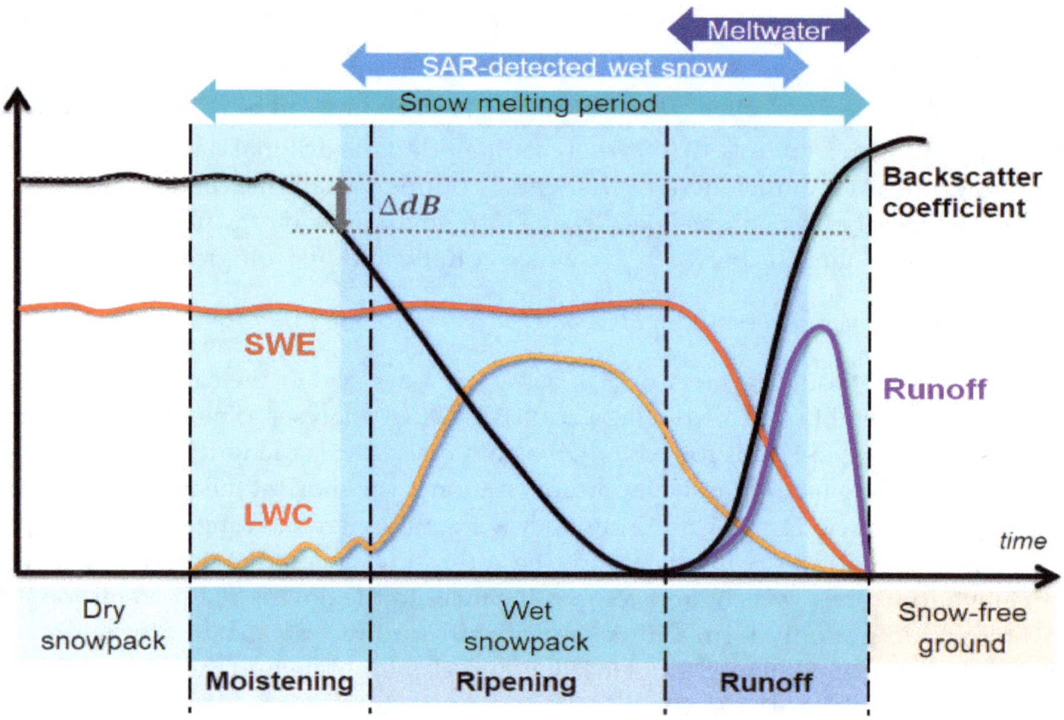

Figure 17. The relationship of SAR-based wet snow detection and the snowmelt processing (modified from [11]). The amount of liquid water content (LWC), snow water-equivalent (SWE), and the released snowmelt runoff together with the SAR signal backscatter coefficient in different snowmelt phases are illustrated.

5.5. The Reason of the Negative LRP of Rainfall in DLM of the Detrended Lake Water Volume

In Sections 4.1.5 and 4.2.3, we only analyzed the LRP of the WSCE%, due to the fact that we focused on depicting the casual effects of increasing the WSCE on the water resource. On the contrary, the other two hydrological factors (mean rainfall and evapotranspiration) are the control variables that are not the investigated targets but help mitigate the omitted variable bias [87] to ensure the robustness of the DLM analysis. Including control variables is indispensable, as there might be correlations between independent variables, which would mislead the estimated lag coefficient and the interested LRP.

Nevertheless, we can still analyze the coefficient of the other two hydrological factors, as shown in Table 1. For evapotranspiration, in both the river and lake cases, the coefficients are negative, which is reasonable, as more evapotranspiration would reduce the water amount. On the contrary, the coefficient of rainfall should be positive, as more rainfall would cause more water volume. However, only the case of the Brahmaputra River shows a positive value. We conclude the main reason causing the negative coefficient value of rainfall for the Lake Urmia case is the detrending processing. It reduces the importance of rainfall, which is already low in comparison to the river's case. In detail, by comparing the rainfall's coefficient in the Lake Urmia case estimated with the original and detrended water volume, we can be observed that the value is positive in the former scenario but becomes negative in the latter one. However, it must be noted that, in both scenarios, the coefficient of the WSCE% and evapotranspiration remain positive and negative, respectively, which agrees with the prior knowledge. It is because the relationship (or importance) between the rainfall and water volume is much weaker than the WSCE% and evapotranspiration-to-water volume in the lake's case (as described in Section 4.1.4). It can also be proven by the much lower maximum $\overline{R}^2$ value of rainfall (0.23) compared to the wet SCE% (0.85) and evapotranspiration (0.61), as shown in Table 1. Furthermore, by comparing with the Brahmaputra River case (0.94), the lake's rainfall-water volume relationship also shows a much weaker linkage.

5.6. Current Limitations and Future Goals

In the present study, we proposed a cloud-free and illumination-independent comprehensive inland water monitoring approach based on freely accessible and sustainable data/services, which guarantees methodological reproducibility and enables users to holistically and continuously monitor the dynamics of any large-scale waterbody. The econometrics' DLM further facilitates the accurate analysis of the lagged causal effect of the snowmelt condition on the water amount. Nevertheless, there are still some limitations in our present study. Hence, future improvements are required.

Firstly, as mentioned above, it is found that the high spatial resolution, cloud-free SAR backscatter-Otsu-based surface area detection method can only be implemented in the lake water depletion period, i.e., dry season, due to the sensitivity of the SAR signal to the surface wetness. On the contrary, the GWP provides continuous and daily observations in all seasons, as shown in Figure 16b, although it has some disadvantages, such as systematic underestimation due to the high spectral reflectance of the lake's bottom when the water is shallow (as shown in Figures 5 and 16) and coarse spatial resolution. Yet, in a future study, we would integrate both multispectral and SAR sensor-based surface areas to achieve a robust, high temporal and spatial resolution continuous surface area mapping strategy.

Secondly, in our study, we used the WSCE% for representing the snowmelt condition. Although it is efficient and straightforward to depict the whole snowmelt dynamics [9,11,76,77], it cannot provide the information of how much snowmelt water is actually generated during the snowmelt period. Instead, the measurement of either SWE or LWC should be used. Unfortunately, although there are many studies aiming at quantifying the SWE or LWC with SAR data, so far, there is no reliable conclusion provided [9]. In addition to the wetness of the snowpack, other factors such as soil moisture, surface roughness, snow grain size, snow density, and temperature would also affect the SAR signal. Consequently, the utilization of either the inversion technique or empirical model is usually required [9]. The employments of multispectral and passive sensors are the alternative; however, they would be affected by either cloud cover and polar darkness or poor spatial resolution, as mentioned in the Introduction. Hence, in the future study, we would explore the SAR-SWE relationship using the inversion technique and compare the results with the passive sensor-based SWE products provided by Copernicus Global Land Service [92] and GlobSnow [92].

In addition, in the current study, we assumed the water balance of Lake Urmia could be explained by the WSCE, rainfall, and evapotranspiration. Since the surface water inflow can be estimated by the lagged effects of the snowmelt and rainfall water, and as Lake Urmia is an endorheic lake, there is no stream outflow. Nevertheless, the influence of groundwater seepage and leakage were not measured in the present study due to (1) the lack of piezometric records around the lake and (2) their influences on the water balance are normally trivial [93]. Thus, the groundwater is commonly ignored in the lake water balance studies [94,95]. However, to enhance the completeness of monitoring the water resource variations of Lake Urmia, the hydrological models such as the MODular three-dimensional finite-difference groundwater flow model (MODFLOW) [96] or a simple groundwater input-output model [97] could be included in the future.

Another improvement that could be considered is using the modified DLM for a hydrological regression analysis. In the present DLM applied in this study, the weight of each lag is set equally. On the contrary, the advanced constrained DLM uses a lag-based smooth function, such as linear declining lag weights or polynomial/spline distributed lag models [85]. These constrains would allow a more accurate estimation of each lag weight of each independent variable, as the potential multicollinearity between different lag of each independent variable might hinder the estimation of each lag coefficient in unconstrained DLM (although the LRP, the main investigated target in the current study, would remain reliable) [88]. Moreover, the autoregressive distributed lag model (ARDL) [98] should also be explored as the past value of the water amount might also influence the current water amount.

The rain-on-snow (ROS) hydrometeorological phenomena should also be investigated for its influence on the snowmelt-led water resource variations. It is found that reactive oxygen species (ROS) events would weaken the snowpack structure and accelerate the melting of the snowpack due to heat transfer, which leads to an earlier and shorter melting period and even causes floods [99–101]. Moreover, in the long run, it can also result in a thinner springtime snowpack with less LWC and SWE and, eventually, reduces the water supply in the dry season [102,103]. Practically, when ROS events take place, they might form an icy crust on the snowpack, as the relatively warm precipitation would melt the surface of the snowpack and seep through the snow, which refreezes during the nighttime [104]. This icy crust would change the surface roughness and the dielectric property and may mislead the SAR-based WSCE mapping results. Hence, the relationship between the WSCE, precipitation, and SAR backscatter coefficient should be examined.

Finally, to utilize the proposed strategy's broad applicability, establishing an automatic processing pipeline is worth investigating. In detail, the whole workflow (as shown in Figure 2) can be divided into five parts, including water level (volume) retrieval, water-covered area extraction, WSCE estimation, watershed delineation, and hydrological factors calculation, as well as an analysis of the causal effects of snowmelt. Each part can be automatized as follows: the most complex first part consists of a user-defined SARvatore task processed with a BRAT command line-based processing script and STL processing scripts; the second part is composed of a manual selection of training samples and automatic applying of the decided Otsu's threshold value for classification; the third part is conducted with the stacking of SAR images and batch ratio/thresholding calculations; the fourth part is the watershed delineation, and hydrological factor calculations can be conducted by built-up Geographic Information System (GIS) tools and written scripts, respectively; at last, the fifth part can be automatized with predefined criteria of selecting the proper lag length for the final DLM analysis. Thus, based on part-wise automatization, the proposed strategy of monitoring large-scale waterbody's dynamics and assessing snowmelt's causal effects can be implemented more efficiently.

6. Conclusions

The significance of inland water resources has been gradually emphasized, especially in times of global warming. Within all waterbody types, lake and river systems are critical not only for human society but, also, for natural ecosystems. In addition to the temperature and rainfall variability, climate change alters the snowmelt phenology and snow distribution, while runoff originating from snowmelt is the primary water supply source for many river and lake systems. Therefore, monitoring the dynamics and variations of water resources, as well as the snowmelt conditions, is necessary to understand cryo-hydrosphere interactions under the influence of climate change. So far, only a few studies addressing this question relying on remote-sensing techniques exist. Most previous studies utilize either in-situ measurements or multispectral sensors, which are limited to small-scale regions or affected by cloud cover, respectively. In the present study, we employed the latest spaceborne Sentinel-1 SAR and Sentinel-3 altimetry data to achieve a high resolution, cloud-free, and illumination-independent comprehensive inland waterbody dynamics (time series surface area and water level) sensing strategy. Moreover, in contrast to previous studies utilizing in-house SAR and altimetry data-processing algorithms, we employed freely available cloud-based services to ensure the broad applicability and transferability.

Based on the altimetry and SAR data, the water level and the water-covered extent (surface area of lakes and the flooded area of rivers) can be successfully estimated. For the surface area mapping, the SAR backscatter coefficient-based Otsu's threshold has proven to be capable of accurately classifying the waterbody boundary by comparing it with the Global WaterPack (GWP) and high-resolution Sentinel-2 imagery. Nevertheless, we found that this approach can only be implemented in the lake water depletion period, i.e., dry season, due to the fact that the rainfall-caused surface wetness over nonwaterbody regions would mislead the backscatter-based classification. For the water level estimation, our results show a high consistency between different passes and even out-performs the well-known altimetry databases owing to the finer along-track resolution of Sentinel-3. Moreover, by fusing both the surface area and water level information, we can achieve comprehensive inland water dynamics monitoring. For our selected study sites, such as Lake Urmia, we could estimate the hypsometry and derive the water volume change. Via utilizing the STL decomposition technique, we can mitigate the "container effect" of the lake and extract the interannual water variation. For the Brahmaputra River, both time series of the water level and the flooded area can be identified and be used for cross-comparison. The synchronized trend observed between them also suggests the high reliability of our derived Sentinel-1 and Sentinel-3 results.

Last, but not least, together with the WSCE mapped with SAR imagery, we can analyze the influence of snowmelt on water resource variations. With the aim to handle the lagged causal effect of snowmelt and mitigate the omitted errors in the regression, the DLM initially developed for econometric applications is employed in the present study. In the lag lengths pre-analysis, we found that the WSCE% has the longest lag length compared to rainfall and evapotranspiration. It is due to the infiltration process of snowmelt water and the fact that the snow-covered area is usually located in the high elevation region, which is more distant to downstream water level-investigated VS. The lag length of the WSCE% was also observed to be nearly twice as long in the Brahmaputra River compared to Lake Urmia owing to the shape differences of their watersheds. Furthermore, the temporal offset between the snowmelt water generation and the SAR-based WSCE detection is also discussed, which is one reason for the long lag length. Eventually, the casual effect of snowmelt conditions on inland water resources is unbiasedly quantified with the DLM: for Lake Urmia, when increasing one percent of the WSCE% during the period of zero to 80 days before, the detrended water volume would increase by 108 cubic meters while hold still other factors; for the Brahmaputra River, when increasing one percent of the WSCE% during the period of 70 to 150 days before, the water level of the investigated VS would increase 0.1 m, while the other independent variables remained constant.

Author Contributions: Conceptualization, Y-L.S.T.; Formal analysis, Y.-L.S.T.; Methodology, Y.-L.S.T. and I.K.; Resources, I.K.; Supervision, A.D. and N.O.; Validation, Y.-L.S.T.; Visualization, Y.-L.S.T.; Writing—original draft, Y.-L.S.T. and I.K.; Writing—review & editing, I.K., A.D. and N.O. All authors have read and agreed to the published version of the manuscript.

Acknowledgments: Support by the German Academic Exchange Service (DAAD) fellowship to Ya-Lun S. Tsai is gratefully acknowledged. Special thanks to Marco Restano (ESA-ESRIN) for supporting the SARvatore service and Yao-Chun Huang (LSE) for guiding the DLM analysis.

References

1. United Nations Environment Programme. *Africa Water Atlas. Division of Early Warning and Assessment (DEWA)*; United Nations Environment Programme (UNEP): Nairobi, Kenya, 2010.
2. FAO. *The State of the World's Land and Water Resources for Food and Agriculture (SOLAW)—Managing Systems at Risk*; Food and Agriculture Organization: Rome, Italy; Earthscan: London, UK, 2011.
3. UNESCO World Water Assessment Programme. The United Nations World Water Development Report 2019: Leaving No One Behind. 2019. Available online: https://en.unesco.org/themes/water-security/wwap/wwdr/2019 (accessed on 7 November 2020).
4. Crétaux, J.-F.; Abarca-del-Río, R.; Berge-Nguyen, M.; Arsen, A.; Drolon, V.; Clos, G.; Maisongrande, P. Lake volume monitoring from space. *Surv. Geophys.* **2016**, *37*, 269–305. [CrossRef]
5. Costanza, R.; Fioramonti, L.; Kubiszewski, I. The UN Sustainable Development Goals and the dynamics of well-being. *Front. Ecol. Environ.* **2016**, *14*, 59. [CrossRef]
6. Bhaduri, A.; Bogardi, J.; Siddiqi, A.; Voigt, H.; Vörösmarty, C.; Pahl-Wostl, C.; Bunn, S.E.; Shrivastava, P.; Lawford, R.; Foster, S.; et al. Achieving Sustainable Development Goals from a Water Perspective. *Front. Environ. Sci.* **2016**, *4*, 64. [CrossRef]

7. Pachauri, R.K.; Allen, M.R.; Barros, V.R.; Broome, J.; Cramer, W.; Christ, R.; Church, J.A.; Clarke, L.; Dahe, Q.; Dasgupta, P. *Climate Change 2014: Synthesis Report. Contribution of Working Groups I, II and III to the Fifth Assessment Report of the Intergovernmental Panel on Climate Change*; IPCC: Cambridge, UK, 2014.
8. UN Water. *Progress on Water-Related Ecosystems: Piloting the Monitoring Methodology and Initial Findings for SDG Indicator 6.6.1*; UN Environment on Behalf of UN Water: Geneva, Switzerland, 2018; ISBN 978-92-807-3712-7.
9. Tsai, Y.-L.S.; Dietz, A.J.; Oppelt, N.; Kuenzer, C. Remote Sensing of Snow Cover Using Spaceborne SAR: A Review. *Remote Sens.* **2019**, *11*, 1456. [CrossRef]
10. Nagler, T.; Rott, H. Retrieval of wet snow by means of multitemporal SAR data. *IEEE Trans. Geosci. Remote Sens.* **2000**, *38*, 754–765. [CrossRef]
11. Marin, C.; Bertoldi, G.; Premier, V.; Callegari, M.; Brida, C.; Hürkamp, K.; Tschiersch, J.; Zebisch, M.; Notarnicola, C. Use of Sentinel-1 radar observations to evaluate snowmelt dynamics in alpine regions. *Cryosphere* **2020**, *14*, 935–956. [CrossRef]
12. McFeeters, S.K. The use of the Normalized Difference Water Index (NDWI) in the delineation of open water features. *Int. J. Remote Sens.* **1996**, *17*, 1425–1432. [CrossRef]
13. Xu, H. Modification of normalised difference water index (NDWI) to enhance open water features in remotely sensed imagery. *Int. J. Remote Sens.* **2006**, *27*, 3025–3033. [CrossRef]
14. Feyisa, G.L.; Meilby, H.; Fensholt, R.; Proud, S.R. Automated Water Extraction Index: A new technique for surface water mapping using Landsat imagery. *Remote Sens. Environ.* **2014**, *140*, 23–35. [CrossRef]
15. Klein, I.; Gessner, U.; Dietz, A.J.; Kuenzer, C. Global WaterPack—A 250 m resolution dataset revealing the daily dynamics of global inland water bodies. *Remote Sens. Environ.* **2017**, *198*, 345–362. [CrossRef]
16. Dirscherl, M.; Dietz, A.J.; Kneisel, C.; Kuenzer, C. Automated Mapping of Antarctic Supraglacial Lakes Using a Machine Learning Approach. *Remote Sens.* **2020**, *12*, 1203. [CrossRef]
17. Alsdorf, D.E.; Rodríguez, E.; Lettenmaier, D.P. Measuring surface water from space. *Rev. Geophys.* **2007**, *45*. [CrossRef]
18. Duan, Z.; Bastiaanssen, W. Estimating water volume variations in lakes and reservoirs from four operational satellite altimetry databases and satellite imagery data. *Remote Sens. Environ.* **2013**, *134*, 403–416. [CrossRef]
19. Schwatke, C.; Dettmering, D.; Bosch, W.; Seitz, F. DAHITI—An innovative approach for estimating water level time series over inland waters using multi-mission satellite altimetry. *Hydrol. Earth Syst. Sci.* **2015**, *19*, 4345–4364. [CrossRef]
20. Birkett, C.M.; Ricko, M.; Beckley, B.D.; Yang, X.; Tetrault, R.L. G-REALM: A lake/reservoir monitoring tool for drought monitoring and water resources management. In Proceedings of the American Geophysical Union, Fall Meeting 2017, New Orleans, LA, USA, 11–15 December 2017. Abstract #H23P-02.
21. Crétaux, J.-F.; Jelinski, W.; Calmant, S.; Kouraev, A.; Vuglinski, V.; Bergé-Nguyen, M.; Gennero, M.-C.; Nino, F.; Del Rio, R.A.; Cazenave, A.; et al. SOLS: A lake database to monitor in the Near Real Time water level and storage variations from remote sensing data. *Adv. Space Res.* **2011**, *47*, 1497–1507. [CrossRef]
22. Yuan, C.; Gong, P.; Zhang, H.; Guo, H.; Pan, B. Monitoring water level changes from retracked Jason-2 altimetry data: A case study in the Yangtze River, China. *Remote Sens. Lett.* **2017**, *8*, 399–408. [CrossRef]
23. Fernandes, M.J.; Lázaro, C.; Nunes, A.L.; Scharroo, R. Atmospheric Corrections for Altimetry Studies over Inland Water. *Remote Sens.* **2014**, *6*, 4952–4997. [CrossRef]
24. Brown, G. The average impulse response of a rough surface and its applications. *IRE Trans. Antennas Propag.* **1977**, *25*, 67–74. [CrossRef]
25. Wang, C.; Chu, Y.; Huang, Z.; Chao, N. Robust, Long-term Lake Level Change from Multiple Satellite Altimeters in Tibet: Observing the Rapid Rise of Ngangzi Co over a New Wetland. *Remote Sens.* **2019**, *11*, 558. [CrossRef]
26. Cretaux, J.-F.; Bergé-Nguyen, M.; Calmant, S.; Jamangulova, N.; Satylkanov, R.; Lyard, F.; Perosanz, F.; Verron, J.; Montazem, A.S.; Le Guilcher, G.; et al. Absolute Calibration or Validation of the Altimeters on the Sentinel-3A and the Jason-3 over Lake Issykkul (Kyrgyzstan). *Remote Sens.* **2018**, *10*, 1679. [CrossRef]
27. Calmant, S.; Seyler, F.; Cretaux, J.F. Monitoring Continental Surface Waters by Satellite Altimetry. *Surv. Geophys.* **2008**, *29*, 247–269. [CrossRef]
28. Uebbing, B.; Kusche, J.; Forootan, E. Waveform Retracking for Improving Level Estimations from TOPEX/Poseidon, Jason-1, and Jason-2 Altimetry Observations Over African Lakes. *IEEE Trans. Geosci. Remote Sens.* **2014**, *53*, 2211–2224. [CrossRef]

29. Gao, Q.; Makhoul, E.; Escorihuela, M.-J.; Zribi, M.; Quintana-Seguí, P.; García, P.; Roca, M. Analysis of Retrackers' Performances and Water Level Retrieval over the Ebro River Basin Using Sentinel-3. *Remote Sens.* **2019**, *11*, 718. [CrossRef]
30. Zarghami, M. Effective watershed management; Case study of Urmia Lake, Iran. *Lake Reserv. Manag.* **2011**, *27*, 87–94. [CrossRef]
31. Djamali, M.; De Beaulieu, J.-L.; Shah-Hosseini, M.; Andrieu-Ponel, V.; Ponel, P.; Amini, A.; Akhani, H.; Leroy, S.A.; Stevens, L.; Lahijani, H.; et al. A late Pleistocene long pollen record from Lake Urmia, NW Iran. *Quat. Res.* **2008**, *69*, 413–420. [CrossRef]
32. UNDP-GEF. Conservation of Iranian Wetlands Project, Department of Environment. In *Integrated Management Plan for Lake Urmia Basin*; UNDP: New York, NY, USA, 2010.
33. Karbassi, A.; Bidhendi, G.N.; Pejman, A.; Bidhendi, M.E. Environmental impacts of desalination on the ecology of Lake Urmia. *J. Great Lakes Res.* **2010**, *36*, 419–424. [CrossRef]
34. Ahmadi, R.; Mohebbi, F.; Hagigi, P.; Esmailly, L.; Salmanzadeh, R. Macro-invertebrates in the Wetlands of the Zarrineh estuary at the south of Urmia Lake (Iran). *Int. J. Environ. Res.* **2011**, *5*, 1047–1052.
35. Ornithology Unit of the Department of the Environment. *The Birds of Lake Orumiyeh and Adjacent Wetlands, Islamic Republic of Iran*; UNEP: Nairobi, Kenya, 2010.
36. Sima, S.; Tajrishy, M. Using satellite data to extract volume–area–elevation relationships for Urmia Lake, Iran. *J. Great Lake Res.* **2013**, *39*, 90–99. [CrossRef]
37. Kabiri, K.; Pradhan, B.; Sharifi, A.; Ghobadi, Y.; Pirasteh, S. Manifestation of remotely sensed data coupled with field measured meteorological data for an assessment of degradation of Urmia Lake, Iran. In Proceedings of the Asia Pacific Conference on Environmental Science and Technology (APEST), Kuala Lumpur, Malaysia, 1–2 February 2012; pp. 395–401.
38. Nouri, H.; Mason, R.J.; Moradi, N. Land suitability evaluation for changing spatial organization in Urmia County towards conservation of Urmia Lake. *Appl. Geogr.* **2017**, *81*, 1–12. [CrossRef]
39. Hassanzadeh, E.; Zarghami, M.; Hassanzadeh, Y. Determining the Main Factors in Declining the Urmia Lake Level by Using System Dynamics Modeling. *Water Resour. Manag.* **2012**, *26*, 129–145. [CrossRef]
40. Delju, A.H.; Ceylan, A.; Piguet, E.; Rebetez, M. Observed climate variability and change in Urmia Lake Basin, Iran. *Theor. Appl. Clim.* **2013**, *111*, 285–296. [CrossRef]
41. Voss, K.A.; Famiglietti, J.S.; Lo, M.; De Linage, C.; Rodell, M.; Swenson, S.C. Groundwater depletion in the Middle East from GRACE with implications for transboundary water management in the Tigris-Euphrates-Western Iran region. *Water Resour. Res.* **2013**, *49*, 904–914. [CrossRef] [PubMed]
42. Eimanifar, A.; Mohebbi, F. Urmia Lake (Northwest Iran): A brief review. *Saline Syst.* **2007**, *3*, 5. [CrossRef] [PubMed]
43. Tourian, M.; Elmi, O.; Chen, Q.; Devaraju, B.; Roohi, S.; Sneeuw, N. A spaceborne multisensor approach to monitor the desiccation of Lake Urmia in Iran. *Remote Sens. Environ.* **2015**, *156*, 349–360. [CrossRef]
44. Sorgeloos, P. Brine Shrimp Artemia in Coastal Saltworks: Hydrobiological Key to Improved Salt Production and Inexpensive Source of Food for Vertically Integrated Aquaculture. In Proceedings of the International Meeting on Saltworks Conversion for Aquaculture, Trapani, Italy, 9–11 May 1986; pp. 9–11.
45. Abbaspour, M.; Javid, A.H.; Mirbagheri, S.A.; Givi, F.A.; Moghimi, P. Investigation of lake drying attributed to climate change. *Int. J. Environ. Sci. Technol.* **2012**, *9*, 257–266. [CrossRef]
46. Pengra, B. The Drying of Iran's Lake Urmia and its Environmental Consequences. UNEP-GRID Sioux Falls: UNEP Global Environmental Alert Service. Available online: https://na.unep.net/geas/getUNEPPageWithArticleIDScript.php?article_id=79 (accessed on 7 November 2020).
47. Alipour, S. Hydrogeochemistry of seasonal variation of Urmia Salt Lake, Iran. *Saline Syst.* **2006**, *2*, 9. [CrossRef]
48. Madani, K. Water management in Iran: What is causing the looming crisis? *J. Environ. Stud. Sci.* **2014**, *4*, 315–328. [CrossRef]
49. Dai, A.; Qian, T.; Trenberth, K.E.; Milliman, J.D. Changes in Continental Freshwater Discharge from 1948 to 2004. *J. Clim.* **2009**, *22*, 2773–2792. [CrossRef]
50. Sarkar, A.; Garg, R.D.; Sharma, N. RS-GIS Based Assessment of River Dynamics of Brahmaputra River in India. *J. Water Resour. Prot.* **2012**, *4*, 63–72. [CrossRef]
51. Immerzeel, W.W.; Van Beek, L.P.H.; Bierkens, M.F.P. Climate change will affect the Asian Water Towers. *Science* **2010**, *328*, 1382–1385. [CrossRef]

52. Dubey, A.K.; Gupta, P.; Dutta, S.; Kumar, B. Evaluation of satellite-altimetry-derived river stage variation for the braided Brahmaputra River. *Int. J. Remote Sens.* **2014**, *35*, 7815–7827. [CrossRef]
53. Samanta, A.; Tripathy, G.R.; Das, R. Temporal Variations in Water Chemistry of the (Lower) Brahmaputra River: Implications to Seasonality in Mineral Weathering. *Geochem. Geophys. Geosyst.* **2019**, *20*, 2769–2785. [CrossRef]
54. Richardson, W.; Thorne, C.R. Multiple thread flow and channel bifurcation in a braided river: Brahmaputra Jamuna River, Bangladesh. *Geomorphology* **2001**, *38*, 185–196. [CrossRef]
55. Finsen, F.; Milzow, C.; Smith, R.; Berry, P.; Bauer-Gottwein, P. Using radar altimetry to update a large-scale hydrological model of the Brahmaputra river basin. *Hydrol. Res.* **2014**, *45*, 148–164. [CrossRef]
56. Huang, Q.; Long, D.; Du, M.; Zeng, C.; Qiao, G.; Li, X.; Hou, A.; Hong, Y. Discharge estimation in high-mountain regions with improved methods using multisource remote sensing: A case study of the Upper Brahmaputra River. *Remote Sens. Environ.* **2018**, *219*, 115–134. [CrossRef]
57. Tsai, Y.-L.S.; Lin, S.-Y.; Kim, J.-R.; Choi, Y. Analysis of the seasonal velocity difference of the Greenland Russell glacier using multi-sensor data. *Terr. Atmos. Ocean. Sci.* **2019**, *30*, 541–562. [CrossRef]
58. Gorelick, N.; Hancher, M.; Dixon, M.; Ilyushchenko, S.; Thau, D.; Moore, R. Google Earth Engine: Planetary-scale geospatial analysis for everyone. *Remote Sens. Environ.* **2017**, *202*, 18–27. [CrossRef]
59. EUMETSAT. Sentinel-3 SRAL Marine User Handbook. 2017. Available online: https://earth.esa.int/eogateway/documents/20142/1564943/Sentinel-3-SRAL-Marine-User-Handbook.pdf (accessed on 7 November 2020).
60. Dinardo, S.; Restano, M.; Ambrózio, A.; Benveniste, J. SAR Altimetry Processing on Demand Service for Cryosat-2 and Sentinel-3 AT ESA G-POD. In Proceedings of the 2016 Conference on Big Data from Space (BiDS'16), Santa Cruz de Tenerife, Spain, 15–17 March 2016; pp. 15–17.
61. Dinardo, S.; Fenoglio-Marc, L.; Buchhaupt, C.; Becker, M.; Scharroo, R.; Fernandes, M.J.; Benveniste, J. Coastal sar and plrm altimetry in german bight and west baltic sea. *Adv. Space Res.* **2018**, *62*, 1371–1404. [CrossRef]
62. Dumont, J.; Rosmorduc, V.; Carrere, L.; Picot, N.; Bronner, E.; Couhert, A.; Guillot, A.; Desai, S.; Bonekamp, H.; Figa, J. Jason-3 Products Handbook. SALP-MU-M-OP-16118-CN. 2016. Available online: https://www.ospo.noaa.gov/Products/documents/hdbk_j3.pdf (accessed on 7 November 2020).
63. Wangchuk, S.; Bolch, T.; Zawadzki, J. Towards automated mapping and monitoring of potentially dangerous glacial lakes in Bhutan Himalaya using Sentinel-1 Synthetic Aperture Radar data. *Int. J. Remote Sens.* **2019**, *40*, 4642–4667. [CrossRef]
64. Henry, J.; Chastanet, P.; Fellah, K.; Desnos, Y. Envisat multi-polarized ASAR data for flood mapping. *Int. J. Remote Sens.* **2006**, *27*, 1921–1929. [CrossRef]
65. Otsu, N. A Threshold Selection Method from Gray-Level Histograms. *IEEE Trans. Syst. Man Cybern.* **1979**, *9*, 62–66. [CrossRef]
66. Greifeneder, F.; Wagner, W.; Sabel, D.; Naeimi, V. Suitability of SAR imagery for automatic flood mapping in the Lower Mekong Basin. *Int. J. Remote Sens.* **2014**, *35*, 2857–2874. [CrossRef]
67. Lehner, B.; Döll, P. Development and validation of a global database of lakes, reservoirs and wetlands. *J. Hydrol.* **2004**, *296*, 1–22. [CrossRef]
68. Nilsson, K.A.; Trout, K.E.; Ross, M. General Model to Represent Multiple Wetland and Lake Stage-Storage Behavior. *J. Hydrol. Eng.* **2010**, *15*, 786–795. [CrossRef]
69. Muala, E.; Mohamed, Y.A.; Duan, Z.; Van Der Zaag, P. Estimation of Reservoir Discharges from Lake Nasser and Roseires Reservoir in the Nile Basin Using Satellite Altimetry and Imagery Data. *Remote Sens.* **2014**, *6*, 7522–7545. [CrossRef]
70. Hayashi, M.; Van Der Kamp, G. Simple equations to represent the volume–area–depth relations of shallow wetlands in small topographic depressions. *J. Hydrol.* **2000**, *237*, 74–85. [CrossRef]
71. Cleveland, R.B.; Cleveland, W.S.; McRae, J.E.; Terpenning, I. STL: A seasonal-trend decomposition. *J. Off. Stat.* **1990**, *6*, 3–73.
72. Fathian, F.; Dehghan, Z.; Eslamian, S. Analysis of water level changes in Lake Urmia based on data characteristics and non-parametric test. *Int. J. Hydrol. Sci. Technol.* **2014**, *4*, 18–38. [CrossRef]
73. Sellinger, C.E.; Stow, C.A.; Lamon, E.C.; Qian, S.S. Recent water level declines in the Lake Michigan– Huron System. *Environ. Sci. Technol.* **2008**, *42*, 367–373. [CrossRef]
74. Hassan, A.A.; Jin, S. Lake level change and total water discharge in East Africa Rift Valley from satellite-based observations. *Glob. Planet. Chang.* **2014**, *117*, 79–90. [CrossRef]

75. Lu, H. Decomposition of vegetation cover into woody and herbaceous components using AVHRR NDVI time series. *Remote Sens. Environ.* **2003**, *86*, 1–18. [CrossRef]
76. Tsai, Y.-L.S.; Dietz, A.J.; Oppelt, N.; Kuenzer, C. Tsai Wet and Dry Snow Detection Using Sentinel-1 SAR Data for Mountainous Areas with a Machine Learning Technique. *Remote Sens.* **2019**, *11*, 895. [CrossRef]
77. Tsai, Y.-L.S.; Dietz, A.J.; Oppelt, N.; Kuenzer, C. Tsai A Combination of PROBA-V/MODIS-based Products with Sentinel-1 SAR Data for Detecting Wet and Dry Snow Cover in Mountainous Areas. *Remote Sens.* **2019**, *11*, 1904. [CrossRef]
78. Seyfried, M.S.; Grant, L.E.; Marks, D.; Winstral, A.; McNamara, J.P. Simulated soil water storage effects on streamflow generation in a mountainous snowmelt environment, Idaho, USA. *Hydrol. Process. Int. J.* **2009**, *23*, 858–873. [CrossRef]
79. Butt, M.J.; Bilal, M. Application of snowmelt runoff model for water resource management. *Hydrol. Process.* **2011**, *25*, 3735–3747. [CrossRef]
80. Abudu, S.; Cui, C.-l.; Saydi, M.; King, J.P. Application of snowmelt runoff model (SRM) in mountainous watersheds: A review. *Water Sci. Eng.* **2012**, *5*, 123–136.
81. Almon, S. The Distributed Lag Between Capital Appropriations and Expenditures. *Econ. J. Econ. Soc.* **1965**, *33*, 178. [CrossRef]
82. Yang, J.; Ou, C.-Q.; Ding, Y.; Zhou, Y.-X.; Chen, P.-Y. Daily temperature and mortality: A study of distributed lag non-linear effect and effect modification in Guangzhou. *Environ. Health* **2012**, *11*, 63. [CrossRef]
83. Allen, M.J.; Sheridan, S.C. Mortality risks during extreme temperature events (ETEs) using a distributed lag non-linear model. *Int. J. Biometeorol.* **2015**, *62*, 57–67. [CrossRef]
84. Gasparrini, A. Distributed lag linear and non-linear models in R: The package dlnm. *J. Stat. Softw.* **2011**, *43*, 1. [CrossRef]
85. Gasparrini, A.; Armstrong, B.; Kenward, M.G. Distributed lag non-linear models. *Stat. Med.* **2010**, *29*, 2224–2234. [CrossRef] [PubMed]
86. Gasparrini, A. Modeling exposure–lag–response associations with distributed lag non-linear models. *Stat. Med.* **2014**, *33*, 881–899. [CrossRef] [PubMed]
87. Stock, J.H.; Watson, M.W. *Introduction to Econometrics*; Addison-Wesley: Boston, FL, USA, 2015.
88. Wooldridge, J.M. (Ed.) *Basic regression analysis with time series data. Introductory Econometrics: A Modern Approach*; South-Western College Publishing: Cincinnati, OH, USA, 2000.
89. Teklehaimanot, H.D.; Lipsitch, M.; Teklehaimanot, A.; Schwartz, J.D. Weather-based prediction of Plasmodium falciparum malaria in epidemic-prone regions of Ethiopia I. Patterns of lagged weather effects reflect biological mechanisms. *Malar. J.* **2004**, *3*, 41. [CrossRef] [PubMed]
90. Zhao, X.; Chen, F.; Feng, Z.; Li, X.; Zhou, X.-H. The temporal lagged association between meteorological factors and malaria in 30 counties in south-west China: A multilevel distributed lag non-linear analysis. *Malar. J.* **2014**, *13*, 57. [CrossRef] [PubMed]
91. Calmant, S.; Da Silva, J.S.; Moreira, D.M.; Seyler, F.; Shum, C.; Cretaux, J.-F.; Gabalda, G. Detection of Envisat RA2/ICE-1 retracked radar altimetry bias over the Amazon basin rivers using GPS. *Adv. Space Res.* **2013**, *51*, 1551–1564. [CrossRef]
92. Takala, M.; Luojus, K.; Pulliainen, J.; Derksen, C.; Lemmetyinen, J.; Kärnä, J.-P.; Koskinen, J.; Bojkov, B. Estimating northern hemisphere snow water equivalent for climate research through assimilation of space-borne radiometer data and ground-based measurements. *Remote Sens. Environ.* **2011**, *115*, 3517–3529. [CrossRef]
93. Lei, Y.; Yao, T.; Bird, B.W.; Yang, K.; Zhai, J.; Sheng, Y. Coherent lake growth on the central Tibetan Plateau since the 1970s: Characterization and attribution. *J. Hydrol.* **2013**, *483*, 61–67. [CrossRef]
94. Kumambala, P.G.; Ervine, A. Water Balance Model of Lake Malawi and its Sensitivity to Climate Change. *Open Hydrol. J.* **2010**, *4*, 152–162. [CrossRef]
95. Kummu, M.; Tes, S.; Yin, S.; Adamson, P.; Józsa, J.; Koponen, J.; Richey, J.; Sarkkula, J. Water balance analysis for the Tonle Sap Lake–floodplain system. *Hydrol. Process.* **2014**, *28*, 1722–1733. [CrossRef]
96. Anderson, M.P.; Woessner, W.W.; Hunt, R.J. *Applied Groundwater Modeling: Simulation of Flow and Advective Transport*; Academic Press: Cambridge, MA, USA, 2015.
97. Dinka, M.O.; Loiskandl, W.; Ndambuki, J.M. Hydrologic modelling for Lake Basaka: Development and application of a conceptual water budget model. *Environ. Monit. Assess.* **2014**, *186*, 5363–5379. [CrossRef]

98. Pesaran, M.H.; Shin, Y.; Smithc, R.J. Bounds testing approaches to the analysis of level relationships. *J. Appl. Econ.* **2001**, *16*, 289–326. [CrossRef]
99. McCabe, G.J.; Clark, M.P.; Hay, L.E. Rain-on-Snow Events in the Western United States. *Bull. Am. Meteorol. Soc.* **2007**, *88*, 319–328. [CrossRef]
100. Merz, R.; Blöschl, G. A process typology of regional floods. *Water Resour. Res.* **2003**, *39*, 12. [CrossRef]
101. Pradhanang, S.M.; Frei, A.; Zion, M.; Schneiderman, E.M.; Steenhuis, T.S.; Pierson, D. Rain-on-snow runoff events in New York. *Hydrol. Process.* **2013**, *27*, 3035–3049. [CrossRef]
102. Cohen, J.; Ye, H.; Jones, J. Trends and variability in rain-on-snow events. *Geophys. Res. Lett.* **2015**, *42*, 7115–7122. [CrossRef]
103. Jeong, D.I.; Sushama, L.; Khaliq, M.N. Attribution of spring snow water equivalent (SWE) changes over the northern hemisphere to anthropogenic effects. *Clim. Dyn.* **2017**, *48*, 3645–3658. [CrossRef]
104. Groisman, P.Y.; Knight, R.W.; Easterling, D.R.; Karl, T.R.; Hegerl, G.C.; Razuvaev, V.N. Trends in Intense Precipitation in the Climate Record. *J. Clim.* **2005**, *18*, 1326–1350. [CrossRef]

3

HY-2A Altimeter Data Initial Assessment and Corresponding Two-Pass Waveform Retracker

Shengjun Zhang [1], Jiancheng Li [2,3], Taoyong Jin [2,3,*] and Defu Che [1]

1 School of Resources and Civil Engineering, Northeastern University, Shenyang 110819, China; zhangshengjun@whu.edu.cn (S.Z.); chedefu@mail.neu.edu.cn (D.C.)
2 MOE Key Laboratory of Geospace Environment and Geodesy, School of Geodesy and Geomatics, Wuhan University, Wuhan 430072, China; jcli@whu.edu.cn
3 Collaborative Innovation Center of Geospatial Technology, Wuhan University, Wuhan 430072, China
* Correspondence: tyjin@sgg.whu.edu.cn

Abstract: The accuracy and resolution of the marine gravity field derived from multisatellite altimeter data sets mainly depend on the corresponding range precision and spatial distribution. Here, we preliminarily investigate the performance of HY-2A altimeter data by analyzing cross-mission sea surface height discrepancies with SARAL/AltiKa and calculating correlation coefficients with respect to tide gauge measurements. We also explore the improved range precision that can be achieved using a two-pass weighted least squares retracker which was proposed for the purpose of optimal gravity field recovery. Firstly, both the exact repetitive mission and the geodetic mission for HY-2A provide new track orientations and different data coverage for recovering the marine gravity field, and these dense geographical distributions are more greatly attributed to the geodetic mission in recent years. Secondly, HY-2A provides reliable sea surface height measurements based on exterior verifications by SARAL/AltiKa geophysical data records and tide gauge measurements, although the accuracy level is slightly lower than SARAL/AltiKa. Another more exciting finding is that the statistics of along-track sea surface heights in one-second intervals show that the two-pass retracking does further improve the range precision by a factor of 1.6 with respect to 20 Hz retracked results in sensor data records. In conclusion, the HY-2A mission can substantially improve the global accuracy and resolution of the marine gravity field and will reveal new tectonic features such as microplates, abyssal hill fabric, and new uncharted seamounts on the ocean floor.

Keywords: HY-2A; waveform retracking; range precision; marine gravity

1. Introduction

Satellite radar altimetry has been proven an effective tool for recovering the marine gravity field relying on its advantages of spatial scale, sustained operation at any time, and relatively low cost with respect to traditional shipboard instruments. Dozens of regional or global marine gravity field models have been published during the past 40 years based on the accumulated multisatellite altimeter data sets, especially those with nonrepetitive ground tracks [1–8]. The recovered marine gravity accuracy mainly depends on the data quality and the dense geographical distribution of satellite altimeters, while the range precision can be further improved by postprocessing techniques (e.g., waveform retracking, path delay correction modeling, and crossover adjustment). By contrast, the spatial distributions (e.g., adjacent intervals and orientations) of ground tracks are objectively determined in accordance with selected orbital parameters for multisatellite altimeter missions.

Therefore, the resolution of a recovered marine gravity field is limited to the altimeter data density, especially along the cross-track directions.

A variety of particular processing techniques, such as the waveform retrackers and low-pass filters, are mostly discussed and adopted in previous marine gravity studies for obtaining improved altimeter measurements [9–11]. Moreover, the altimeter data sets from geodetic missions (GMs) provide high-density coverage that is crucial for recovering high-resolution marine gravity fields and thus play a more important role than data from exact repetitive missions (ERMs) [12], and the ERM data sets show remarkable advantages in oceanic operational applications. In addition, the ERM data sets should also be used for the construction of gravity fields as these repeated profiles have improved signal-to-noise by averaging nonunique, repetitive cycles. The diverse orientations of ground tracks are satisfied by four series of altimeter satellites with different orbit inclinations for 66°, 92°, 98.5° and 108°, respectively. The near polar orbital altimeter has a larger global coverage scope which is closer to the north and south poles, but has poor performances in obtaining the east components of vertical deflection at low-latitude zones. By contrast, the designed inclinations with bigger difference to 90° provide altimeter measurements with complementary features and all these different orbital inclinations make the solution of vertical deflection components more stable in marine gravity recovery studies [3].

HY-2A, China's first satellite for exploring the oceanic dynamic environment, was successfully launched on 16 August 2011 and the corresponding data sets were operationally distributed by China's National Satellite Ocean Application Service (NSOAS) since October 2011. The major payload of the HY-2A satellite is a dual-frequency radar altimeter operating at both Ku and C band for acquiring geophysical parameters such as sea surface height (SSH), significant wave height (SWH), and wind speed (WS) over sea surface. Firstly, HY-2A provides different orientations of ground tracks since its planned orbital inclination is 99.34° and the planned repetitive period is almost 14 days during the early stage of designed life, which are obviously different from other altimeter satellites [13]. Secondly, the geodetic mission phase (~168 days) with drifting orbits planned during the end stage of design life provides another gratifying aspect of denser spatial coverage for marine gravity recovery studies (Table 1). In summary, HY-2A has a great potential for enhancing the resolution and accuracy of the marine gravity field as long as the accuracy of SSH observations maintains an acceptable level for about several centimeters. Fortunately, the previous studies have verified that HY-2A has a similar performance to Jason-2 in foundation of cross calibration and crossover analysis, indicating that SSH observations have an accuracy of 6~8 cm or so [14–17], and meet the normal requirement of oceanic scientific researches and applications. However, most of the previous studies only focus on Interim Geophysical Data Record (IGDR) data, but the suitable waveform retrackers that can bring further improvement for SSH accuracy were poorly discussed before. Besides, HY-2A has a significantly lower application rate in the altimetry community compared to other typical missions, even lower than the afterward missions such as SARAL/AltiKa, Jason-3, and Sentinel-3A.

Under this circumstance, this paper will focus on two issues: (1) Whether the performance of HY-2A is reliable. The validation will be processed by comparing with the in situ time series of tide gauge measurements and by calculating the multimission crossover differences with contemporary missions; (2) Whether the range precision of HY-2A can be further improved by application of a two-pass weighted least squares retracker which was proposed by Sandwell and Smith (2005) for the purpose of optimal gravity field recovery. Previous studies have also shown that this two-pass approach successfully improves the range precision by a factor of ~1.5 for most conventional altimeters operating on both Ku-band and Ka-band [18–20]. If the HY-2A range precision also benefits from this two-pass retracking method, then there can be a potential dramatic enhancement in global marine gravity recovery relying on its geodetic mapping orbit.

Table 1. HY-2A mission specifications.

	HY-2A
Mission duration	16 August 2011–present
Size	8.56 m × 4.55 m × 3.185 m
Mass	≤1575 kg
Orbit type	Sun-synchronous orbit
Altitude	971~973 km
Inclination	99.34°
Repeat period	~14 days, ~168 days
Mode	Pulse-limited
Footprint size	2~10 km
Frequency	13.58 GHz. 5.2 GHz
Chirp bandwidth(Ku)	320 MHz/80 MHz/20 MHz
Chirp bandwidth(C)	160 MHz

2. HY-2A Data Description

The HY-2A, China's first oceanic satellite for exploring the dynamic environment with a radar altimeter on board, has been running operationally in orbit for more than 6 years since it was launched from Taiyuan on 16 August 2011. This satellite repeats its ground track as planned with a repetitive period of ~14 days for the first 4.5 years or so, while a geodetic mission with drifted ground tracks for each subcycle is operating currently since March 2016. All the level 2 products of HY-2A are administrated and distributed by NSOAS. Fortunately, we have successfully obtained most of the HY-2A Sensor Data Records (SDRs) including 20 Hz waveforms since January 2014 to April 2017 by official application from NSOAS (http://www.nsoas.gov.cn/). The specific information for the acquired HY-2A data sets is listed in Table 2.

Table 2. The general information for each cycle of acquired HY-2A altimeter data sets ('E' and 'G' indicates HY-2A data from ERM and GM, respectively).

Cycle	Time Scope	Data Quantity	Cycle	Time Scope	Data Quantity
E059	2013.12.21–2014.01.04	386 passes	E074	2014.07.19–2014.07.28	273 passes
E060	2014.01.04–2014.01.18	386 passes	E075	2014.08.03–2014.08.16	336 passes
E061	2014.01.18–2014.02.01	386 passes	E109	2015.11.21–2015.12.04	201 passes
E062	2014.02.01–2014.02.15	386 passes	E110	2015.12.07–2015.12.19	301 passes
E063	2014.02.15–2014.03.01	386 passes	E111	2015.12.19–2016.01.02	386 passes
E064	2014.03.01–2014.03.15	386 passes	E112	2016.01.02–2016.01.16	324 passes
E065	2014.03.15–2014.03.29	363 passes	E113	2016.01.16–2016.01.28	223 passes
E066	2014.03.29–2014.04.12	386 passes	E114	2016.01.30–2016.02.13	308 passes
E067	2014.04.12–2014.04.26	386 passes	E115	2016.02.13–2016.02.27	386 passes
E068	2014.04.26–2014.05.10	386 passes	E116	2016.02.27–2016.03.12	307 passes
E069	2014.05.10–2014.05.24	334 passes	E117	2016.03.12–2016.03.15	108 passes
E070	2014.05.24–2014.06.07	386 passes		Geodetic Mission	
E071	2014.06.07–2014.06.21	386 passes	G001	2016.03.24–2016.09.08	4542 passes
E072	2014.06.21–2014.07.05	376 passes	G002	2016.09.08–2017.02.23	4630 passes
E073	2014.07.05–2014.07.19	386 passes	G003	2017.02.23–2017.05.10	2030 passes

This paper focuses on the performance of HY-2A data, either the retracked measurements in level 2 products or the corresponding results of the two-pass waveform retracker. The spatial coverage is another crucial factor for evaluating altimetry data except for the common focus on range precision. Accordingly, the geographical distributions of HY-2A ground tracks were firstly analyzed over southeastern China seas by selecting the Cycle_E115 (from 13 February 2016 to 27 February 2016, 'E' represents ERM) and Cycle_G001 (from 24 March 2016 to 8 September 2016, 'G' represents GM) of HY-2A data. As shown in Figure 1, the HY-2A ERM data with a repetitive period of 14 days show significantly different ground tracks (red lines) over research regions compared with other typical altimeter missions, such as Jason-1/10 days (yellow lines), Jason-1 tandem with drifted intermediate orbit/10 days (orange lines), Geosat/17 days (purple lines), Envisat/35 days (blue lines), Envisat/30 days (green lines) and CryoSat-2/369 days (olive lines). The insert maps on the right side show magnified distributions within the black rectangular boxes for these indistinguishable tracks. The blue lines also represent the ground tracks of 35-day repeated orbit for ERS-1, ERS-2, and SARAL/AltiKa, while T/P, Jason-2, and Jason-3 have similar repetitive periods and geographical tracks with Jason-1 and its tandem mission. As shown in Figure 2, HY-2A GM data indeed has denser ground tracks (yellow lines) compared with ERM data (red lines) over the same research areas (subgraph), while a magnification of the local region within a blue rectangular box is also executed for giving better contrast effects.

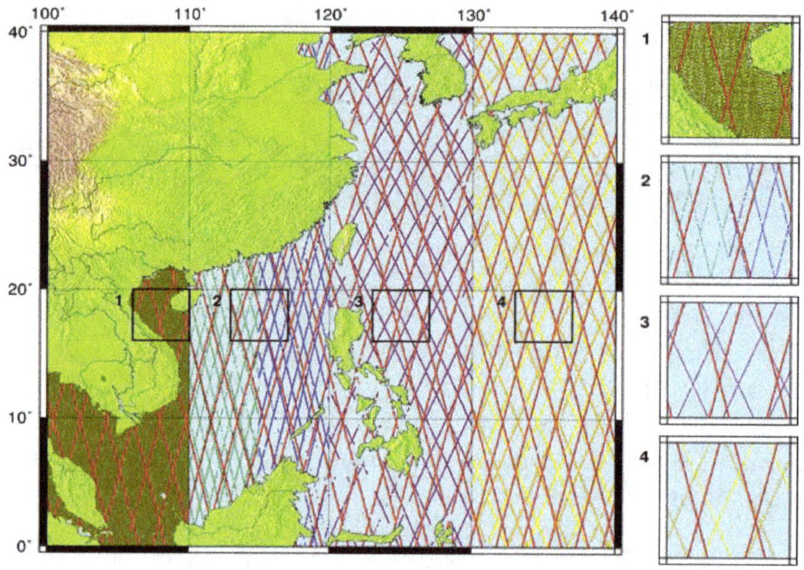

Figure 1. Distribution of multisatellite altimetry data over Research Region (HY-2A/ERM: red; Jason-1/ERM: yellow; Jason-1 Tandem/ERM: orange; Geosat/ERM: purple; Envisat/35 days: blue; Envisat/30 days: green; CryoSat-2/369 days: olive).

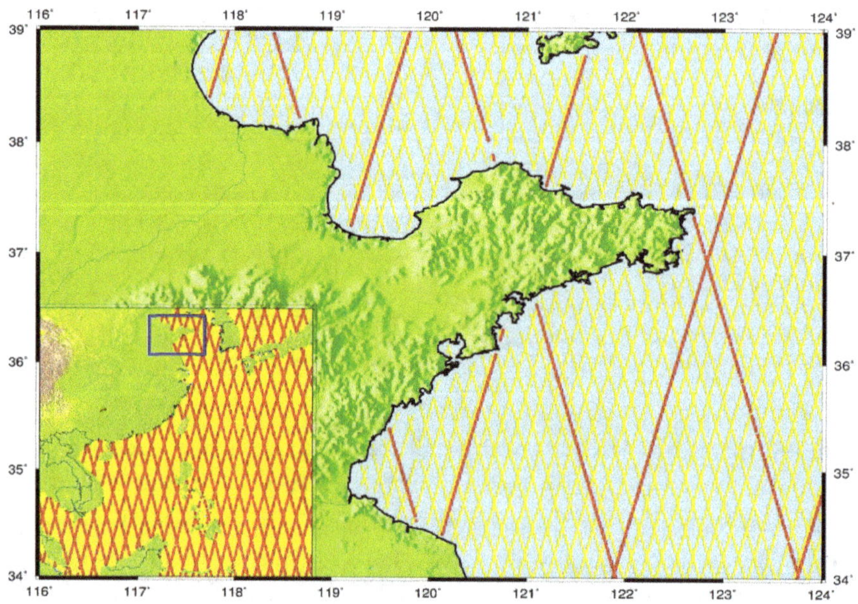

Figure 2. Distribution of HY-2A data over locally magnified region and research area (ERM: red; GM: yellow).

According to Sandwell and Smith's previous study [3], two or more satellites with different orbital inclinations and repetitive periods provide more stable computation of vertical deflections due to their diverse track orientations and intersections. Figure 1 proves that HY-2A ERM data is a great supplement to providing diverse track orientations of multisatellite materials for recovering marine gravity fields, while Figure 2 further verifies that the latest operation of HY-2A geodetic mission has brought denser data coverage. Under the premise of not considering the factor of range precision temporarily, there is enough reason to believe that HY-2A may potentially bring improvement of resolution for multisatellite, altimeter-derived marine gravity fields. Consequently, another crucial factor that directly relates to marine gravity recovery is the range precision of HY-2A and this study accordingly focuses on evaluating the performance of HY-2A measurements originally provided in level 2 products, which will be discussed in Section 3. Additionally, the two-pass retracker which was verified to be commonly effective for pulse-limited altimeter missions will be further investigated for HY-2A on evaluating whether the purpose of lower noise level is achieved. The procedures of estimating parameters and assessing performance are separately discussed in Sections 4 and 5.

3. Initial Assessment of HY-2A Measurement

3.1. Multi-Mission Crossover Analysis

Measurements from radar altimetry missions are classically validated at regional to global scales using relative calibration approaches based on statistical analysis at crossover points intra- and intermissions. Honestly, the smaller difference for intramission situation indicates better stability and internal conforming accuracy for onboard instruments while those crossover discrepancies between multimissions are more reliable for evaluating the range precision between altimetry measurements. Accordingly, we planned to firstly assess the HY-2A performance by analyzing crossover discrepancies with another typical high-precision mission. Meanwhile, the crossover difference for intramission situation is also considered for evaluating inner coincidence.

The satellite mission SARAL/AltiKa (hereafter referred as SARAL), which was launched in March 2013 by the Centre National d'Etudes Spatiales (CNES) and the Indian Space Research Organization (ISRO), shows outstanding performance in providing high-precision measurements due to the following improvements. SARAL operates at Ka-band frequency (35.75 GHz) and has a relatively small pulse-limited footprint and beam width in contrast to traditional Ku-band altimeters, which should decrease the contamination from bright off-nadir reflection. Furthermore, the relatively large bandwidth of SARAL allows a better vertical range sampling (~31 cm) than for Ku-band radar altimeters (~47 cm) which should strongly improve the determination of range [21–23]. Consequently, the SARAL geophysical data record (GDR) from cycle 1 to 12 was used as calibration data in this study, which is a homogeneous data set corresponding to more than one year (14 March 2013 to 8 May 2014).

Returning to the theory of crossover methodology, the validity requires that SSH measurements be entirely corrected from all ocean variability between the intersected passes of altimeters, while these SSH variations are caused by ocean tide, solid earth tide, barometer tide and wind effects. Therefore, the relevant environmental and geophysical corrections of the altimeter range measurements have been applied to calculating SSH. These corrections include dry and wet tropospheric path delay, ionospheric correction, sea state bias, ocean tide, solid earth tide, pole tide, high frequencies wind effect, and inverted barometer correction. All these corrections are provided in GDR products for HY-2A and SARAL, respectively, and no extra updating procedure is applied in this study. The specific information is listed in Table 3. Besides, the incompletely constrained time-varying effects of ocean surface will inevitably affect the crossover discrepancies for either intra- and intermissions. As a result, crossover differences with and without time limit need to be respectively considered during the statistical procedure. We selected the HY-2A data sets from E059 to E068 among the cross-mission validation based on the chosen SARAL data in order to ensure the overlap of measuring time.

Table 3. List of parameters used to estimate the SSH for both HY-2A and SARAL missions.

Contrastive Parameters	HY-2A	SARAL
Cycle range	59–68	1–12
Time scope	2013.12.21–2014.5.10	2013.3.14–2014.5.8
Dry troposphere correction	ECMWF	ECMWF
Wet troposphere correction	Radiometer	Radiometer
Ionospheric correction	Dual-frequency	GIM
Sea state bias	NSOAS empirical solution	NOAA empirical solution
Ocean tide	GOT00.2	GOT4.8
Solid earth tide	Cartwright and Tayler tables	Cartwright and Tayler tables
Pole tide	Wahr [24]	Wahr [24]
Inverted barometer correction	ECMWF	ECMWF
High-frequency fluctuations	Mog2D model	Mog2D model
Mean sea surface	NSOAS gridding solution	MSS_CNES_CLS_2011

Based on the procedures mentioned before, along-track SSH measurements are calculated for both HY-2A and SARAL missions. The HY-2A/HY-2A and SARAL/SARAL crossover points are respectively determined by fitting ground tracks of sample data, while the crossovers are defined as the positions where each satellite crosses its own ground position. In addition, the crossover points for intermissions are defined as the positions where HY-2A and SARAL cross the same sea surface position. To sum up, the geographical distributions of HY-2A data (red lines), SARAL data (blue lines), HY-2A/HY-2A (blue dots), SARAL/SARAL (red dots), and HY-2A/SARAL (green dots) crossovers are shown in Figure 3.

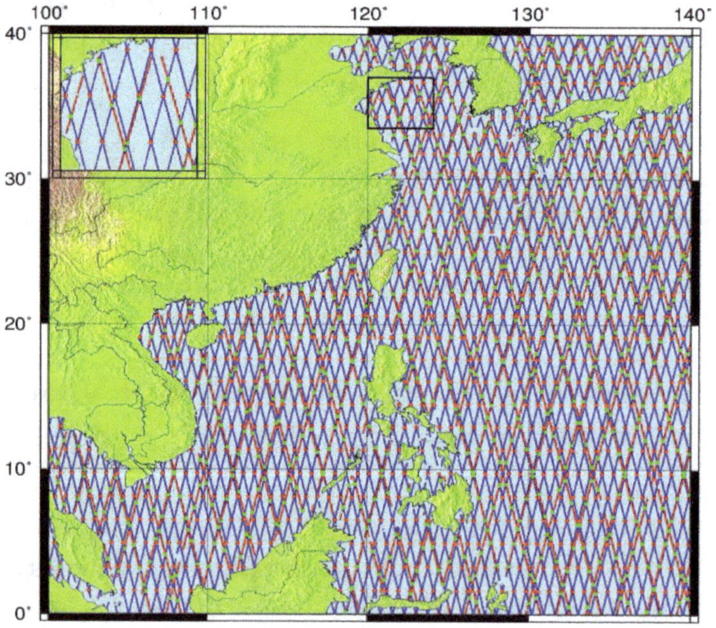

Figure 3. Distribution of HY-2A and SARAL altimetry data and corresponding crossover points intra- and intermissions (Red line: HY-2A data; Blue line: SARAL data; Blue dots: crossover points for HY-2A itself; Red dots: crossover points for SARAL itself; Green dots: crossover points for HY-2A and SARAL intermissions).

Following the calculated positions of crossover points, the minimum, maximum, mean, and root-mean-square (RMS) values of crossover discrepancies are computed between corresponding ascending and descending passes over ocean surfaces without and with time limits of 1, 2, 5 and 10 days, respectively. The specific results are listed in Table 4, which indicate that the measurements within the shortest time limits have the best coincidence under each situation. The differences between GOT00 and GOT4.8 might be as large as a centimeter level and influential in comparing SSH between HY-2A and SARAL. Consequently, we use the GOT-e 4.7 model (http://web.cecs.pdx.edu/~zaron/pub/GOT-e.html), which contains GOT-compatible grids and provides more valid values to cover the shoreline, for updating corresponding ocean tide corrections for both missions. The intermission crossover analysis is thus executed twice and the statistical information is listed in Table 5. HY-2A has a valid but slightly worse performance than SARAL according to crossover analysis for either intramission or intermission situations. Besides the close measuring time, the uniform GOT-e model also reduces the SSH discrepancies for intermissions. A preliminary conclusion is drawn that the accuracy of HY-2A measurements is around ~10 cm while the inherent uncertainty of SARAL is taken into account. This validation initially implies that HY-2A is capable of investigating marine gravity anomalies due to the new data coverage and reliable range precision.

Table 4. Statistical information of crossover differences under intramission situations. Unit: m.

Time Limit	HY-2A					SARAL				
	Num	Min	Max	Mean	RMS	Num	Min	Max	Mean	RMS
–	5136	−0.995	0.998	0.083	0.232	32146	−0.976	0.893	0.016	0.142
≤10 days	816	−0.979	0.841	0.096	0.182	2937	−0.607	0.732	0.017	0.082
≤5 days	442	−0.931	0.841	0.090	0.177	1468	−0.255	0.297	0.016	0.071
≤2 days	242	−0.899	0.841	0.086	0.170	557	−0.232	0.279	0.014	0.062
≤1 day	143	−0.664	0.608	0.095	0.148	371	−0.196	0.279	0.015	0.061

Table 5. Statistical information of crossover differences under intermission situation ('HY' and 'SA' represent HY-2A and SARAL, 'OTC' for ocean tide correction, while '_A' and '_D' denote ascending and descending passes, respectively). Unit: m.

Mission	Time Limit	OTC in GDR					Updated OTC with GOT-e				
		Num	Min	Max	Mean	RMS	Num	Min	Max	Mean	RMS
HY_A & SA_D	–	13,784	−0.998	0.958	0.191	0.250	13,216	−0.999	0.980	0.070	0.199
	≤1 day	131	−0.804	0.543	0.242	0.140	127	−0.942	0.406	0.097	0.139
HY_D & SA_A	–	11,590	−0.999	0.917	0.100	0.206	11,302	−0.999	0.782	−0.029	0.190
	≤1 day	100	−0.868	0.641	0.092	0.151	99	−0.999	0.321	−0.046	0.155
HY_A & SA_A	–	418	−0.993	0.650	0.209	0.226	396	−0.993	0.515	0.081	0.181
	≤1 day	5	0.041	0.439	0.222	0.162	5	−0.097	0.312	0.087	0.165
HY_D & SA_D	–	359	−0.992	0.951	0.124	0.224	349	−0.717	0.802	−0.0002	0.201
	≤1 day	3	0.047	0.419	0.175	0.211	3	−0.078	0.294	0.046	0.210
HY & SA	–	26,151	−0.999	0.958	0.150	0.235	25,263	−0.999	0.980	0.024	0.201
	≤1 day	239	−0.868	0.641	0.178	0.184	234	−0.999	0.406	0.035	0.175

3.2. Validation with Tide Gauge Measurements

Besides the classical cross-mission validation, another well-known method is that sea surface heights from radar altimetry missions and in situ sea level measurements from tide gauges serve as each other's buddy check. Consequently, we explored the performance of HY-2A observations in this section by comparing with time series of in situ sea level at tide gauge stations. Simultaneously, the 12 cycles of SARAL data were also evaluated during this procedure as a comparison. Nevertheless, the HY-2A time series were extended to 16 August 2014 (E075) because we wanted to involve the time series for as long as possible. This validation needed time overlap between altimetric measurements and in situ data series rather than between the two altimetric missions, while a significant data interruption happened to HY-2A after Cycle_E075.

Radar altimeters provide SSH observations, which need to be referenced and corrected from geophysical signals to provide sea level anomalies (SLA) compared with in situ measurements [25]. Therefore, the item of mean sea surface provided in level 2 products needs to be subtracted on the basis of calculating SSH as described in Section 3.1, after which along-track SLA series are obtained. We used valid-only satellite altimetry measurements relying on the quality flags and recommended data editing criteria in handbooks for both missions. The SLA series were then averaged onto a regular $1^\circ \times 1^\circ$ grid, with a temporal resampling corresponding to data intervals at half a month [26].

The University of Hawaii Sea Level Center (UHSLC) offers tide gauge data with two levels of quality control (QC), where more rigorous QC requires more effort and a greater time lag. The Research Quality Data (RQD) sets received thorough QC and were considered to be the final science-ready data set, while the RQD were released 1–2 years after data was received from the data originators by UHSLC. Consequently, this RQD obtained via ftp://ftp.soest.hawaii.edu/uhslc/rqds was used for evaluating the performance of HY-2A measurements in this study. Although the spatial coverage of tide gauges is limited to coastal areas or isolated islands in the open ocean, the tide gauges will sample the ocean every hour, while at a given point, the satellite altimetry sampling is higher than ten days. The RQD is maintained by Pat Caldwell [27] and archived as the NOAA Joint Archive for Mean Sea Level (JAMSL) and only available from the UHSLC at hourly and daily resolution in order to avoid confusion with monthly data provided by the Permanent Service for Mean Sea Level (PSMSL). Consequently, the daily RQD was used for validation and a preprocessing procedure was applied to high-frequency in situ measurements for resampling a time series at the low frequency of altimetry data in 15-day intervals.

Based on the acquired global RQD sets and time span for two altimeter missions, 292 tide gauge stations (red dots in Figure 4) were probably for validation as only these in situ data series span across the period from 2013 to 2015, while 6 stations (yellow stars in Figure 4) among them were initially chosen for validation. All these 6 stations are located at the isolated islands over the open ocean and their geographical distributions represent the most situations over open oceans. The specific information for the chosen tide gauges is listed in Table 6.

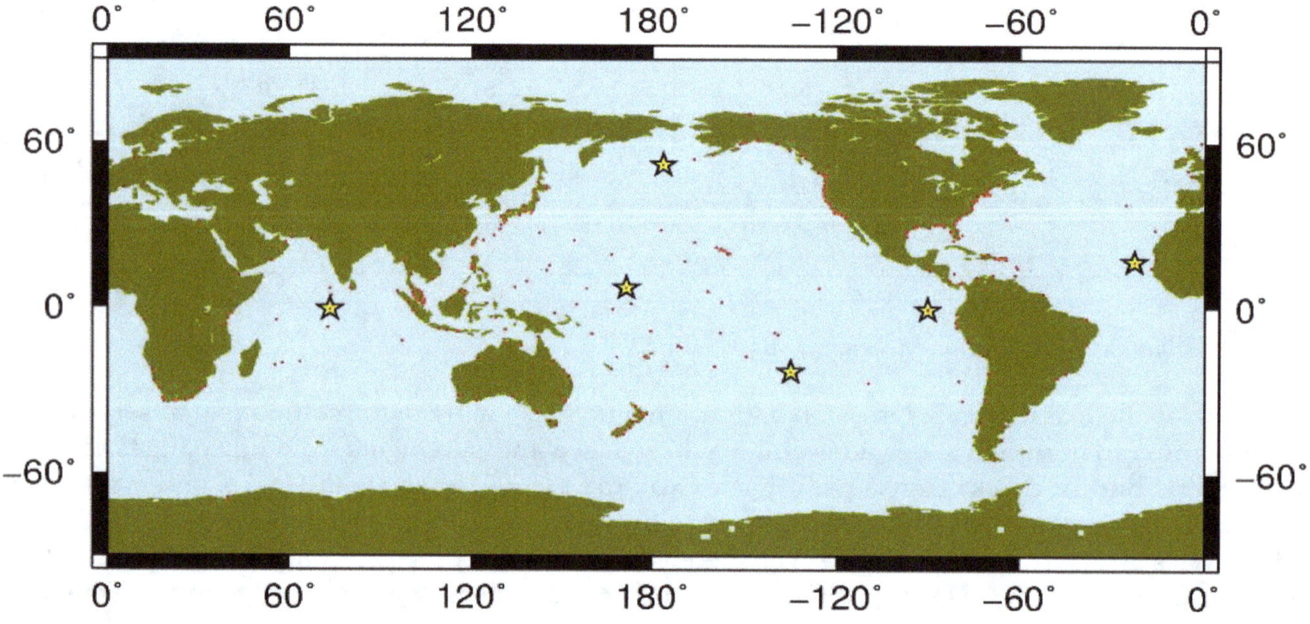

Figure 4. The 6 tide gauge stations (yellow stars) considered in this study. Red dots denote 292 tide gauge stations that span across the period from 2013 to 2015.

Table 6. Specific information of tide gauges in this study and corresponding correlation parameter (CI: Completeness index or percentage of data span without missing data; QC-YEARS: years which have received quality control).

Station	Latitude	Longitude	QC-YEARS	CI	Country	Contributor	Correlation Coefficient	
							SARAL	HY-2A
Majuro-B	07°–07°N	171°–22°E	1993–2016	98	Rep. Marshall I.	Nat. Tidal Ctr., BOM	0.8079	0.9572
Gan	00°–41°S	073°–09°E	1987–2015	93	Rep. of Maldives	UH Sea Level Center	0.8224	0.7606
Palmeira C.Verde	16°–45°N	022°–59°W	2000–2015	92	Portugal	UH Sea Level Cente	0.6278	0.5135
Adak Alaska	51°–52°N	176°–38°W	1950–2016	92	USA	National Ocean Service	0.4677	0.6054
Rikitea	23°–08°S	134°–57°W	1969–2015	93	French Polynesia	UH Sea Level Center	0.8859	0.4982
Santa Cruz	00°–45°S	090°–19°W	1978–2015	96	Ecuador	UH Sea Level Center	0.9016	0.9281
Los Angeles, CA	33°–43°N	118°–16°W	1923–2016	99	USA	National Ocean Service	0.6589	0.6276
Cape May, NJ	38°–58°N	074°–58°W	1965–2016	91	USA	National Ocean Service	0.8171	0.7386
Ko Lak	11°–48°N	099°–49°E	1985–2017	95	Thailand	Naval Hydro. Dept.	0.9157	0.9407
Hong Kong-B	22°–18°N	114°–13°E	1986–2016	99	China	HK Observatory	0.8045	0.1467

After resampling the tide gauge time series, a set of correlation coefficients between the altimetry grids and the in situ records were computed. The maximum correlation is found at four 1° × 1° grid points and corresponding satellite altimetry time series are then extracted for plotting collation maps. The maps are shown in Figure 5 and corresponding correlation coefficients for altimetry and in situ series are listed in Table 6. The average values of correlation coefficients are respectively, 0.71 and 0.75 for HY-2A and SARAL. Both the missions provide valid performance due to our numerical test and the agreement is slightly better for in situ/SARAL than for in situ/HY-2A.

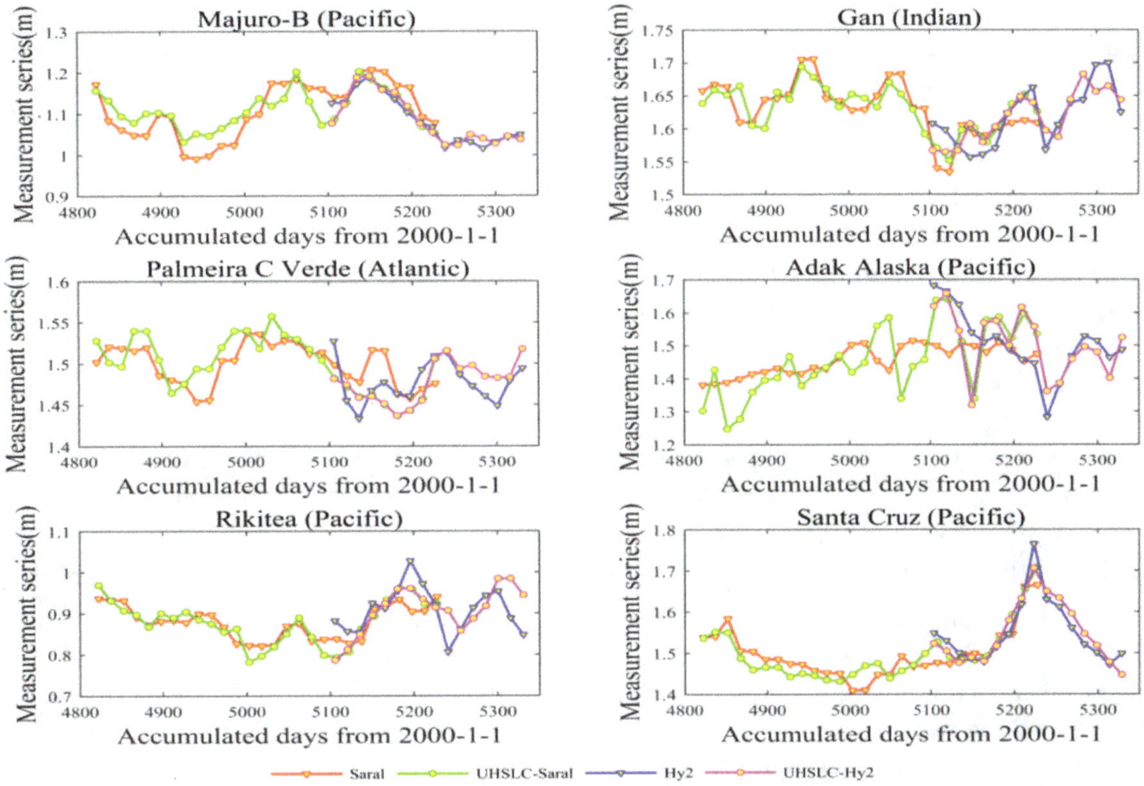

Figure 5. Comparison of HY-2A and SARAL SLA series with tide gauge sea level series at 6 stations (Red: SARAL; Green: TG for SARAL; Blue: HY-2A; Purple: TG for HY-2A).

We also calculated the correlation coefficients for the tide gauge sites used for validation in the recent published studies [28,29]. Four UHSLC sites (Los Angeles, CA; Cape May, NJ; Ko Lak; Hong Kong) with required data series were selected due to the time-scope repetition with HY-2A and SARAL. The specific information for these sites is also listed in Table 6. Slightly larger average values of correlation coefficients are obtained for HY-2A (0.76) and SARAL (0.79), respectively, while the Hong Kong station is excluded for statistics due to the serious data loss which occurred at surrounding areas for HY-2A. We can draw a similar conclusion that the HY-2A altimeter provides valid sea surface height measurements but at a slightly lower accuracy level than SARAL. Although HY-2A provides noisier measurements than SARAL according to cross-mission analysis and verification with in situ tide gauge measurements, we also found the reliable side of HY-2A data and the explorer procedure about the two-pass waveform retracker for the purpose of gravity recovery is accordingly reasonable and necessary.

4. Two-Pass Retracker for HY-2A

4.1. Theory of Two-Pass Retracker

Another major objective of our analysis is to evaluate whether the precision of HY-2A range measurements are further improved by using a two-pass waveform retracker which was proposed by Sandwell and Smith [9] initially and has already proven to be effective for other pulse-limited altimeter missions [19,20]. It is well known that waveform retracking is an effective and mature technique for solving limitations of data contamination by distribution of land or islands and diverse algorithms are developed either on empirical statistics or fitting functions for raw waveforms. Over the open ocean, the typical shapes of raw waveforms are well matched by the Brown [30] model with five parameters: arrival time, rise time, amplitude, square of the antenna mispointing angle, and thermal noise. The most concerning parameter for applications in marine gravity recovery is the arrival time, which provides an estimate of distance between the satellite and the instantaneous sea surface. In addition, the rise time and amplitude should also be set as variables for each raw waveform under any situations, while the thermal noise and square of the antenna mispointing angle can be treated as constant parameters for common waveforms obtained over open oceans.

Owing to the relationship between the errors in retracked estimates of arrival time and rise time [31], one effective approach to improving the precision of the arrival time is to fix the rise time parameter to a predetermined value based on surrounding waveforms. This can be accomplished with the two-pass retracking method where the waveforms are first fitted by the three-parameter Brown model (arrival time, rise time, and amplitude) with invariable constants describing thermal noise and antenna mispointing angles. The rise time parameter is then smoothed along-track, before retracking the waveforms a second time using a two-parameter Brown model (arrival time and amplitude) with the rise time being fixed to the along-track smoothed value [9]. Either the former three parameters or the latter two parameters will be solved during the weighted least squaring procedure, while the determination of the constants for describing thermal noise and antenna mispointing angles is critical before the two-pass model fitting procedure. The parameters and procedures involved in the two-pass retracker are deduced from the principle and power model of radar echoes over ocean surfaces.

The returned signal of conventional pulse-limited radar altimeters is proportional to the illuminated area scaled by the off-nadir roll-off of antenna gain pattern [30]. We are only interested in the slopes of SSH when using vertical deflection as an intermediate variable for recovering marine gravity field, and the constant offset with long-wave characteristics can be ignored accordingly. We also consider the case where the antenna bore sight is normally pointed to nadir. Under these simplifying assumptions, the power in the returned waveform can be modeled as following Equation (1), where A is the amplitude, t_0 is the arrival time, t is time for each sample gate, σ is the effective pulse length, and α is the trailing edge decay factor [32]. The effective pulse length σ can be calculated from two factors $\sigma^2 = {\sigma_h}^2 + {\sigma_p}^2$. The σ_h is the broadening of the pulse due to the interaction of ocean waves $\sigma_h = SWH/2c$, where c is the transferring speed of radar signals. The $\sigma_p = 0.513\tau_p$ is the standard deviation of the length of the outgoing pulse where $\tau_p = 1/B$ and B represents the chirp bandwidth for HY-2A (Ku: 320 MHz/80 MHz/20 MHz; C: 160 MHz). Three bandwidths for chirp signals are adopted in Model Compatible Tracker (MCT): the 320 MHz signal is arranged for oceanic measurement while

80 MHz is for the coastal zone, and the 20 MHz is for land or ice coverage. The trailing edge decay factor α mostly depends on the square of off-nadir pointing angle and can be treated as a constant value for HY-2A. Therefore, the fitting model has only three parameters $[A, t_0, \sigma_h]$ to be solved by using the weighted least squares approach, while the constant value α should be predetermined.

$$M(t) = \frac{A}{2}[1 + erf(\frac{t - t_0}{\sqrt{2}\sigma})]exp[-\alpha(t - t_0)] \quad (1)$$

$$X^2 = \sum_{i=1}^{N} [\frac{P_i - M(t_i - t_0 \cdot \sigma_h \cdot A)}{W_i}] \quad (2)$$

$$W_i = \frac{P_i + P_0}{\sqrt{K}} \quad (3)$$

The chi-square measure of the misfit between the returned power samples P_i and the model estimate $M(t_i)$ for the corresponding time t_i is formulated as Equation (2), where N is the number of gates (time samples) in the waveform, and W_i is the uncertainty assigned to each sample and given by Equation (3), where K is the number of statistically independent return echoes that were averaged to produce the waveform data records and P_0 is the power offset value reflecting the thermal noise level of the instrument. $K = 96$ is empirically determined for HY-2A as this mission collects ocean data at a pulse repetition frequency (PRF) of 1~4 kHz and similarly provides 20 Hz products in contrast to other altimeter missions [19]. P_0 will be given further discussion for selecting the optimal option in Section 4.2 although the results of the weighted least squares approach are largely insensitive to the numerical values for both K and P_0 [9].

In summary, the three parameters $[A, t_0, \sigma_h]$ can be firstly calculated through an iterative weighted least squares approach once the numerical values of α and P_0 are given previously. Besides, a threshold retracker is also introduced to provide an initial estimate of t_0 to accelerate the iterative process, while the corresponding threshold value is defined with respect to the cumulative sum of waveform power and should be predetermined as well. Secondly, the rise time parameter about SWH is smoothed along-track using a low-pass filter having a 0.5 gain at a full wavelength of 90 km [9,19]. The final step of the two-pass retracker is to process the raw waveforms again by fixing the rise time parameter to the smoothed value and only solving for the parameters of arrival time and amplitude. Based on the theory and procedures of the two-pass retracker mentioned before, we will discuss how to determine the corresponding parameters in the next section.

4.2. Parameter Determination

The typical shapes of HY-2A waveforms over the open ocean are also well matched by the Brown model similar to other Ku-band missions. By contrast, both HY-2A and Jason-1 waveforms have equally spaced gates or samples although the counts are 128 and 104, respectively. Both sets of waveforms have a low noise power level prior to the sharp arrival of the main pulse, which starts to show up around gate 30~32. Besides, HY-2A waveforms have relatively steeper trailing edge decays and corresponding parameters describing waveform tail should be adjusted. The special deformed waveforms over sea ice-covered areas, large lakes, or land areas are not discussed in this paper as we focus on the performance of the two-pass retracker for optimal gravity field recovery over open ocean areas. Consequently, we initially assume that the two-pass retracker is still effective for this China altimetry mission and accordingly determine the adjusted parameters for fitting HY-2A echoes.

As discussed in the theory section, we began the analysis for the two-pass retracker with five unknown parameters $[A, t_0, \sigma_h]$ and $[\alpha, P_0]$. The first three parameters need to be calculated by using the weighted least squares approach, while the latter items should be predetermined once the objective waveform samples are sufficient. In addition, a threshold value needs to be predetermined for providing an initial estimate of t_0 to accelerate the iterative process. The determination of $[\alpha, P_0]$ and the threshold value was discussed on the basis of HY-2A waveform samples from Cycle_E115.

Two descending passes labeled 222 and 260, respectively, were further selected for determining optimal parameters. Moreover, the screening criteria of valid results during the retracking procedure were also involved.

Firstly, α is a crucial parameter for fitting the waveforms, and the two-pass retracker is more sensitive to α than P_0 and threshold value [20]. The factor α depends mostly on the square of off-nadir pointing angle with a small order of magnitude and can be treated as a constant value for HY-2A. Based on the sampled passes, we gathered a number of successfully fitted waveform records and mean misfit values between original powers and modeled results respectively when using several typical values of α within the range of 0.005 and 0.015. Then we repeated this evaluation by sweeping through values of α at steps of 1.0×10^{-4} within range of 0.009 and 0.012. A best-fit α value of 0.0105 was approximately determined by considering both the valid records and misfit values for HY-2A. Meanwhile, the best-fit α values for different altimeter missions determined by previous studies are listed in Table 7 for summarization [18–20].

Table 7. The summarized best-fit α values for different satellite altimeter missions.

Mission	**Geosat**	**ERS-1**	**Envisat**	**T/P Jason-1**	**CryoSat-2 LRM**	**CryoSat-2 SAR/SIN**	**SARAL**	**HY-2A**
α value	0.006	0.022	0.09	0.0058	0.013	0.00744	0.0351	0.0105

Secondly, we investigated the relationship between waveform fitting results and the parameter P_0 describing the background noise level caused by temperature-dependent thermal noise in the receiver of the satellite altimeter. The magnitude of P_0 depends on the engineering characteristics of the altimeter and varies from the automatic gain control settings. Generally, P_0 can be either treated as a constant or empirically estimated through mean values or sum values of echo power for the first several sample gates [19,33]. 5 is used in this study for avoiding the false estimation due to complicated waveforms. Considering that the model-fitting results using the weighted least squares approach are largely insensitive to the numerical values of P_0, we treated P_0 as a constant similarly following the previous studies about ERS-1, Jason-1, and SARAL [19,20]. Besides, the power value for each gate is scaled by an arbitrary value of 3.3×10^6 to convert the original integers to floating point numbers and this scale value is completely irrelevant to the two-pass retracking procedure. Fortunately, the inheritance numerical value of P_0 from Jason-1 (5500) is still reliable for HY-2A waveforms under the selected scale value (3.3×10^6) because the ratios between the background noise level and the maximum power of echoes are similar for both missions.

Thirdly, a threshold value with respect to the cumulative sum of waveform power needs to be predetermined for providing an initial estimate of t_0 to accelerate the iterative process. Actually, the two-pass retracking results through iterative fitting are largely insensitive to the threshold value according to our experiments. Accordingly, we determined the optimal threshold value by minimizing the difference with respect to the results of the threshold retracker developed by Davis [34] for an exact initial objective gate, which leads to a valid threshold value of 0.015 for HY-2A.

Further discussion is needed about seriously deformed waveforms that will get unreliable results after the retracking procedure once the predetermined parameters $[\alpha, P_0]$ are given. Accordingly, we developed a waveform editing criteria based on estimated parameters $[A, \sigma_h]$ as well as the fits to the waveforms executed. The editing threshold was established by constructing histograms of amplitude, chi-square misfit, and SWH versus standard deviation of the arrival time parameter. For HY-2A, valid waveforms should have model amplitudes within the range of 40,000–80,000, while the chi-squared misfit measurement should not exceed 800. The majority values of chi-squared misfit measurement are distributed within the range of 350~800 for valid retracked results. We also removed waveforms having SWH outside of the range 0.3–10 m for excluding observations over extremely unusual sea state conditions. Considering the robustness of the threshold retracker, we adopted the estimated results by the predetermined threshold instead when the HY-2A waveform records failed the editing criteria of the two-pass retracker.

While the predetermined parameters, threshold value, and waveform editing criteria were given, the results of the least squares fit of model waveforms and original waveform shapes are shown in Figure 6 for typical values of chi-squared misfit measurements. After examining thousands of waveforms over a range of sea states, we generally found that the fit to the HY-2A data is visually good, which suggested that the predetermined parameters were reliable. In accordance with the analysis by Garcia et al. [19], we also found a significant improvement (~5%) in the 20-Hz range precision if the least squares analysis was performed on three adjacent waveforms simultaneously, assigning the two outlier waveforms 1/2 the weight of the central waveform. Further enlargement of this three-waveform window did not reduce the range noise but increased processing time.

Once the two-pass retracker with three parameters $[A, t_0, \sigma_h]$ to be solved in the first step was optimized for selected constant parameters $[\alpha, P_0]$ and threshold value, we retracked the complete Cycle_E115 with the full three-parameter model and corresponding waveform editing criteria. Then the rise time parameter was low-pass filtered with its 0.5 gain at a full wavelength of 90 km. The final step was to retrack the waveforms by fixing the rise time parameter to the smoothed value and only solving the parameter of the arrival time and amplitude.

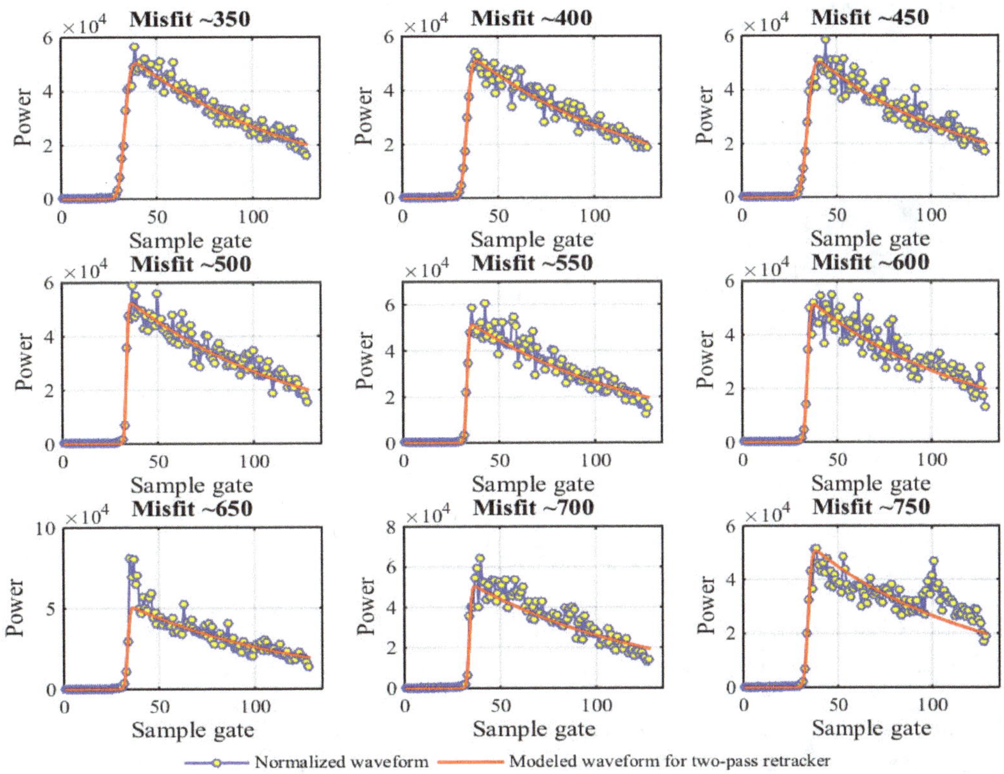

Figure 6. Least squares fit of model waveform of HY-2A for predetermined parameters (Blue: normalized waveform in SDR; Red: Modeled waveform of two-pass retracker).

5. Discussion

To assess the noise level of HY-2A data by the two-pass retracker, we performed a statistical analysis on the retracked SSH values (altitude minus range). All the corrections for path delay and geophysical environment were considered using items supplied by level 2 products. The noise was estimated as the standard deviation of the SSH with respect to the mean difference from the EIGEN6C4 model computed over a 1-second interval. The EIGEN6C4 model was used to account for areas of very steep geoid gradient where there could be a significant height variation (>20 mm) over the

7-km distance that the ground track moves in one second. This involved the provided 20 estimates corresponding to the HY-2A sampling rate along ground tracks.

Figure 7 shows the noise levels for the HY-2A sample pass labeled 260 in Cycle_E115. The noise level in the solved arrival time from the waveform leading edge should increase along with enhancing SWH and we plotted the noise level versus SWH thusly. Figure 7 also shows three independent estimates of noise level: the initial estimates derived from the retracked range measurement in SDR (red dots), the estimates from 3-parameter retracking (green dots), and ultimate 2-parameter retracking (blue dots), respectively. Additionally, the solid smoothed curves in the subgraph below with three corresponding colors are median averages of these estimates in 0.5-m SWH bins. The noise levels of three-parameter solutions and retracked results in SDR are at a similar level, while the latter is slightly superior. One more important finding is that the noise level of two-parameter retracked data is significantly lower than that for three-parameter solutions. Under the typical conditions of 2-m SWH, mean values of standard deviation for SDR, 3-parameter, and 2-parameter results are, respectively, 66.3, 67.1 and 40.9 mm for HY-2A, which is comparable to typical altimeter missions. The values for 3-parameter and 2-parameter retracked Jason-1 are, respectively, 75.9 and 46.4 mm, while CryoSat-2 LRM results are 64.7 and 42.7 mm [19]. This also demonstrates that the 2-parameter retracking decreases the noise level by a factor of 1.6 with respect to 3-parameter retracking, which is very close to the expected noise reduction of 1.57 based on the Monte Carlo simulation [9].

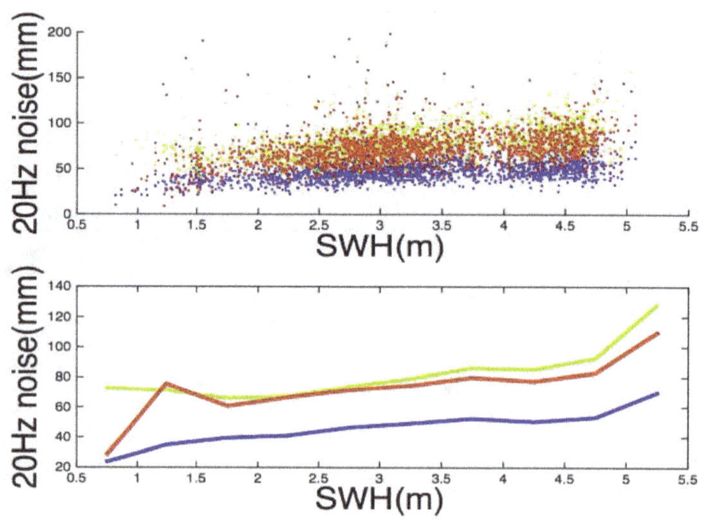

Figure 7. Standard deviation of retracked height with respect to EIGEN6C4 for HY-2A (Red—sensor data record; Green—3-parameter retracking; Blue—2-parameter retracking).

Spectral analysis is another effective way of analyzing the signal and noise ratio in the altimeter observations. On the basis of the Welch method [35], the power spectral density of along-track SSH for 3-parameter and 2-parameter retracked HY-2A data were respectively estimated using MATLAB and shown in Figure 8a. Both spectra show a rapid increase in power for wavelengths greater than 10 km and both the 3- and 2-parameter show a white noise level at wavelengths less than about 5 km. To highlight the differences between the two spectra, we calculated the spectra of the difference series between them as shown in Figure 8b. The increase in the difference spectra for wavelengths shorter than 90 km simply reflects the wavelength over which the SWH is smoothed between the two steps of iterative retracking procedures. At longer wavelengths, both the 3- and 2-parameter retrackers provide the same height measurement because the profiles contain the same SWH signal. The "hump" in the difference spectra between the wavelengths of 90 and ~5 km is also seen in spectra of two-pass retracked data from other pulse-limited altimeter missions [19,20]. This also demonstrates the band of wavelengths where two-pass retracking provides the most benefit [9].

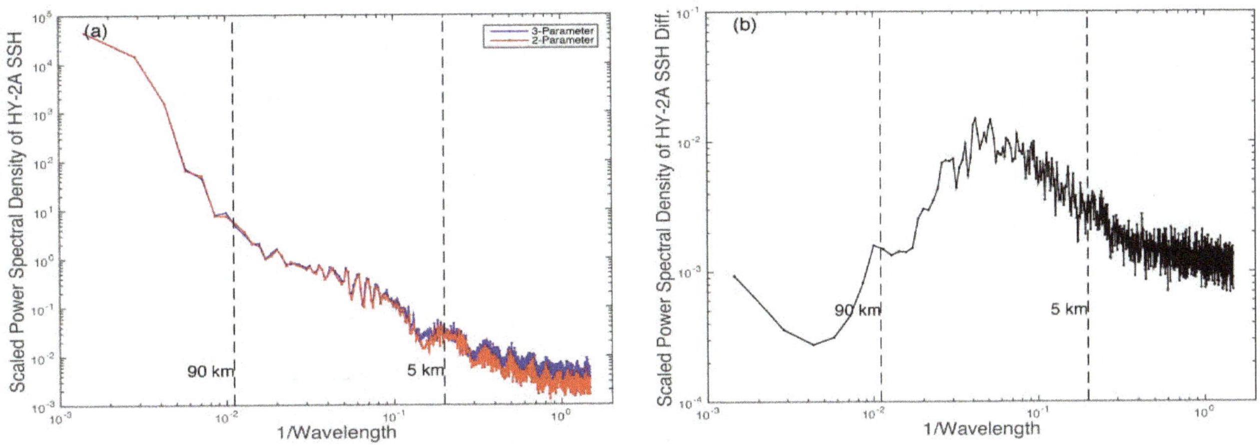

Figure 8. (**a**) Power spectral density of along-track retracked height from 3-parameter (blue) and 2-parameter (red) model for HY-2A data; (**b**) Power spectral density of along-track retracked height difference between 3-parameter and 2-parameter model for HY-2A. Wavelength range of spectral hump within 5–90 km is marked by vertical dashed lines.

The discrepancy at crossover points is another important index factor for assessing the accuracy of inner coincidence for altimeter SSH observations as mentioned before. Based on the HY-2A SDR data from Cycle_E115 over the southeastern China seas (100°~140°E, 0°~40°N), the statistic values of discrepancy at crossover points for three retracked results (SDR; 3-parameter and 2-parameter) are listed in Table 8, while the corrections for path delay and geophysical environment are also considered using items supplied by level 2 products. In addition, the corresponding results for Jason-1 GM data from Cycle500 are similarly calculated for comparison.

Table 8. Statistics of the SSH discrepancy at crossover points over southeastern China seas for HY-2A (Cycle_E115) and Jason-1 (Cycle500) sample data Unit: m.

Mission	Retracker Description	Crossover Points	Min	Max	MEAN	RMS
HY-2A	No retracker	125	−0.984	0.958	−0.287	0.412
	SDR—MLE4	125	−0.985	0.833	−0.280	0.391
	3-parameter	125	−0.975	0.916	−0.290	0.389
	2-parameter	125	−0.918	0.630	−0.272	0.377
Jason-1 GM	No retracker	126	−1.996	1.031	0.026	0.439
	SDR—MLE3	126	−0.713	0.547	0.048	0.191
	3-parameter	126	−0.531	1.286	0.072	0.230
	2-parameter	126	−0.446	1.006	0.066	0.208

Table 8 indicates that all the three retracked results show improvement with respect to results without any retracker, and more remarkable for Jason-1. Moreover, the ultimate retracked results derived from the 2-parameter model are slightly superior to that of the 3-parameter model for both missions, which proves that the second iterative fitting procedure with smoothed SWH values as priority information does bring improvement for range precisions. In addition, the larger discrepancies may lead to an initial conclusion that the precision of HY-2A SSH is slightly lower than Jason-1 data, which is consistent with the previous conclusion in Section 3. Nevertheless, the accuracy improvement by the two-pass retracker was verified for both missions and fortunately HY-2A can potentially calculate sea surface slopes at a similar accuracy level with respect to Jason-1 GM data due to the statistics of standard deviation over a 1-second interval.

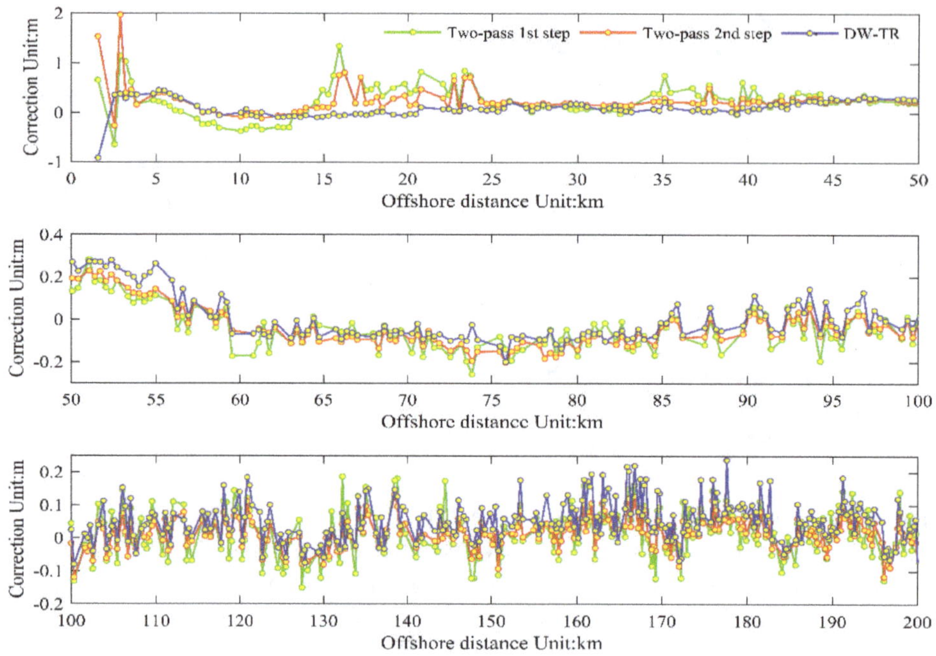

Figure 9. Comparison of range correction series for two-pass retracker and DW-TR retracker (Green: Two-pass retracker 1st step; Red: Two-pass retracker, 2nd step; Blue: DW-TR results).

We also introduced the DW-TR retracker proposed by Huang et al. [28] for comparison based on the same sequence of HY-2A waveforms. The sample data (Cycle_E075 Pass_001) is located at the Gulf of Alaska and contains 535 waveforms with an offshore distance of less than ~200 km. The range corrections for the sample data with respect to corresponding offshore distances are plotted in Figure 9. The comparison shows that the correlation coefficient is large (0.92) for typical waveforms with offshore distance larger than 50 km, and it dramatically declines to ~0.10 for contaminated waveforms with offshore distances less than 50 km. The two-pass retracker pays more attention to optimal gravity recovery over the open ocean, while the DW-TR retracker focuses mostly on the coastal applications. As a result, we can initially draw the conclusion that the two-pass waveform retracker is valid over the open ocean.

Moreover, the 2nd-step of the two-pass retracker (red lines) got a closer result with DW-TR retracker than the 1st step. It's hard to evaluate which retracker is more accurate relying on the current comparison. However, the accuracy improvement between the two steps of the retracker is detectable if we take the DW-TR result as reference. This improvement is also verified three other ways described before, STD calculation, power spectral density estimation, and crossover analysis. It's reasonable to draw the conclusion that the two-pass retracker can bring range precision enhancement for HY-2A.

6. Conclusions

The validation activities represent an essential part of describing HY-2A altimetry data quality assessment and allow a quick feedback to operational teams and users in the altimetry community. Based on the investigations performed in this study, HY-2A firstly provides unique data coverage over ocean regions within 80°N and 80°S according to its specific orbital parameters with respect to other altimetric missions. Secondly, HY-2A provides reliable measurements in level 2 products based on the analysis of the multimission crossover discrepancies and the exterior verification with tide gauge measurement series. Thirdly, the two-pass retracker does bring improvement of HY-2A range precision and significant decline of noise level by a factor of 1.6, similar to previous pulse-limited altimeter missions.

In spite of the declined accuracy level of SSH observations with respect to SARAL based on our analysis, the along-track variances over a 1-second interval for retracked HY-2A results stay at a comparable level with respect to other missions (e.g., Jason-1/GM, CryoSat-2/LRM), which is an indicator for calculating reliable sea surface slopes. Additionally, the new data coverage due to HY2A's individual orbital parameters may provide detective signals over somewhere unreached previously. Therefore, the accumulation of HY-2A data, especially for the GM phase recently, provides a great opportunity to explore the finer structure of marine gravity fields.

Almost 5 years ago, marine gravity fields having accuracies of 3~5 mGal (e.g., S&S V18 [18] and DNSC08 [11]) were based primarily on dense track coverage from 18 months of Geosat data from geodetic missions collected in the early 1990s and 12 months of European Remote-Sensing Satellite-1 (ERS-1/GM) data collected in 1995~1996. Over the last five years, the latest global mainstream marine gravity model such as DTU13 [36] and V23.1 [37] have got precision breakthrough at 1~3 mGal which mainly benefits from new missions with nonrepetitive ground tracks such as Jason-1/GM and CryoSat-2/369d. Moreover, both the orbits of SARAL and HY-2A satellites began to drift by up to 10 km from their initial 35-day track and 14-day track, respectively, and resulted in several cycles of altimeter data for new ocean coverage. Consequently, there are enough reasons to believe that the objective of 1-mGal accuracy and 1-minute resolution for the global marine gravity field is accessible in the recent future. Besides, the next big step in gravity field improvement may come with the higher range precision and finer resolution provided by the planned SWOT mission, which is scheduled for launch in 2020 [38].

Acknowledgments: The multi-satellite altimeter data were provided by CNES and European Space Agency, while the HY-2A data and tide-gauge data were respectively distributed by NSOAS and UHSLC. The authors also thank Wang Haihong for providing DW-TR results for HY-2A sample waveforms. The manuscript was considerably improved through constructive comments from three anonymous reviewers and the assigned academic editor, which are gratefully acknowledged. This work was supported by the National Natural Science Foundation of China (No. 41721003, 41210006), by the National Key Research and Development Program of China (No. 2016YFB0501702), by the 973 Project of China (No. 2013CB73301, 2013CB73302), and by the DAAD Thematic Network Project under Grant 57173947.

Author Contributions: Jiancheng Li and Taoyong Jin conceived and designed the experiments. Shengjun Zhang performed the experiments and wrote the first draft, and all authors analyzed the data and wrote the final draft.

References

1. Rapp, R.H. Geos 3 data processing for the recovery of geoid undulations and gravity anomalies. *J. Geophys. Res.* **1979**, *84*, 3784–3792. [CrossRef]
2. Haxby, W.F.; Karner, G.D.; Labrecque, J.L.; Weissel, J.K. Digital images of combined oceanic and continental data sets and their use in tectonic studies. *EOS Trans. Am. Geophys. Union* **1983**, *64*, 995–1004. [CrossRef]
3. Sandwell, D.T.; Smith, W.H.F. Marine gravity anomaly from Geosat and ERS1 satellite altimetry. *J. Geophys. Res.* **1997**, *102*, 10039–10054. [CrossRef]
4. Andersen, O.B.; Knudsen, P. Global marine gravity field from the ERS-1 and Geosat geodetic mission altimetry. *J. Geophys. Res.* **1998**, *103*, 8129–8137. [CrossRef]
5. Hwang, C. Inverse Vening Meinesz formula and deflection-geoid formula: Applications to the predictions of gravity and geoid over the South China Sea. *J. Geodesy* **1998**, *72*, 304–312. [CrossRef]
6. Sandwell, D.T.; Garcia, E.S.; Soofi, K.; Wessel, P.; Chandler, M.; Smith, W.H. Towards 1-mGal accuracy in global marine gravity from Cryosat-2, Envisat and Jason-1. *Lead. Edge* **2013**, *32*, 892–898. [CrossRef]
7. Andersen, O.B.; Jain, M.; Knudsen, P. The Impact of Using Jason-1 and Cryosat-2 Geodetic Mission Altimetry for Gravity Field Modeling. *IAG 150 Years Symp.* **2016**, *143*, 205–210.
8. Zhang, S.; Sandwell, D.T.; Jin, T.; Li, D. Inversion of marine gravity anomalies over southeastern China seas from multi-satellite altimeter vertical deflections. *J. Appl. Geophys.* **2017**, *137*, 128–137. [CrossRef]
9. Sandwell, D.T.; Smith, W.H.F. Retracking ERS-1 altimeter waveforms for optimal gravity field recovery. *Geophys. J. Int.* **2005**, *163*, 79–89. [CrossRef]

10. Hwang, C.; Guo, J.; Deng, X.; Hsu, H.Y.; Liu, Y. Coastal gravity anomalies from retracked Geosat GM altimetry: Improvement, limitation and the role of airborne gravity data. *J. Geodesy* **2006**, *80*, 204–216. [CrossRef]
11. Andersen, O.B.; Knudsen, P.; Berry, P. The DNSC08GRA global marine gravity field from double retracked satellite altimetry. *J. Geodesy* **2010**, *84*, 191–199. [CrossRef]
12. Sandwell, D.T. Antarctic marine gravity field from high-density satellite altimetry. *Geophys. J. Int.* **1992**, *109*, 437–448. [CrossRef]
13. Zhao, G.; Zhou, X. Precise orbit determination of Haiyang-2 using satellite laser ranging. *Chin. Sci. Bull.* **2013**, *58*, 589–597. [CrossRef]
14. Legeais, J.-F.; Ablain, M.; Faugère, Y.; Mertz, F.; Soussi, B.; Vincent, P. HY-2A and DUACS (Data Unification and Altimeter Combination System) altimeter products. In Proceedings of the OSTST 2012 (Ocean Surface Topography Science Team) Meeting, Venice, Italy, 22–29 September 2012.
15. Bao, L.; Gao, P.; Peng, H.; Jia, Y.; Shum, C.K.; Lin, M.; Guo, Q. First accuracy assessment of the HY-2A altimeter sea surface height observations: Cross-calibration results. *Adv. Space Res.* **2015**, *55*, 90–105. [CrossRef]
16. Peng, H.; Lin, M.; Mu, B. Global statistical evaluation and performance analysis of HY-2A satellite radar altimeter data. *Haiyang Xuebao* **2015**, *37*, 54–66. (In Chinese)
17. Yang, L.; Zhou, X.; Lin, M.; Lei, N.; Mu, B.; Zhu, L. Global statistical assessment of HY-2A altimeter IGDR data. *Prog. Geophys.* **2016**, *31*, 629–636. (In Chinese) [CrossRef]
18. Sandwell, D.T.; Smith, W.H.F. Global marine gravity from retracked Geosat and ERS-1 altimetry: Ridge segmentation versus spreading rate. *J. Geophys. Res.* **2009**, *114*, B014411. [CrossRef]
19. Garcia, E.S.; Sandwell, D.T.; Smith, W.H.F. Retracking CryoSat-2, Envisat, and Jason-1 radar altimetry waveforms for improved gravity field recovery. *Geophys. J. Int.* **2014**, *196*, 1402–1422. [CrossRef]
20. Zhang, S.; Sandwell, D.T. Retracking of SARAL/AltiKa Radar Altimetry Waveforms for Optimal Gravity Field Recovery. *Mar. Geodesy* **2017**, *40*, 40–56. [CrossRef]
21. Prandi, P.S.; Philipps, S.; Pignot, V.; Picot, N. SARAL/AltiKa global statistical assessment and cross-calibration with Jason-2. *Mar. Geodesy* **2015**, *38*, 297–312. [CrossRef]
22. Vu, P.; Frappart, F.; Darrozes, L. Multi-Satellite Altimeter Validation along the French Atlantic Coast in the Southern Bay of Biscay from ERS-2 to SARAL. *Remote Sens.* **2018**, *10*, 93. [CrossRef]
23. Salameh, E.; Frappart, F.; Marieu, V. Monitoring Sea Level and Topography of Coastal Lagoons Using Satellite Radar Altimetry: The Example of the Arcachon Bay in the Bay of Biscay. *Remote Sens.* **2018**, *10*, 297. [CrossRef]
24. Wahr, L.W. Deformation of the earth induced by polar motion. *J. Geophys. Res. (Solid Earth)* **1985**, *90*, 9363–9368. [CrossRef]
25. Valladeau, G.; Legeais, J.F.; Ablain, M.; Guinehut, S.; Picot, N. Comparing altimetry with tide gauges and Argo profiling floats for data quality assessment and mean sea level studies. *Mar. Geodesy* **2012**, *35*, 42–60. [CrossRef]
26. CalVal In-Situ Altimetry/tide Gauges. *Validation of Altimeter Data by Comparison with Tide Gauge Measurements*; CLS: Ramonville St-Agne, France, 2014; CLS.DOS/NT/13-285, Issue 1.1, Nomenclature: SALP-RP-MA-EA-22294-CLS.
27. Caldwell, P.C.; Merrifield, M.A.; Thompson, P.R. *Sea Level Measured by Tide Gauges from Global Oceans—The Joint Archive for Sea Level Holdings (NCEI Accession 0019568)*; Version 5.5; NOAA National Centers for Environmental Information: Silver Spring, MD, USA, 2015.
28. Huang, Z.; Wang, H.; Luo, Z.; Shum, C.K.; Tseng, K.-H.; Zhong, B. Improving Jason-2 Sea Surface Heights within 10 km Offshore by Retracking Decontaminated Waveforms. *Remote Sens.* **2017**, *9*, 1077. [CrossRef]
29. Xu, X.; Birol, F.; Cazenave, A. Evaluation of Coastal Sea Level Offshore Hong Kong from Jason-2 Altimetry. *Remote Sens.* **2018**, *10*, 282. [CrossRef]
30. Brown, G. The average impulse response of a rough surface and its applications. *IEEE Trans. Antennas Propag.* **1977**, *25*, 67–74. [CrossRef]
31. Maus, S.; Green, C.M.; Fairhead, J.D. Improved ocean-geoid resolution from retracked ERS-1 satellite altimeter waveforms. *Geophys. J. Int.* **1998**, *134*, 243–253. [CrossRef]
32. Rodriguez, E. Altimetry for non-Gaussian oceans: Height biases and estimation of parameters. *J. Geophys. Res. Oceans* **1988**, *93*, 14107–14120. [CrossRef]

33. Martin, T.V.; Zwally, H.J.; Brenner, A.C.; Bindschadler, R.A. Analysis and retracking of continental ice sheet radar altimeter waveforms. *J. Geophys Res.* **1983**, *88*, 1608–1616. [CrossRef]
34. Davis, C.H. A surface and volume scattering retracking algorithm for ice sheet satellite altimetry. *IEEE Trans. Geosci. Remote Sens.* **1993**, *31*, 811–818. [CrossRef]
35. Welch, P.D. The use of fast Fourier transforms for the estimation of power spectra: A method based on time averaging over short, modified periodograms. *IEEE Trans. Audio Electroacoust.* **1967**, *15*, 70–73. [CrossRef]
36. Andersen, O.B.; Knudsen, P.; Kenyon, S.; Holmes, S. Global and arctic marine gravity field from recent satellite altimetry (DTU13). In Proceedings of the 76th EAGE Conference and Exhibition, Amsterdam, The Netherlands, 16–19 June 2014.
37. Sandwell, D.T.; Müller, R.D.; Smith, W.H.F.; Garcia, E.; Francis, R. New global marine gravity model from CryoSat-2 and Jason-1 reveals buried tectonic structure. *Science* **2014**, *346*, 65–67. [CrossRef] [PubMed]
38. Fu, L.L.; Ubelmann, C. On the transition from profile altimeter to swath altimeter for observing global ocean surface topography. *J. Atmos. Ocean. Technol.* **2013**, *31*, 560–568. [CrossRef]

4

Calibrating the SAR SSH of Sentinel-3A and CryoSat-2 over the Corsica Facilities

Pascal Bonnefond [1,*], Olivier Laurain [2], Pierre Exertier [2], François Boy [3], Thierry Guinle [3], Nicolas Picot [3], Sylvie Labroue [4], Matthias Raynal [4], Craig Donlon [5], Pierre Féménias [6], Tommaso Parrinello [6] and Salvatore Dinardo [7]

1 SYRTE, Observatoire de Paris, PSL Research University, CNRS, Sorbonne Universités, UPMC University Paris 06, LNE, 75014 Paris, France
2 Géoazur, Observatoire de la Côte d'Azur, 06905 Sophia-Antipolis CEDEX, France; Olivier.Laurain@oca.eu (O.L.); Pierre.Exertier@oca.eu (P.E.)
3 Centre National d'Etudes Spatiales (CNES), 31401 Toulouse CEDEX, France; Francois.Boy@cnes.fr (F.B.); Thierry.Guinle@cnes.fr (T.G.); Nicolas.Picot@cnes.fr (N.P.)
4 Collecte Localisation Satellites (CLS), 31520 Ramonville Saint-Agne, France; Sylvie.Labroue@cls.fr (S.L.); mraynal@cls.fr (M.R.)
5 European Space Agency/European Space Research and Technology Centre (ESA/ESTEC), 2201 AZ Noordwijk, The Netherlands; Craig.Donlon@esa.int
6 European Space Agency/European Space Research Institute (ESA/ESRIN), 00044 Frascatti, Italy; Pierre.Femenias@esa.int (P.F.); Tommaso.Parrinello@esa.int (T.P.)
7 HeSpace, 64293 Darmstadt, Germany; Salvatore.Dinardo@eumetsat.int
* Correspondence: Pascal.Bonnefond@obspm.fr

Abstract: Initially developed to monitor the performance of TOPEX/Poseidon and to follow the Jason legacy satellite altimeters at Senetosa Cape, Corsica, this calibration/validation site has been extended to include a new location at Ajaccio. This addition enables the site to monitor Envisat and ERS missions, CryoSat-2 and, more recently, the SARAL/AltiKa mission and Sentinel-3A satellites. Sentinel-3A and CryoSat-2 carry altimeters that use a synthetic aperture radar (SAR) mode that is different to the conventional pulse-bandwidth limited altimeters often termed "low resolution mode" (LRM). The aim of this study is to characterize the sea surface height (SSH) bias of the new SAR altimeter instruments and to demonstrate the improvement of data quality close to the coast. Moreover, some passes of Sentinel-3A and CryoSat-2 overfly both Senetosa and Ajaccio with only a few seconds time difference, allowing us to evaluate the reliability and homogeneity of both ground sites in term of geodetic datum. The Sentinel-3A and CryoSat-2 SSH biases for the SAR mode are respectively +22 ± 7 mm and −73 ± 5 mm (for CryoSat-2 baseline C products). The results show that the stability of the SAR SSH bias time series is better than standard LRM altimetry. Moreover, compared to standard LRM data, for which the measurements closer than ~10 km from the coast were generally unusable, SAR mode altimeters provide measurements that are reliable at less than few hundred meters from the coast.

Keywords: altimetry; SAR; calibration; validation

1. Introduction

The Corsica geodetic facilities that are located both at Senetosa Cape for the TOPEX/Poseidon (T/P) and Jason satellite ground tracks and near Ajaccio for the Envisat satellite ground tracks have been developed to calibrate successive satellite altimeters in an absolute sense. Since 1998, the successful calibration process used to calibrate many oceanographic satellite altimeter missions has been regularly updated, see details in [1–3].

One of the main issues in comparing offshore altimetric sea surface height (SSH) measurements together with in-situ tide gauge data is taking into account the geoid slope (of several cm/km) over distances of tens of kilometers between both locations. As discussed in [4], other factors including ocean dynamics and tidal differences, impact the comparisons with independent coastal in situ measurements, but are negligible in such areas and within the short distance in our study (less than 20 km). A specific Global Positioning System (GPS) campaign was conducted in 1999 in order to determine a geoid map of about 20 km long and 5.4 km wide centered on the T/P and Jason satellite's ground track #085 at the Senetosa Cape site [4]. In addition, a second campaign was conducted in 2005 in the Ajaccio area under the Envisat ground track #130 (Figure 1) [5,6]. Both local geoids are the key datum for the absolute calibration of the satellite altimeters. Indeed, any distortion or systematic errors between offshore and tide gauge locations will cause a systematic error in the absolute SSH bias. From this point of view, the calibration process also requires accurate vertical offsets to be determined for each individual tide gauge in use. In order to achieve and maintain accuracy at the level of just a few millimeters or even less, geodetic campaigns must be carried out regularly in order to check or to eventually update such offsets.

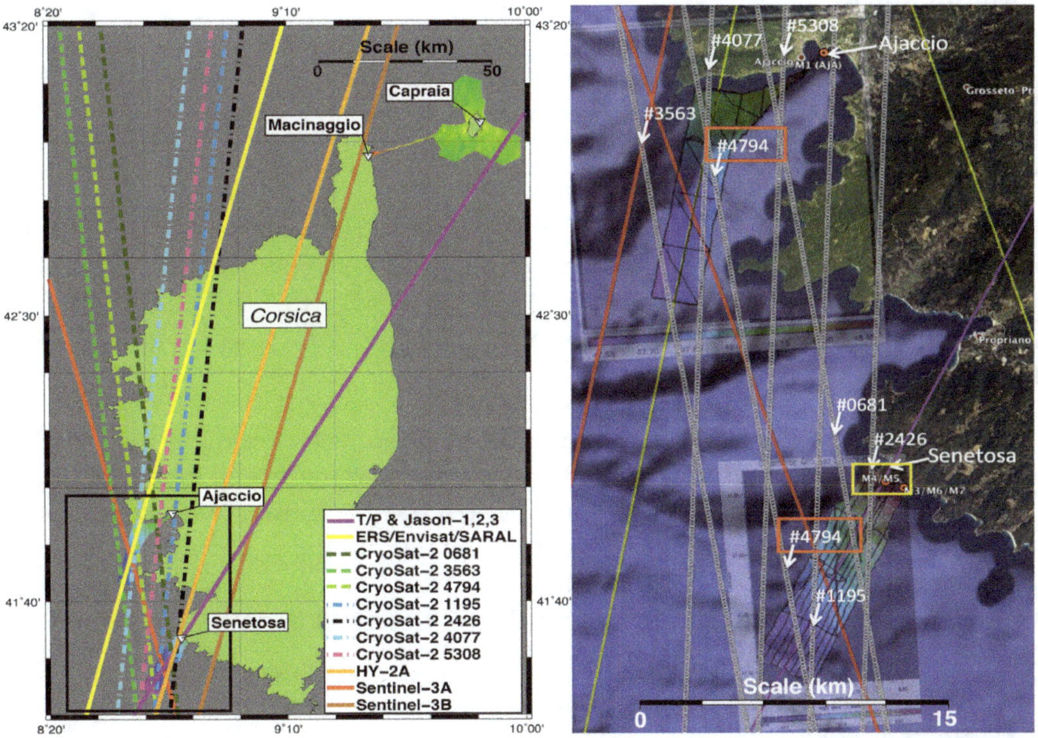

Figure 1. Left: General configuration of the Corsica calibration site with all of the satellite altimeter missions ground tracks that have been monitored since 1998 (the black frame on the left bottom corresponds to the zoom on the right map). **Right**: zoom on Ajaccio and Senetosa sites with Sentinel-3A (red) and CryoSat-2 (grey) tracks. The red rectangles highlight the CryoSat-2 pass #4794 that crosses the Senetosa and Ajaccio sites (see Section 0). The yellow rectangle shows the area that is discussed in Section 0 for pass #2426 (see Figure 8c).

The present paper focuses on the calibration of both the Sentinel-3A and CryoSat-2 missions using the same facilities and similar data processing. Sentinel-3A has a 27-day repeat cycle, where CryoSat-2 has a repeat cycle of 369 days and a 30-day sub-cycle; this repetition allows them to perform the absolute calibration over the Senetosa and Ajaccio area on a monthly basis for both satellites. Due to the configuration of Sentinel-3A repeat ground track and some CryoSat-2 passes, each altimeter overflies both the Senetosa and Ajaccio sites with a time delay of about five seconds corresponding to a distance of about 37 km (Figure 1). As a consequence, it allows us to determine two SSH biases for each mission that should be equivalent considering that, most of the time, the sea state conditions at both sites are almost identical due to the short time delay and the short distance between sites. From the analysis of all available calibration passes, systematic differences can be related to errors in the geodetic references or in the situ data. For T/P-Jason and ERS-Envisat-SARAL/AltiKa this was difficult to determine as each altimeter mission overflew only a single site, Senetosa or Ajaccio respectively. Since July 2016, the SARAL/AltiKa mission is in a drifting orbit phase allowing a similar study to the one presented here that is discussed in [7].

In this paper we first present the data and the methodology (Section 0). In the results section, Section 0, we present a re-processing of the Ajaccio tide gauge data following a re-analysis of the Global Positioning System (GPS) surveys that are used to monitor the location and any vertical movement of the tide gauges (Section 0). This resulted in an updated datum and a −30 mm vertical correction to the tide-gauge-derived SSH that is used in the calibration process of all the missions calibrated using the Ajaccio tide gauge data. The two following sections, 0 and 0, are then dedicated to the analysis of the Sentinel-3A [8] and CryoSat-2 [9] SSH biases, respectively.

2. Data and Methodology

2.1. Data

The data sets used for Sentinel-3 and CryoSat-2 are described below.

Sentinel-3A:

- Data sources: S3-PDGS (payload data ground segment) reprocessing 2017, Non-Time Critical (NTC) products: cycle 5–19, processing baseline 2.15 [10]. The processing baseline (PB) 2.15 consists of the following processors: SRAL L1 IPF V6.11, MWR L1 IPF V6.04, SRAL L2 IPF V6.07.
- Sentinel-3A SRAL (Ku/C Radar Altimeter) processing for SAR and Pseudo LRM (PLRM) and applied corrections
    - Standard processing: dual-frequency ionosphere measurement and wet troposphere radiometry. In this study, we also compared the dual frequency ionosphere measurements with the global ionosphere maps model (GIM), together with a comparison of the wet troposphere radiometer measurements with the wet troposphere model from European Centre for Medium-Range Weather Forecasts (ECMWF)). The chosen wet correction for the model corresponds to the field that is computed at sea level (MOD_WET_TROPO_COR_ZERO_ALTITUDE)
    - Dry troposphere
    - Sea state bias (SSB)
    - Solid, loading and pole tides
- In situ data
    - Ajaccio: Service Hydrographique et Océanographique de la Marine (SHOM) radar tide gauge data in real time
    - Senetosa: pressure tide gauges (data retrieved end of July 2017)

CryoSat-2:

- Data sources: from SARvatore service at ESA G-POD (https://gpod.eo.esa.int/services/CRYOSAT_SAR/) using baseline B products with two different retrackers. It is important to note that this product does not correspond to the current baseline C products processed by the CryoSat-2 payload data ground segment (PDGS), the main difference being the correction of a known range bias of 673 mm:

    - SAMOSA2 retracker (default): The SAMOSA (SAR Altimetry MOde Studies and Applications) model [11] has been implemented in its first-order formulation (SAMOSA2) in a retracker as in [12]. This retracker uses a look-up table to mitigate the effect of the model's approximation of the squared PTR (point target response) with a Gaussian curve, it implements a stack masking for the Doppler beams padded to zero and has in principle no enhancement in the coastal zone.
    - SAMOSA+ retracker (the SAMOSA2 model tailored for inland water, sea ice and the coastal zone domain): in the case of SAMOSA+, the open ocean SAMOSA [11] model and retracker [13] has been implemented in the retracking scheme with two significant additions [12]: The first one concerns the selection of the first-guess epoch. This is not selected as the position of the waveform peak but as the position of the moving correlation peak in 20 consecutive waveforms (after aligning them for tracker shift). The rationale behind this choice is to attempt to mitigate the typical off-ranging effect in coastal data. The second concerns the treatment of land-contaminated waveforms. In case waveforms are not contaminated by land, the SAMOSA model was used with the mean square slope set to zero, i.e., as in [13], whereas in the case of land contaminated waveforms, we used a dual step retracking: in the first step, the SWH is still estimated as in [13], while in the second step, the SWH was set to zero and the third free parameter in the retracking becomes the mean square slope. The output of this second step is the range and the amplitude Pu (retracked waveform amplitude, see Equation (22) in [12]).

- Cryosat-2 SIRAL (synthetic aperture interferometric radar altimeter) data processing for SAR and applied corrections
    - Dry troposphere
    - Wet troposphere from ECMWF model
    - Ionosphere from GIM
    - Sea state bias = 3.5% of SWH
    - Solid, loading and pole tides
- In situ data
    - Ajaccio: SHOM radar tide gauge data in real time
    - Senetosa: pressure tide gauges (data retrieved end of July 2017)

2.2. Methodology

The main difference in the processing in this study, compared to previous studies [1–3,6], is the difference in the footprint area in over which the geoid height is interpolated. In the case of LRM (low resolution mode, e.g., Jason missions), the footprint is defined as a circle using the significant wave height to define the diameter [14], while in the case of SAR (synthetic aperture radar, e.g., Sentinel-3A) the footprint is defined as a rectangle with a width of 300 m along-track and a length across-track that is also a function of SWH with the same formula as LRM (see Figure 5 or Figure 8 for an illustration).

Our standard processing of the SSH bias uses (when available) the wet tropospheric correction derived from the radiometer in addition to the ionospheric correction derived from the dual frequency of the altimeter. We performed a comparison with numerical weather prediction (NWP) model corrections that are available in the products for the wet troposphere and the ionosphere corrections, from ECMWF and GIM respectively (see Section 0).

3. Results

3.1. Solving the Vertical Offset in the Ajaccio Tide-Gauge-Derived SSH

In [3] we showed that there was an anomalous SSH bias of -30.5 ± 4.5 mm between the Ajaccio tide gauge and our GPS-based sea level measurements. We thus retrieved the GPS data from recent surveys performed in 2013, 2014 and 2015 and computed new height differences between the antenna reference point (ARP) of the permanent receiver (AJAC) and the fundamental marker close to the tide gauge (G). This processing was done in baseline mode with the GAMIT software [15] and the results are shown in Table 1. The standard errors of the individual determinations range from 0.2 to 1.1 mm mainly due to the number of days of observation. It provides a much more coherent weighted mean value of 50.0371 m, rather than the 2005 reference, with a standard error of the weighted mean of 0.4 mm (Table 1), thus changing our historical reference by 18.3 mm.

Additionally, during the deployment of the Ajaccio tide gauge in 2012, the Service Hydrographique et Océanographique de la Marine (SHOM) performed SSH observations with a contact gauge and determined a 13.0 mm bias (standard deviation of 3.6 mm) which was introduced as a "calibration offset" into the tide gauge system. We decided to remove this quantity from the instrument and by adding it to the new correction (of 18.3 mm); the total systematic error of the tide gauge SHH has been established to be 31.4 mm. This is in very good agreement with the SSH bias observed between the Ajaccio tide gauge and our GPS-based sea level measurements (of -30.5 ± 4.5 mm [3]). We then changed our datum to a rounded offset of -30 mm in the tide gauge SSH. The results presented in this paper take this correction into account.

Another geodetic survey is planned for next year to confirm the height difference between the ARP of the permanent receiver (AJAC) and the fundamental marker close to the tide gauge (G), but we are already very confident in the current result given the 0.4 mm standard error.

Table 1. Height differences between permanent Global Positioning System (GPS) (AJAC) and tide gauge (G) at Ajaccio (in m).

Year	Reference Used	AJAC-G	Number of Days
2005	50.0188 ± 0.0100		
2013		50.0386 ± 0.0011	2
2014		50.0360 ± 0.0010	3
2015		50.0370 ± 0.0002	7
Average		50.0371 ± 0.0004	

3.2. Sentinel-3A SSH Calibration and Corrections Validation

The Sentinel-3A ascending pass #741 overflies the Senetosa site and ~5 s later the Ajaccio site. For the first time this allows us to compare the SSH biases, which can be independently determined at both locations, and then to assess geodetic references together with in-situ measurements. Launched on February 16th 2016, Sentinel-3A was in a ground-segment ramp-up phase during the first months with different versions of the IPF (instrument processing facilities) making the different cycles difficult to compare. Thanks to a reprocessing campaign [10] completed in September 2017 by the respective ESA and EUMETSAT agencies for cycles 5–19 (June 2016 to July 2017) with the same Processing Baseline 2.15, it is now possible to analyze a homogeneous set of data using the Non-Time Critical (NTC) product. We thus analyzed and compared the two modes which are used to derive the sea surface height from the radar measurements: (i) the SAR (synthetic aperture radar) mode and (ii) the PLRM (pseudo LRM) mode that mimics the classic LRM. A complete description of the products used and the processing options is given in Section 0.

Concerning the ionospheric correction, both the dual-frequency altimeter derived and the GIM model were averaged over a distance of ~140 km centered on the geoid area of Senetosa and Ajaccio (slightly shifted by ~14 km for Senetosa to avoid measurements over Sardinia). The results presented in Figure 2 show good consistency between the instrument and GIM model in terms of bias (+6 mm) and stability (5–6 mm). Moreover, the results from either the Senetosa or Ajaccio sites show good consistency.

Concerning the wet tropospheric correction deduced from the satellite radiometer, we have reduced the land contamination (as the radiometer has a large footprint) by filtering to keep the non-contaminated signal within our area of study. The selected corrections are linearly interpolated with an outlier rejection at 3-σ up to the edges of the study area. They are then considered as a constant for both Senetosa and Ajaccio processing. The Senetosa interpolation area was used for both the Senetosa and Ajaccio sites (Figure 3) because in this area the radiometer is almost not contaminated by land. The permanent GPS receiver at Senetosa and Ajaccio also allows us to compute the wet tropospheric corrections at these locations. Recently, Collecte Localisation Satellites (CLS) have developed an algorithm to reduce the land contamination in coastal conditions (available for cycle five to 15) [16]. The three wet tropospheric corrections were compared and the differences are shown in Figure 2. We note that the radiometer wet tropospheric correction provided in the products appears to be biased on average (Senetosa and Ajaccio) by −17.5 mm compared to the model and −20.5 mm compared to the GPS. The differences between the two sites are very small (3–5 mm). The ~−20 mm bias is probably due to a remaining land contamination in the radiometer footprint. The analysis of the new radiometer correction developed by CLS to limit this effect shows better agreement with the correction derived from GPS. On the same time series (cycle 5–15), the bias is not only reduced by 15 mm in average but the standard deviation is also improved by 2 mm in average (Table 2).

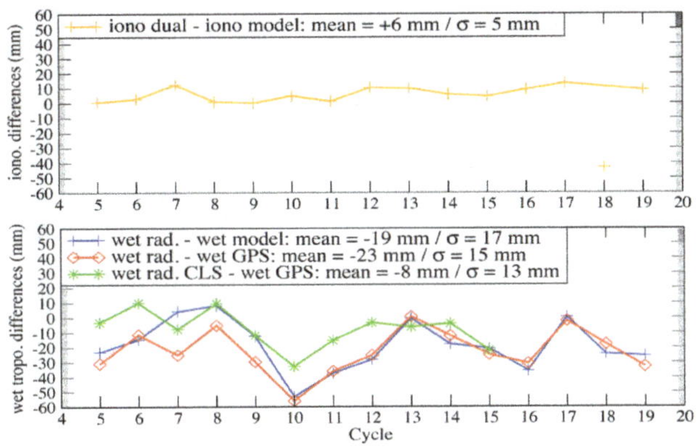

Figure 2. Sentinel-3A corrections differences at the Senetosa (**left**) and Ajaccio (**right**) calibration sites. Ionospheric correction differences in the upper panels. Wet tropospheric correction differences in the lower panels.

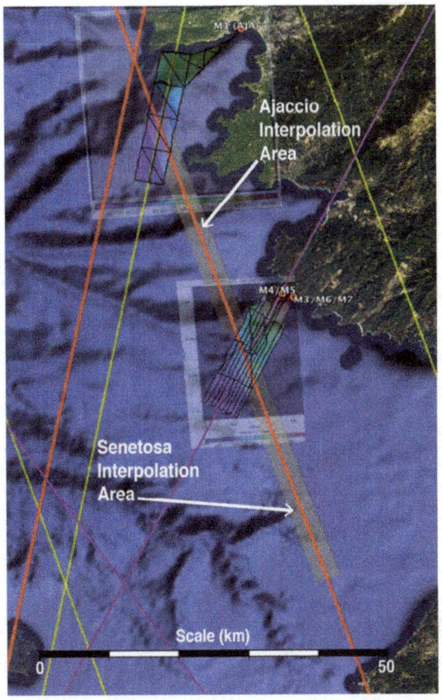

Figure 3. Map of the Sentinel-3A configuration: the shaded areas correspond to the locations where the wet tropospheric correction (radiometer and model) is interpolated.

A second study of the wet tropospheric correction was conducted using the Ajaccio interpolation area (Figure 3) that should be more sensitive to land contamination compared to Senetosa. All the other parameters are identical to the standard processing performed for Ajaccio. Figure 4 shows that, in this case, the wet tropospheric correction is biased by −79 mm (light blue) using the Ajaccio GPS wet tropospheric correction as a reference. The 59 mm increase in the difference between using the Ajaccio and Senetosa interpolation areas illustrates the challenge to fully remove land contamination in the radiometer footprint in such complex coastal areas. In this case, the CLS correction clearly provides a great improvement even though a residual bias of −14 mm still exists (orange). In addition, the differences between the wet tropospheric corrections deduced from both the Senetosa and Ajaccio GPS data (~37 km separation distance and only five seconds time lag for a typical satellite overpass) are given. It is worth noting that, if the corrections are equivalent (+1 mm) on average, the dispersion of 12 mm certainly reflects different meteorological conditions at both sites at the same time. We conclude and confirm that it is very important to use GPS measurements as close as possible to satellite radiometer data when computing wet tropospheric correction terms.

Table 2. Wet tropospheric correction differences on the common cycles (5–15) (in mm).

	Radiometer-GPS	
	Mean	**σ**
Senetosa		
Standard product	−23	16
CLS	−8	13
Ajaccio		
Standard product	−20	17
CLS	−5	16

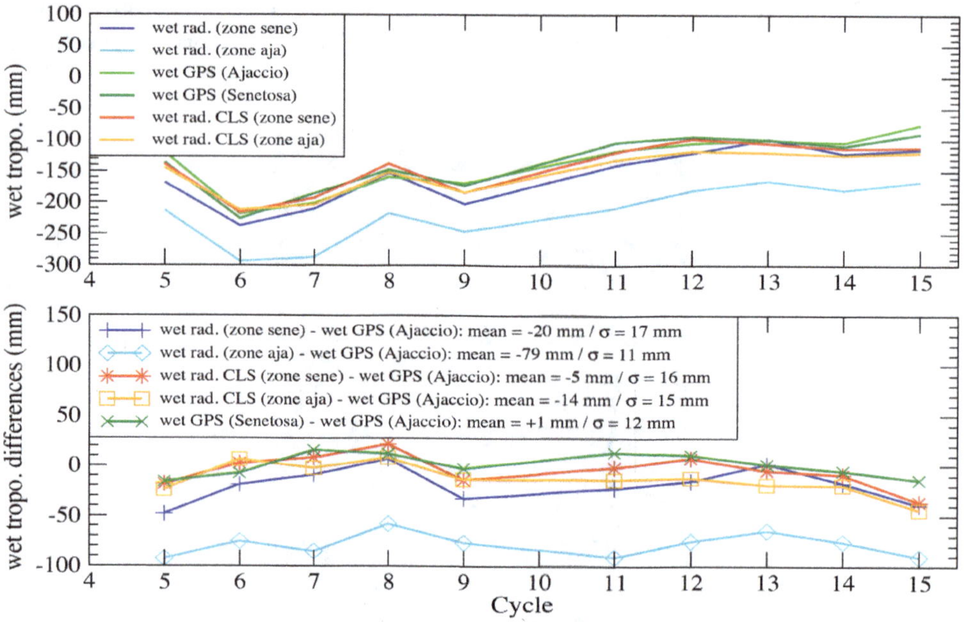

Figure 4. Sentinel-3A wet tropospheric correction differences at Ajaccio (using the Ajaccio interpolation area). Wet tropospheric correction in the upper panel: derived from satellite microwave radiometer using the Senetosa interpolation area (dark blue), radiometer using the Ajaccio interpolation area (light blue), GPS from Ajaccio (light green), GPS from Senetosa (dark green), radiometer (from Collecte Localisation Satellites (CLS)) using the Senetosa interpolation area (red) and radiometer (from CLS) using the Ajaccio interpolation area (orange). Wet tropospheric correction differences in the lower panel: reference is always the wet tropospheric correction from the Ajaccio GPS and same color code is used for the corrections being compared.

Finally, we also compared the corrections derived from the satellite instruments (radiometer and altimeter) using either SAR or PLRM processing. We found very small differences (<1 mm) that are caused by range differences between these two processing schemes. Finally, we compared the sea state bias corrections for SAR and PLRM that, for the moment, use the same correction tables: there are few millimeter differences originating from significant wave height (SWH) and wind speed differences, as the waveform shapes are different for SAR and PLRM.

The SSH calibration was performed independently at Ajaccio and Senetosa using respectively a radar tide gauge and three pressure tide gauges. Figure 5 illustrates the process used that is detailed in [3]. We recall that by convention the SSH bias corresponds to the difference ($SSH_{altimetry} - SSH_{tide\ gauge}$). This process was first carried out without taking into account the Ajaccio tide gauge SSH offset (see Section 0); a 26 mm offset clearly appears in the averaged results (−5 mm for Ajaccio and +21 mm for Senetosa) highlighting that the Ajaccio offset is clearly detectable by our absolute SSH calibration process. As a result, the Ajaccio tide gauge SSH time series were corrected by −30 mm, definitively. In Figure 5, the insert in green illustrates the improvement brought by SAR processed data. Indeed, when overflying the small Sanguinaires islands, only 2–3 data look bad and can be easily detected and rejected using a simple 3-σ criteria (circled crosses). For comparison with LRM in the same area, all the Envisat data that were at a distance less than 10 km from the coast were eliminated [6]. However, it is important to note that the areas used to perform the calibration (blue boxes in Figure 5) are far enough (>10 km) to avoid any land contamination for the altimeter, except maybe in PLRM mode in the Ajaccio area.

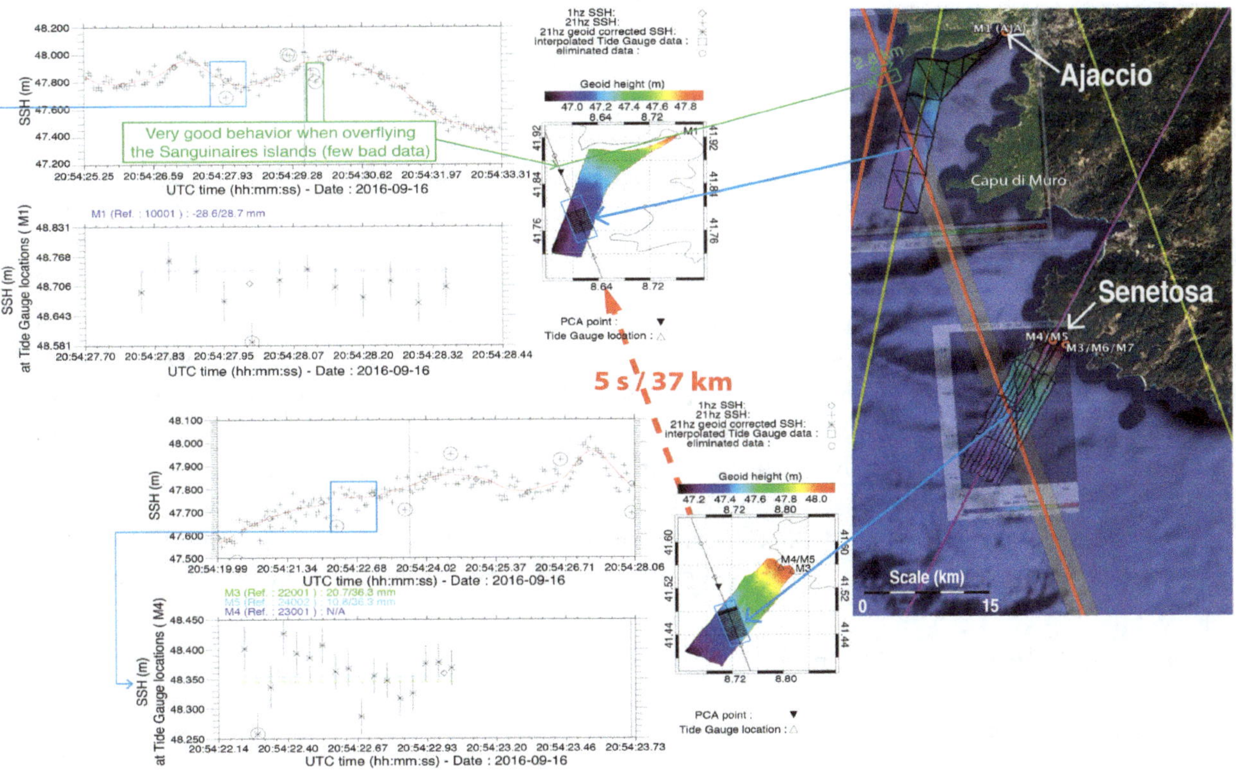

Figure 5. Example of sea surface height (SSH) bias processing for Sentinel-3A in synthetic aperture radar (SAR) mode (pass 741, cycle 8). For each independent processing (Ajaccio, top/Senetosa, bottom) the upper plot shows ±4 s of altimeter SSH series around the point of closest approach (PCA). The lower plots show the selected SSH (inside the geoid area) that are corrected from the geoid height differences between the SSH location and the tide gauge locations.

The Figure 6 upper plots shows the SSH bias time series for SAR and PLRM respectively at Senetosa (left) and Ajaccio (right), while the lower plots show the differences (SAR-PLRM) either for "orbit-range" only or total SSH bias ("orbit-range + corrections"). The Ajaccio time series looks slightly noisier for SAR and much more so for PLRM compared to the Senetosa time series. This can be explained by more possible land contamination in the radar footprint in the Ajaccio area (e.g., the Capu di Muro and Sanguinaires islands, see Figure 5). The very good consistency of the "orbit-range" and "SSH bias" differences indicates that the correction differences between SAR and PLRM are very small, as explained previously. They are mainly due to SSB differences because SWH and wind speed estimations are different between SAR and PLRM processing. For the moment, the SSB model in the product is the same for SAR and PLRM. An improvement in terms of consistency is foreseen by the further tuning of SSB models for both SAR and PLRM, instead of using the same model. Thus, the main difference is due to range estimation that is different in SAR and PLRM processing.

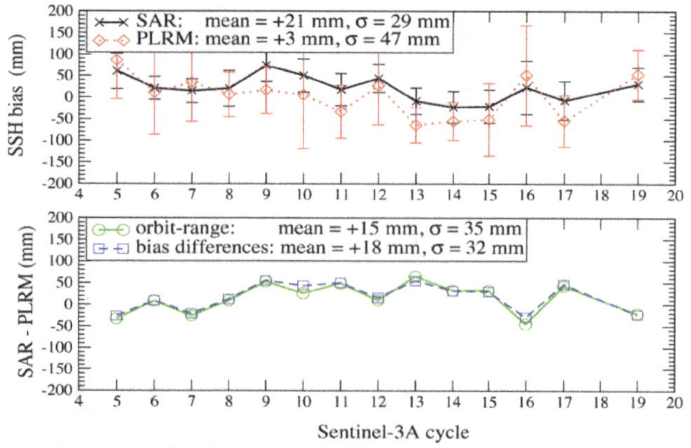

Figure 6. Sentinel-3A SSH bias time series from Senetosa (**left**) and Ajaccio (**right**). The upper panels show the time series for SAR (black crosses) and Pseudo LRM (PLRM) (red diamonds). The lower panels show the differences between SAR and PLRM for the "orbit-range" (green circles) and the total SSH bias (blue squares).

For each cycle, we have computed the averaged SSH bias from both Senetosa and Ajaccio independently (see the time series in Figure 7). First, it shows that the SSH bias for SAR is very stable with a standard deviation of 24 mm. Concerning the PLRM, as expected due to data processing issues (reconstruction of pseudo LRM waveforms from original SAR measurements that make the PLRM measurements noisier than real LRM), the standard deviation of 42 mm is higher than the classical LRM missions (e.g., 38 and 37 mm for Jason-1 and Jason-2 respectively [2]). Second, the 13 mm difference in the SSH bias between SAR and PLRM comes mainly from the differences in the altimeter range. From the entire data set (13 cycles), our best estimate of the SSH bias for Sentinel-3A is then:

- Sentinel-3A SAR: +22 $\pm$ 7 mm
- Sentinel-3A PLRM: +9 $\pm$ 12 mm

Recent studies using the transponder installed in Crete show that the SAR range bias is 0 $\pm$ 12 mm [17] and 8 $\pm$ 12 mm [18], so our results are in very good agreement because most of the SSH bias (+22 mm) originates from the radiometer wet tropospheric correction.

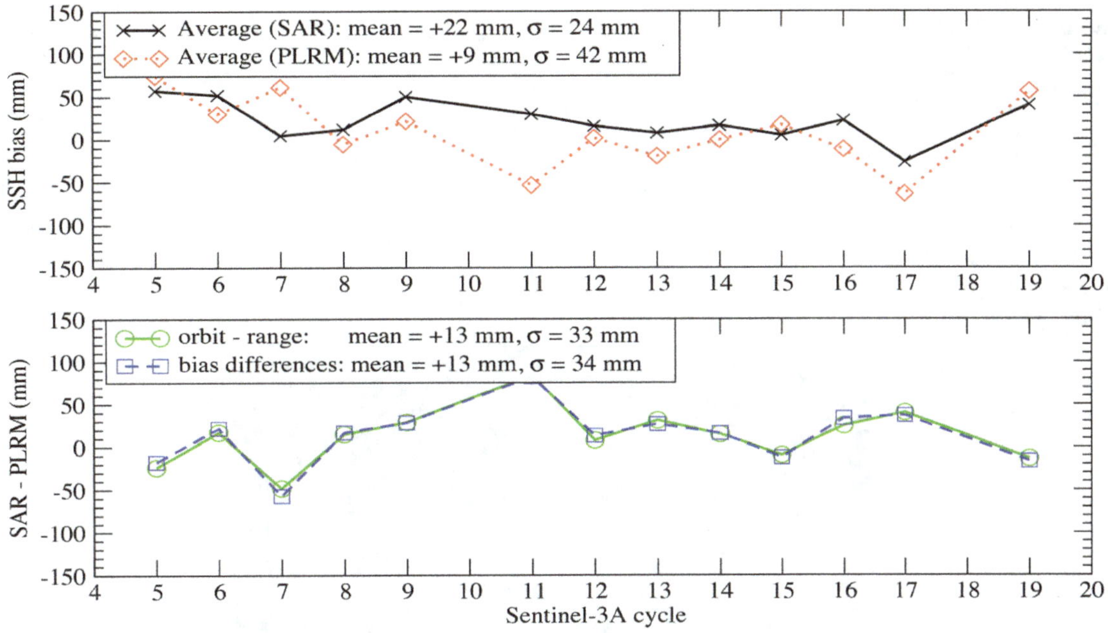

Figure 7. Average of Senetosa and Ajaccio Sentinel-3A SSH biases for both SAR and PLRM (upper panel). The lower panel shows the differences between SAR and PLRM for the "orbit-range" (green circles) and the total SSH bias (blue squares).

3.3. CryoSat-2 SSH Calibration

Some of the CryoSat-2 altimeter passes overfly the Senetosa or Ajaccio facilities (or both). These passes are listed on Figure 1 (with their relative orbit number). In this study, we analyzed all these overflights with the same methodology as used for Sentinel-3A. In terms of our data set, we used the SARvatore service [19], which is part of the European Space Agency (ESA) grid processing on demand (G-POD) service. G-POD allows us to select an area and then process all of the L1a (FBR, full bit rate) data available to provide L2 products. It is possible to choose different options, but we have left all the options as default values. G-POD uses the SAMOSA2 retracker as default. However, another retracking approach tailored for coastal zones [11,12] was also available (SAMOSA+, SAR altimetry mode studies and applications) and we generated a separate set of data using this option. The different processing schemes were then used to analyze their impact on the SSH bias: SAMOSA2 for one set and SAMOSA+ for the other. The details on the data sets are given Section 0.

The configurations of the overflights are different from one CryoSat-2 ground track pattern to another (see Figure 1 right) but most of them have data very close to the coast (a few hundred meters, see Table 3). Descending pass #2426 is particularly interesting because the closest data is at ~1400 m from the M4/M5 tide gauges at Senetosa. Figure 8 illustrates the quality of the data for cycle 1 and the improvement for some data when using SAMOSA+ retracking. For the data before the point of closest approach (PCA), the SSH from SAMOSA+ decreases and appears more homogeneous with the rest of the time series (marked by green arrows in Figure 8a,b): one measurement at only 400 m across-track from the coast has been clearly improved even with about 37% of the footprint on land (marked by red arrows in Figure 8). However, most importantly, all the data immediately after the coastline (in time after the PCA, because it is a descending track) are of very good quality (marked by blue arrows in Figure 8) and do not look contaminated by land effects even with the classic SAMOSA2 retracker. This is clearly linked to the small area of the footprint for SAR in comparison with LRM. In this example, the footprint is a rectangle with a width of 300 m along-track and a length across-track of 3000 m (see white rectangle area in Figure 8c).

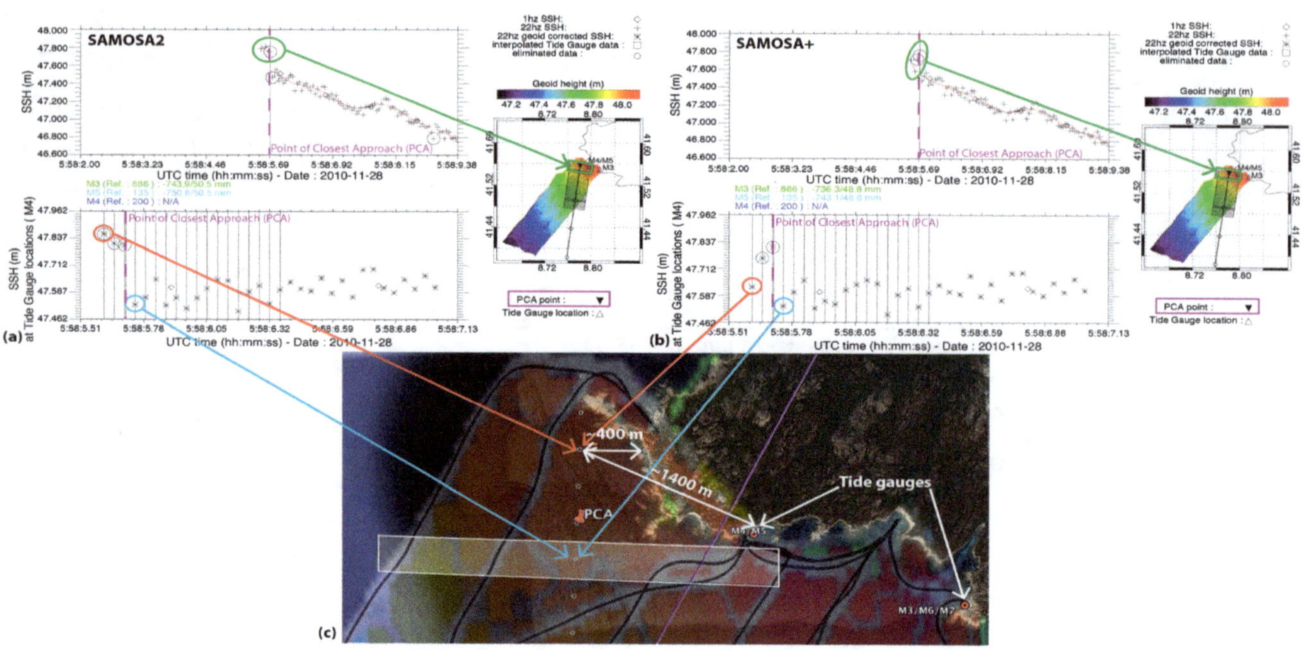

Figure 8. Example of retracking improvement for pass #2426 cycle 001. The upper panels of (**a**,**b**) show the altimeter SSH (crosses) without geoid correction, while the lower ones show the altimeter SSH after geoid correction (stars with error bars). An automatic outlier detection is performed and the rejected data are circled. (**a**) Corresponds to the G-POD standard retracking (SAMOSA2) while (**b**) corresponds to the improved retracking (SAMOSA+ is the SAMOSA2 model tailored for inland water, sea ice and coastal zone domains). (**c**) Shows a zoom of the data locations (circles) very close to the coast; the white rectangle represents the SAR footprint for one data; the black lines correspond to the Catamaran-GPS tracks [4] that were used to map the Senetosa geoid (colored map in background).

Table 3 provides the details of the SSH calibration for each pass and for the two retrackings used, and Figure 9 illustrates the time series of these SSH biases. Two passes in the Ajaccio area were not used because too few data were in the geoid area (#3563 and #5308, see Figure 1). The standard deviations of the individual mean SSH bias were respectively 14.6 mm and 14.0 mm for SAMOSA2 and SAMOSA+. This shows a small improvement for SAMOSA+, but if the same computation is done only for the "coastal" passes (#4077, #4794 and #2426), the standard deviations are this time respectively 15.2 mm and 12.5 mm. SAMOSA+ clearly improved the SSH when very close to the coast, avoiding some outliers even in the time series (see data circled in red in Figure 9 for pass #4794 at Ajaccio). The mean SSH bias for Senetosa and Ajaccio was −746 mm for both SAMOSA2 and SAMOSA+, but the overall standard deviation of the time series was improved by ~13 mm with SAMOSA+ ($\sqrt{(29^2–26^2)}$). The CryoSat-2 SSH bias is then:

- CryoSat-2 SAMOSA2: −746 ± 5.5 mm (−73 ± 5.5 mm for baseline C)
- CryoSat-2 SAMOSA+: −746 ± 5.0 mm (−73 ± 5.0 mm for baseline C)

We were unable to find in the literature another value of the CryoSat-2 absolute SSH bias (using sea level in situ measurements), but our results look very coherent with the updated Svalbard transponder result [20]: 722 ± 6 mm (standard deviation of 29 mm, baseline B, before correcting the 673 mm known range bias). It is worth noting that the 24 mm remaining difference between these two determinations can be due to the geophysical corrections, and notably that the SSB is not affecting the transponder processing.

For pass #4794, overflying both the Senetosa and Ajaccio facilities, we computed the averaged time series in the same way as for the Sentinel-3A pass shown in Figure 7. This is presented in Figure 10, where the average SSH bias from the Ajaccio and Senetosa overflights (5 s time lag and 37 km distance) showed a very low standard deviation (15 mm for SAMOSA+ and 17 mm for SAMOSA2) and a bias very close to the global average: −740 mm for SAMOSA+ and −741 mm for SAMOSA2 vs. −746 mm for the global average (for both SAMOSA+ and SAMOSA2). For comparison, with a similar configuration, the standard deviation over 13 cycles for Sentinel-3A was 24 mm.

Table 3. Details of the SSH biases for each pass and for the two retrackers (SAMOSA2 and SAMOSA+).

Passes	SAMOSA2			SAMOSA+			Comment
	Mean (mm)	σ (mm)	Number of Cycles	Mean (mm)	σ (mm)	Number of Cycles	
Ajaccio							
3563	NA	NA	NA	NA	NA	NA	Coastal *
4077	−746	32	5	−743	30	5	Coastal *
4794	−727	36	5	−731	24	5	Coastal *
5308	NA	NA	NA	NA	NA	NA	Coastal *
Average	−736	33	10	−737	26	10	
Senetosa							
0681	−766	17	5	−768	14	4	d > 3 km
1195	−733	23	3	−733	23	3	d > 8 km
2426	−757	36	4	−756	35	4	Coastal *
4794	−742	26	6	−746	23	6	d > 13 km
Average	−751	27	18	−751	26	17	
Total	−746	29	28	−746	26	27	

* Coastal means that we take the measurements up to the coast; for the others the distance "d" is the closest measurement to the coast.

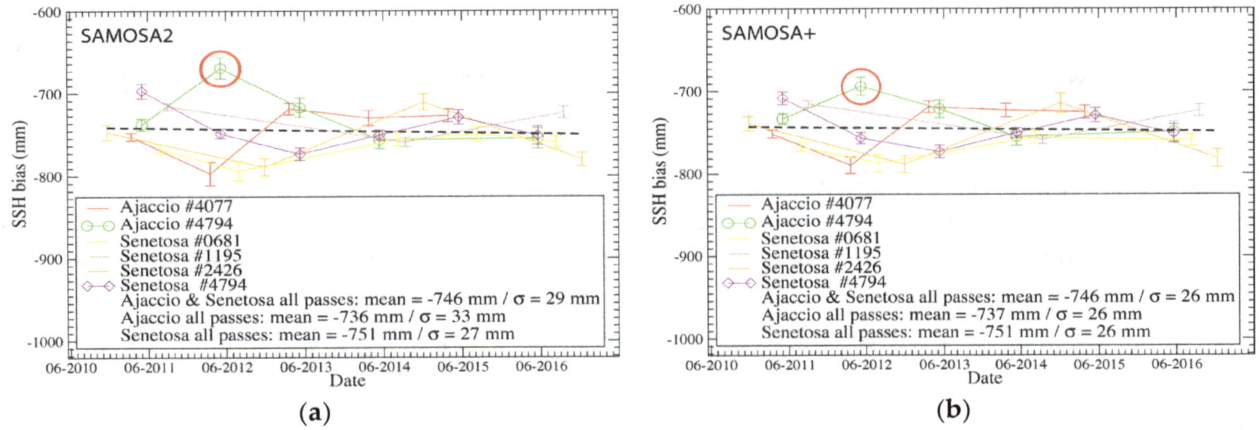

Figure 9. Time series of the CryoSat-2 SSH biases for each pass and the averages from Ajaccio, Senetosa and both (dash black line): (**a**) corresponds to the G-POD standard retracking (SAMOSA2) while (**b**) corresponds to the improved retracking (SAMOSA+) tuned for coastal, inland waters and ice.

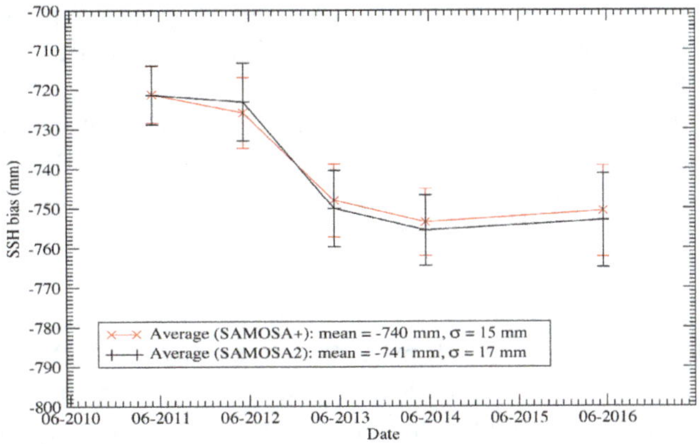

Figure 10. Example of pass #4794 with a configuration « à la » Sentinel-3A from which the SSH biases from Ajaccio and Senetosa were averaged (using SAMOSA+).

4. Discussion and Conclusions

The configuration of the Sentinel-3A and CryoSat-2 passes allowed us to cross-compare the results obtained independently from both the Senetosa and Ajaccio sites. Our results confirm previous results that were obtained from SARAL/AltiKa together with GPS-based sea level measurements [3]. The SSH measured by the Ajaccio tide gauge is biased by ~30 mm and our present study shows the origin of this offset. By taking into account this offset, we compared results from the Senetosa and Ajaccio sites for both the Sentinel-3A and CryoSat-2 altimeters. For the Sentinel-3A the site-difference between Senetosa and Ajaccio was −4 mm (21–25 mm, see Figure 6) while for CryoSat-2 the difference was −14 mm ((−751)–(−737) mm, see Figure 9 from SAMOSA+ results). These comparisons show that no significant systematic error remains in the geodetic datum or in situ measurements. This is a very important result that gives us confidence also for the historical calibrations performed independently at each site. However, differences still exist at the millimeter level between the SSH bias obtained at the Senetosa and Ajaccio sites for Sentinel-3A and CryoSat-2 (respectively −4 and −14 mm). This probably reflects some remaining errors in the altimeter measurement system (coastal contamination, sea state bias, ...) and/or in the local geodetic datum of both the Senetosa and Ajaccio sites. Finally, our analysis of the wet tropospheric corrections available for the Sentinel-3A calibration showed that the effect of land contamination on the radiometer was ~−20 mm, which is actually the main part of the SSH bias (+22 ± 7 mm). In terms of the range itself, we can consider that SAR measurement system on Sentinel-3A as "unbiased". For CryoSat-2, it is important to recall that the product used from G-POD is based on baseline B, so to be comparable to the current baseline C, one must remove the known range bias of 673 mm to the SSH biases determined in this study: when using baseline C products, the CryoSat-2 SSH bias is −73 ± 5 mm. Last but not least, the SSH biases for both Sentinel-3A and CryoSat-2 were compared to the range biases determined with transponders located respectively in Crete and Svalbard, and the remaining differences (respectively 22 mm and 24 mm) between these two determinations can be due to geodetic datum errors or geophysical corrections and, notably, the sea state bias that is not affecting transponder processing. This is probably the first time in altimetry that the results from transponders and in situ SSH calibration agree at this level of accuracy.

Acknowledgments: This study was conducted and financed thanks to Centre National d'Etudes Spatiales (CNES), Centre National de la Recherche Scientifique (CNRS), and the French Ministry of Research.

Author Contributions: Pascal Bonnefond and Olivier Laurain designed the study, analyzed the data and wrote the paper. All the other co-authors helped in paper writing and analysis.

References

1. Bonnefond, P.; Exertier, P.; Laurain, O.; Menard, Y.; Orsoni, A.; Jan, G.; Jeansou, E. Absolute Calibration of Jason-1 and TOPEX/Poseidon Altimeters in Corsica Special Issue on Jason-1 Calibration/Validation. *Mar. Geod.* **2003**, *26*, 261–284. [CrossRef]
2. Bonnefond, P.; Exertier, P.; Laurain, O.; Jan, G. Absolute Calibration of Jason-1 and Jason-2 Altimeters in Corsica during the Formation Flight Phase. *Mar. Geodesy* **2010**, *33*, 80–90. [CrossRef]
3. Bonnefond, P.; Laurain, O.; Exertier, P.; Guillot, A.; Picot, N.; Cancet, M.; Lyard, F. SARAL/AltiKa absolute calibration from the multi-mission Corsica facilities. *Mar. Geodesy* **2015**, *38*, 171–192. [CrossRef]
4. Bonnefond, P.; Haines, B.; Watson, C. In Situ Calibration and Validation: A Link from Coastal to Open-ocean altimetry, chapter 11. In *Coastal Altimetry*; Vignudelli, S., Kostianoy, A., Cipollini, P., Benveniste, J., Eds.; Springer: Berlin/Heidelberg, Germany, 2011; pp. 259–296. ISBN 978-3-642-12795-3.
5. Bonnefond, P.; Exertier, P.; Laurain, O.; Menard, Y.; Orsoni, A.; Jeansou, E.; Haines, B.; Kubitschek, D.; Born, G. Leveling Sea Surface using a GPS catamaran. *Mar. Geodesy* **2003**, *26*, 319–334. [CrossRef]
6. Bonnefond, P.; Exertier, P.; Laurain, O.; Lyard, F.; Bijac, S.; Cancet, M.; Chimot, J.; Féménias, P. Corsica: An experiment for long-term Altimeter calibration and sea level monitoring. In Proceedings of the ESA Living Planet Symposium, Bergen, Norway, 28 June–2 July 2010.

7. Bonnefond, P.; Verron, J.; Babu, K.N.; Bergé-Nguyen, M.; Cancet, M.; Chaudhary, A.; Crétaux, J.F.; Frappart, F.; Haines, B.J.; Laurain, O.; et al. The Benefits of the Ka-Band as Evidenced from the SARAL/AltiKa Altimetric Mission: Quality Assessment and Unique Characteristics of AltiKa Data. *Remote Sens.* **2017**, *10*, 83. [CrossRef]
8. Donlon, C.; Berruti, B.; Buongiorno, A.; Ferreira, M.-H.; Féménias, P.; Frericka, J.; Goryl, P.; Kleina, U.; Laur, H.; Mavrocordatos, C.; et al. The Global Monitoring for Environment and Security (GMES) Sentinel-3 mission. *Remote Sens. Environ.* **2012**, *120*, 37–57. [CrossRef]
9. Bouffard, J.; Webb, E.; Scagliola, M.; Garcia-Mondéjar, A.; Baker, S.; Brockley, D.; Gaudelli, J.; Muir, A.; Hall, A.; Mannan, R.; et al. CryoSat Instrument Performance and Ice Product Quality Status. *Adv. Space Res.* **2017**. [CrossRef]
10. Sentinel-3A SRAL Reprocessed Dataset Available on CODA from 16 October. Available online: https://www.eumetsat.int/website/home/News/DAT_3648215.html (accessed on 6 November 2017).
11. Ray, C.; Martin-Puig, C.; Clarizia, M.P.; Ruffini, G.; Dinardo, S.; Gommenginger, C.; Benveniste, J. SAR altimeter backscattered waveform model. *IEEE Trans. Geosci. Remote Sens.* **2015**, *53*, 911–919. [CrossRef]
12. Dinardo, S.; Fenoglio-Marc, L.; Buchhaupt, C.; Becker, M.; Scharroo, R.; Fernandez, J.M.; Benveniste, J. Coastal SAR and PLRM Altimetry in German Bight and West Baltic Sea. *Adv. Space Res.* **2017**, in press. [CrossRef]
13. Fenoglio-Marc, L.; Dinardo, S.; Scharroo, R.; Roland, A.; Dutour Sikiric, M.; Lucas, B.; Becker, M.; Benveniste, J.; Weiss, R. The German Bight: A validation of CryoSat-2 altimeter data in SAR mode. *Adv. Space Res.* **2015**, *55*, 2641–2656. [CrossRef]
14. Chelton, D.B.; Walsh, E.J.; MacArthur, J.L. Pulse compression and sea level tracking in satellite altimetry. *J. Atmos. Ocean. Technol.* **1989**, *6*, 407–428. [CrossRef]
15. King, R.; Bock, Y. Documentation for the GAMIT GPS Analysis Software. Release 10, Massachusetts Institute of Technology, Cambridge. 2000. Available online: http://www-gpsg.mit.edu/~simon/gtgk/GAMIT.pdf (accessed on 5 October 2017).
16. Picard, B.; Fréry, M.; Pardé, M.; Bonnefond, P.; Laurain, O.; Crétaux, J.-F. A Wet Tropospheric Correction Dedicated to Hydrological and Coastal Applications. In Proceedings of the Ocean Surface Topography Science Team Meeting, Miami, FL, USA, 23–27 October 2017; Available online: https://meetings.aviso.altimetry.fr/fileadmin/user_upload/tx_ausyclsseminar/files/OSTST-2017_PICARD_NewCoastalNN_POSTER.pdf (accessed on 6 November 2017).
17. Mertikas, S.; Donlon, C.; Mavrocordatos, C.; Féménias, P.; Galanakis, D.; Tziavos, I.; Boy, F.; Vergos, G.; Andersen, O.; Frantzis, X.; et al. Multi-mission calibrations results at the Permanent Facility for Altimetry Calibration in west Crete, Greece attaining Fiducial Reference Measurement Standards. In Proceedings of the Ocean Surface Topography Science Team Meeting, Miami, FL, USA, 23–27 October 2017; Available online: https://meetings.aviso.altimetry.fr/fileadmin/user_upload/tx_ausyclsseminar/files/OSTST_CalVal_20_Oct_2017_Mertikas_AAA.pdf (accessed on 6 November 2017).
18. Garcia-Mondejar, A.; Mertikas, S.; Galanakis, D.; Labroue, S.; Bruniquel, J.; Quartly, G.; Féménias, P.; Mavrocordatos, C.; Wood, J.; Garcia, G.; et al. Sentinel-3 Transponder Calibration Results. In Proceedings of the Ocean Surface Topography Science Team Meeting, Miami, FL, USA, 23–27 October 2017; Available online: https://meetings.aviso.altimetry.fr/fileadmin/user_upload/tx_ausyclsseminar/files/S3_OSTST_Poster_20171012_V8.pdf (accessed on 6 November 2017).
19. Dinardo, S. Guidelines for the SAR (Delay-Doppler) L1b Processing. 2013. Available online: http://wiki.services.eoportal.org/tiki-download_wiki_attachment.php?attId=2540 (accessed on 5 October 2017).
20. Garcia-Mondejar, A.; Fornari, M.; Bouffard, J.; Féménias, P.; Roca, M. CryoSat-2: Range, Datation and Interferometer Calibration with Svalbard Transponder. *Adv. Space Res.* **2017**, under review.

5

The Benefits of the Ka-Band as Evidenced from the SARAL/AltiKa Altimetric Mission: Scientific Applications

Jacques Verron [1,*], Pascal Bonnefond [2], Lofti Aouf [3], Florence Birol [4], Suchandra A. Bhowmick [5], Stéphane Calmant [4], Taina Conchy [6], Jean-François Crétaux [4], Gérald Dibarboure [7], A. K. Dubey [5], Yannice Faugère [8], Kevin Guerreiro [4], P. K. Gupta [5], Mathieu Hamon [9], Fatma Jebri [4], Raj Kumar [5], Rosemary Morrow [4], Ananda Pascual [10], Marie-Isabelle Pujol [8], Elisabeth Rémy [9], Frédérique Rémy [4], Walter H. F. Smith [11], Jean Tournadre [12] and Oscar Vergara [4,8]

1 Institut des Géosciences de l'Environnement (IGE), CNRS, 38041 Grenoble, France
2 SYRTE, Observatoire de Paris, PSL Research University, CNRS, Sorbonne Universités, UPMC Univ. Paris 06, LNE, 75014 Paris, France; pascal.bonnefond@obspm.fr
3 Météo-France, 31057 Toulouse, France; lotfi.aouf@meteo.fr
4 Laboratoire d'Etudes en Géophysique et Océanographie Spatiales (LEGOS), 31400 Toulouse, France; florence.birol@legos.obs-mip.fr (F.B.); stephane.calmant@ird.fr (S.C.); jean-francois.cretaux@legos.obs-mip.fr (J.-F.C.); guerrei@legos.obs-mip.fr (K.G.); fatma.jebry@gmail.com (F.J.); rosemary.morrow@legos.obs-mip.fr (R.M.); remy.omp@free.fr (F.R.); oscar.vergara@legos.obs-mip.fr (O.V.)
5 Space Applications Centre (ISRO), Ahmedabad 380015, India; suchandra@sac.isro.gov.in (S.A.B.); a_dubey@sac.isro.gov.in (A.K.D.); pkgupta@sac.isro.gov.in (P.K.G.); rajkumar_62@rediffmail.com (R.K.)
6 Universidade do Estado de Amazonas, Manaus 69020, Brazil; taina.conchy@inpa.gov.br
7 Centre National d'Etudes Spatiales (CNES), 31400 Toulouse, France; gerald.dibarboure@cnes.fr
8 Collecte Localisation Satellites (CLS), 31520 Ramonville Saint-Agne, France; yannice.faugere@cls.fr (Y.F.); mpujol@cls.fr (M.-I.P.)
9 Mercator Océan, 31520 Ramonville Saint-Agne, France; mathieu.hamon@mercator-ocean.fr (M.H.); elisabeth.remy@mercator-ocean.fr (E.R.)
10 Institut Mediterrani d'Estudis Avançats (IMEDEA) (CSIC-UIB), 07190 Esporles, Illes Balears, Spain; ananda.pascual@imedea.uib-csic.es
11 Laboratory for Satellite Altimetry, NOAA, College Park, MD 20740-3818, USA; walter.hf.smith@noaa.gov
12 Laboratoire d'Océanographie Physique et Spatiale (LOPS), 29280 Plouzané, France; jean.tournadre@ifremer.fr
* Correspondence: jacques.verron@univ-grenoble-alpes.fr

Abstract: The India–France SARAL/AltiKa mission is the first Ka-band altimetric mission dedicated primarily to oceanography. The mission objectives were firstly the observation of the oceanic mesoscales but also global and regional sea level monitoring, including the coastal zone, data assimilation, and operational oceanography. SARAL/AltiKa proved also to be a great opportunity for inland waters applications, for observing ice sheet or icebergs, as well as for geodetic investigations. The mission ended its nominal phase after three years in orbit and began a new phase (drifting orbit) in July 2016. The objective of this paper is to highlight some of the most remarkable achievements of the SARAL/AltiKa mission in terms of scientific applications. Compared to the standard Ku-band altimetry measurements, the Ka-band provides substantial improvements in terms of spatial resolution and data accuracy. We show here that this leads to remarkable advances in terms of observation of the mesoscale and coastal ocean, waves, river water levels, ice sheets, icebergs, fine scale bathymetry features as well as for the many related applications.

Keywords: altimetry; Ka-band; oceanography; hydrology; ice; geodesy

1. Introduction

The India–France SARAL/AltiKa mission is the first Ka-band altimetric mission dedicated to oceanography. SARAL was launched on 25 February 2013. Today, the SARAL/AltiKa mission is in its fifth year, which means that it has reached its design life. The mission will continue as long as the satellite and the ground segment continue to operate. The lifetime requirement of the AltiKa payload was just three years, which has clearly been exceeded. The instruments are still in perfect conditions.

The SARAL satellite is composed of a spacecraft bus developed by the Indian Space Agency (ISRO), and a payload developed by the French Space Agency (CNES). As its full name (SARAL stands for "Satellite with ARgos and ALtiKa") indicates, two missions are onboard the SARAL satellite: The AltiKa altimeter component and ARGOS-3, the new generation of ARGOS instrument. For the scientific applications discussed in this paper, we focus on AltiKa. The altimeter payload is composed of the AltiKa altimeter-radiometer in Ka-band, a Doris system for precise orbit determination (POD), and a Laser Retro-reflector Array instrument used for precise calibration of other POD instruments. AltiKa is a nadir-looking altimeter in line with Poseidon-3 on-board Jason-3 and other existing altimeters, but operates in a single frequency band: the Ka-band. It is the reduced ionosphere effects in Ka-band that makes it possible to use a mono-frequency altimeter. The altimeter shares the antenna with a bi-frequency radiometer required to correct the altimeter range for the wet troposphere path delay. The SARAL satellite orbit is almost polar, sunsynchronous and with a 35-day repeat cycle. This is the same orbit as Envisat with the initial motivation to continue the time series and to benefit from the existing mean sea surface. Since March 2015, technical issues have been encountered on the reaction wheels and this has led to the decision of relaxing the orbit. Then, after 4 July 2016, SARAL/AltiKa left its repetitive orbit by beginning a new phase named "SARAL Drifting Phase" (SARAL-DP). The data processing as well as data latency have not been changed. From this date, SARAL satellite does fly free of station keeping maneuvers which means there is a drift between successive ground tracks which are no more repetitive. The altimetry payload is operated nominally on the drifting orbit. The cycle duration of 35 days is no longer kept but the orbit has been chosen to preserve anyway subcycles within the range of 15 to 17 days quite relevant for mesoscale sampling.

For further information on SARAL/AltiKa, one can refer to introductory papers [1,2] and more widely to the Special Issue of Marine Geodesy dedicated to this satellite [3].

Since the beginning of the SARAL/AltiKa mission, performances proved to be compliant with nominal specifications with an overall observed performance for the sea surface height (SSH) RMS of 3.4 cm, lower than the mission requirement of 4 cm (see also [4]). In general, SARAL/AltiKa performances appear to be quite similar, and often better, than the Ku-band reference altimetric satellites such as Jason-2. Before launch, concerns were raised about the sensitivity of the Ka-band to rain events, leading to missing and invalid measurements. However, in practice, the SARAL/AltiKa data return is remarkably high with few missing data. Over the ocean, the data coverage is greater than 99.5%, which, once again, exceeds the mission requirements. SSH differences at crossovers is the main metric to assess the overall performance of satellite altimetry missions. In this regard, the results obtained from the SARAL/AltiKa mission are of the same order, or even better, as those obtained for Jason-2 mission. The same result is obtained for the sea level anomalies (SLA): SARAL/AltiKa data quality is as good (or even slightly better) as Jason-2. Another key factor for altimetry applications is the spectral content of the data. AltiKa altimeter provides a signal-to-noise ratio which has never been obtained before. Compared to Jason-2 and CryoSat-2 (SAR mode) data, the spectral content of SARAL/AltiKa observations is largely improved for all wavelengths below 70 km. This is explained by the much lower white noise, due to the excellent Ka-band retracking performances and also to the 40 Hz measurement rate (higher than the standard 20 Hz on Jason-class and CryoSat-2 missions). In conclusion, the mission performances widely confirm the nominal expectations in terms of accuracy, data quality and data availability.

The mission objectives are the observation of the oceanic mesoscale dynamics, including the coastal zone, but also the global and regional sea level monitoring, data assimilation and operational oceanography. It also includes the monitoring of the level of the main continental water bodies (lakes, rivers, enclosed seas), of large-scale sea level variations, the observation of polar oceans (as a result of the high inclination of its orbit), the analysis and forecast of wave and wind fields, the study of continental ice and sea ice, the access to low rains climatology and the marine biogeochemistry (mostly through improved mesoscale access).

Beyond the excellent quality of SARAL/AltiKa measurements, there are some specificities that emerge from the Ka-band altimetry. Different scientific results illustrate these specificities. The objectives of this paper are to give precise evidence of the main features of the SARAL/AltiKa that are associated with the Ka-band. In a companion paper [4], we looked at the quality assessment and specificities of AltiKa data. In the present paper, we focus more on the scientific aspects. Rather naturally, the paper is organized following the main scientific applications of SARAL/AltiKa: oceanography, hydrology, ice and geodesy. Some examples are given to illustrate some salient features of today's SARAL/AltiKa data with regard to standard altimetry: data accuracy allowing significant advances in the field of ocean mesoscale studies and for operational applications, in coastal areas, over ice, for inland waters and finally for geodesy.

2. The Ocean

As previously mentioned, SARAL/AltiKa's main scientific objective is to provide data products to the oceanographic community with the aim to improve our knowledge of the ocean mesoscale variability, mainly associated to eddies, meandering currents, fronts, filaments and squirts. The mesoscale variability refers to ocean signals with space scales of 50 km to 500 km and time scales from a few days to a few months. The associated kinetic energy is at least an order of magnitude larger than that of the mean circulation. It is now well recognized that the mesoscale processes are a key component of the ocean circulation, shaping the mean currents, controlling the exchanges with the atmosphere and the marine biogeochemistry, and clearly playing a major role in the way the ocean participate to the climatic system.

The sea level measurements of SARAL/AltiKa are used in theoretical and observational studies, modelling, data assimilation, etc. Because mesoscale and fine scale dynamical processes are particularly important in the coastal ocean, the observational capabilities of SARAL/AltiKa are also expected to bring an important contribution to the study of coastal dynamics. It is particularly important for many downstream applications including operational oceanography which is fed by a large amounts of in-situ and space-based data. Similar to previous altimeters, SARAL/AltiKa not only provide sea level measurements, but also ocean wave observations through the significant wave height (SWH) parameter. Ocean waves play a key role in ocean–atmosphere exchanges at the sea interface. It is essential to have the most accurate knowledge possible of the sea state for climate and meteorological studies.

2.1. *Observability of the Fine-Scale Ocean Dynamics*

Technological advances in all of the recent satellite altimeter missions have improved their signal-to-noise ratio, allowing us to observe finer-scale ocean processes with along-track altimeter data. By calculating wavenumber spectra from along-track SSH anomalies from the recent missions, we have been able to estimate the background noise levels of each mission, and the corresponding observable ocean scales, where the signal-to-noise is greater than 1. This statistical estimate of the altimetric noise and the observable scales has been performed over the global oceans using Jason-2 (Ku-band), SARAL/AltiKa (Ka-band) and CryoSat-2 in low-resolution mode (LRM) by [5], and from Jason-2, SARAL/AltiKa and Sentinel-3 in Synthetic Aperture Radar (SAR) mode by [6]. Regional analyses have also been performed over the western Mediterranean Sea by [7] using Jason-2, SARAL/AltiKa and CryoSat-2 SAR mode.

An example of the globally-averaged sea level anomaly (SLA) wavenumber spectra are shown in Figure 1 from three missions with different technologies : Jason-2, SARAL/AltiKa and Sentinel-3 in

SAR mode [6]. All missions observe similar SLA spectral levels at scales larger than 100 km wavelength. At shorter scales, the SLA spectra are dominated by the error level of each missions. Jason-2 has a higher mean noise level with a strong "bump" from 10 to 50 km wavelength due to surface roughness inhomogeneities within its relatively large footprint, mainly from surface waves. SARAL/AltiKa has the smallest noise at very short scales, but is also affected by the spectral "bump" at scales of 3–20 km wavelength (note that standard data processing is applied here; with more sophisticated treatment, these spectral bumps can be reduced without changing the hierarchy of satellite data quality for small ocean scales [4]). Sentinel-3 in SAR mode has a significant reduction in the spectral bump and the smallest error level at 10 km wavelength, with a different "red" noise spectrum. Both SARAL/AltiKa and Sentinel-3 SAR data observe similar SSH variability down to scales of 40 km wavelength.

Each altimeter's error field shows distinct seasonal and geographical variations, with higher noise in winter due to the rougher sea state [5]. We have recently extended this work to analyse SARAL/AltiKa over a longer three-year period when both missions were on a repeat groundtrack with low mean sea surface errors. SARAL/AltiKa has relatively low 1 Hz noise, but still exhibits strong seasonal variations in noise (Figure 2). The 1 Hz noise is higher in each hemisphere in the region of highest winter waves, but also increases in the boundary currents due to wave-current interactions, and in the tropics due to rain cells or calm water patches, all leading to inhomogeneities within the footprint. In the western Mediterranean Sea, similar seasonal patterns were found [7] (not shown).

These statistical noise levels impact on the ocean scales we can observe. In the global ocean, SARAL/AltiKa can observe scales of 30 km in the strong boundary currents and the tropics, reaching 70–80 km in the quieter eastern basin where the signal-to-noise ratio is low [5]. Conversely, the higher noise of Jason-2 limits the observation scales: we can only resolve processes of 40–50 km in the western boundary currents, and 80–100 km in the eastern basins. Note that these spectral wavelength limits correspond to individual feature diameters of half that size, i.e., with SARAL/AltiKa we can detect strong western boundary eddies down to 15 km diameter (30 km in wavelength). In the Mediterranean Sea, where the Rossby radius is small, the difference between SARAL/AltiKa and Jason-class altimeters is also striking [7]. In winter, when the mixed layers are deepest and the submesoscale is energetic, all of the altimeter missions can observe wavelengths down to 40–50 km. In summer, when the submesoscales are weaker, SARAL/AltiKa with its lower noise can detect ocean scales down to 35 km wavelength, whereas the higher noise from Jason-2 blocks the observation of scales less than 50–55 km wavelength.

The noise levels impacting on the observation of fine-scale dynamics depend on many factors, including surface roughness from different wind–wave–swell conditions, impact from rain or other geophysical corrections [5,8], and errors in the geoid or mean sea surface (MSS). Even with the new generation of MSS products, when a satellite drifts off the long-term mean repeat track with its accurate MSS, the error in SLA increases. Figure 3 shows the evolution of the SLA standard deviation, calculated with respect to different MSS products, during a period when the SARAL/AltiKa mission drifted from its long-term mean groundtrack (cycles 23–25, marked in gray, from [9]). The data have been filtered to retain only short wavelengths less than 200 km, which have the largest gridded MSS errors. The standard deviation for Jason-2 (in black) is calculated from the precise alongtrack MSS, whereas for SARAL/AltiKa it is calculated with respect to different MSS models. Figure 3 highlights the lower standard deviation of SARAL/AltiKa SLA compared to Jason, but even when the recent MSS models are used with the most geodetic data included (MSS CNES_CLS15), the error increases during the drifting orbit phase. Work is in progress to better understand and estimate these different geophysical signals, in order to reduce their effects on the SLA noise.

Although SARAL/AltiKa and Jason-2 show a similar pattern in seasonal and geographical variations of their noise, these studies quantify that the low background noise levels of SARAL/AltiKa enable this mission to make the most precise estimates of the fine-scale ocean dynamics, in the open ocean and regional and coastal seas. This is encouraging for further studies with SARAL/AltiKa , but also in the preparation for the future Ka-band wide-swath altimeter mission from NASA/CNES, the Surface Water Ocean Topography (SWOT) mission, whose lower noise should enable us to observed even smaller ocean scales.

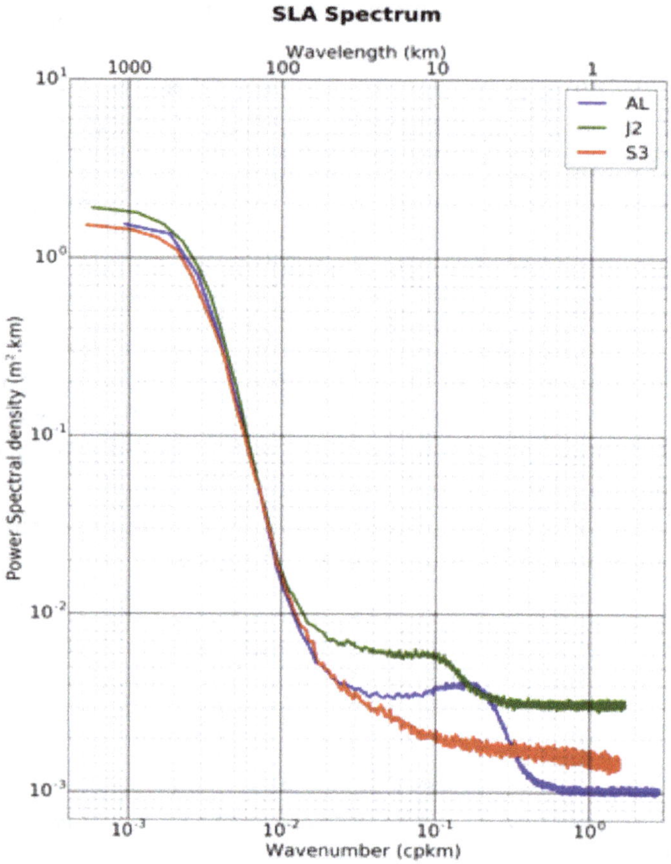

Figure 1. Globally-averaged wavenumber spectra of SSH anomalies (in $m^2/cpkm$) from the highest resolution along-track data for three different missions: Jason-2 20 Hz (J2 in green); SARAL/AltiKa 40 Hz (AL in blue); and Sentinel-3 20 Hz SAR (S3 in red). The spectral "bump" in Jason-2 and SARAL/AltiKa data is due to surface roughness inhomogeneities within the footprint, mainly from surface waves.

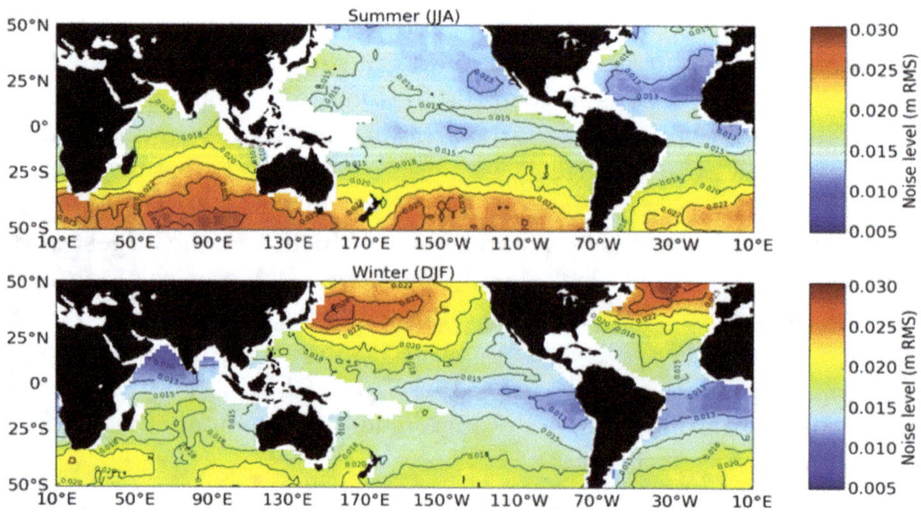

Figure 2. Geographical distribution of the 1 Hz noise level over the global oceans in: (**upper panel**) June–August; and (**lower panel**) December–February. The noise is estimated from wavenumber spectra averaged over $15° \times 15°$ boxes, as a constant value fit over the wavelengths 14–30 km.

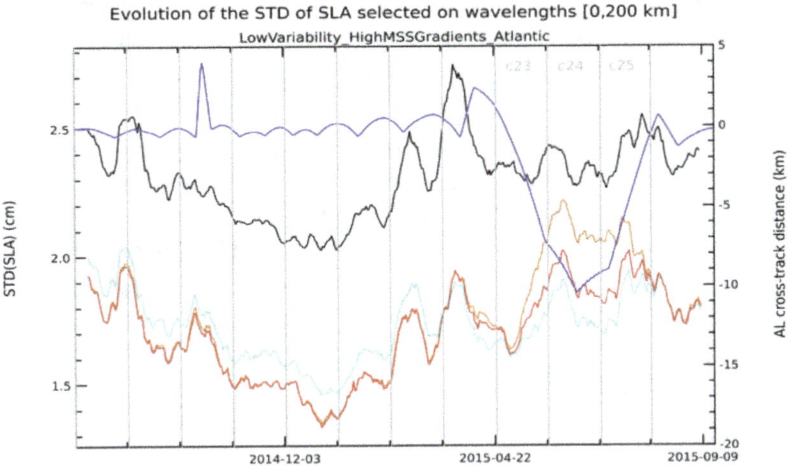

Figure 3. Temporal evolution of the standard deviation of the SLA, along Jason-2 tracks (calculated with respect to its alongtrack precise MSS in black) and along SARAL/AltiKa tracks (calculated with respect to different gridded MSS products: MSS CNES_CLS15 (in red), CNES_CLS11 (in orange) and DTU15 (in cyan). The SLA was high-pass filtered to keep only wavelengths shorter than 200 km. Statistics are computed over a low variability and intense bathymetric gradients region in the northeast Atlantic [20°S, 0°N; 30°W, 0°E]. The blue line shows the cross-track distance (in km; right axis) of the SARAL/AltiKa tracks compared to their long-term repeat groundtrack; this distance increases sharply during cycles C23, C24, and C25, marked in gray from 22 April 2015 onwards.

2.2. *Observability of the Coastal Ocean Dynamics*

Satellite altimetry represents a mature technology in the open ocean but its exploitation remains an issue in the coastal ocean for different reasons. Land contamination in the altimeter footprint lead to a signal which is much more complex to analyze than offshore and often lead to the rejection of the altimetry measurements during the standard processing step. Geophysical corrections are significantly less accurate in the coastal area. The radiometer measurements used to derive the wet-troposphere correction becomes perturbed by land 25–50 km from the coast (depending on the instrument). The coastal ocean dynamics have also shorter space and time scales than the scales of open ocean dynamics which are mostly resolved by altimetry. Consequently, coastal altimetry data are largely discarded in standard products distributed to users and then largely unexploited. However, having as many accurate observations of the coastal ocean as possible is crucial for improving our understanding and monitoring of physical ocean processes that impact a wide range of applications (coastal area management, risk prevention, fisheries, marine safety and shipping, pollution, etc). In the context of climate change, we need to observe and understand how the coastal sea level variations, in particular extreme events such as storm surges, but also those associated to much lower time scales, impact the coastal morphology (via erosion). In the last years, a lot of efforts have been dedicated in the coastal altimetry community in order to extend the observational capabilities of altimeters always closer to the coastline (e.g., http://www.coastalal.eu). The launch of the SARAL/AltiKa altimetry mission, with an instrument able to provide accurate sea level measurements at a higher spatial resolution and closer to the coastline, is a major step forward.

2.2.1. Observability of the Coastal OceanDynamics in the Central Mediterranean Sea

In a first illustration of the benefits of SARAL/AltiKa , we analyse if the improved altimetric capabilities of SARAL/AltiKa that has been shown in different studies (e.g., [5,7]) can lead to the detection of finer ocean scales in the Central Mediterranean Sea. In this area, the observation of the different current branches by altimetry is particularly challenging due to their short width and high space-time variability [10]. Figure 4 (left) shows an example of the wavenumber spectra calculated over the study area from both Jason-2 and SARAL/AltiKa along-track SSH. This very local result obtained from a data set including coastal altimetry observations (unlike the studies mentioned above) confirms what has already been obtained elsewhere. At wavelengths >80 km, SARAL/AltiKa and Jason-2 SLA spectra are very similar. Below 80 km, the spectral energy of SARAL/AltiKa decreases significantly more than for Jason-2. Using the same spectral analysis approach than [5,7] for different altimetry tracks crossing the Central Mediterranean Sea, we deduce that the mesoscale observational limit occurs around 46 km wavelength (for a 12–35 km estimated noise level) for Jason-2 and around 21 km wavelengths (for a 12–18 km estimated noise level) for SARAL/AltiKa. These numbers are slightly lower than the results obtained by [7] further north in the northwestern Mediterranean Sea, confirming the strong dependency of the observational capability of altimetry missions to the level of mesoscale energy. These results have been used to define for both altimetry missions the optimal spatial filtering of alongtrack SLA: a low-pass Lanczos filter with a 21/45 km cutoff frequency for SARAL/Jason-2.

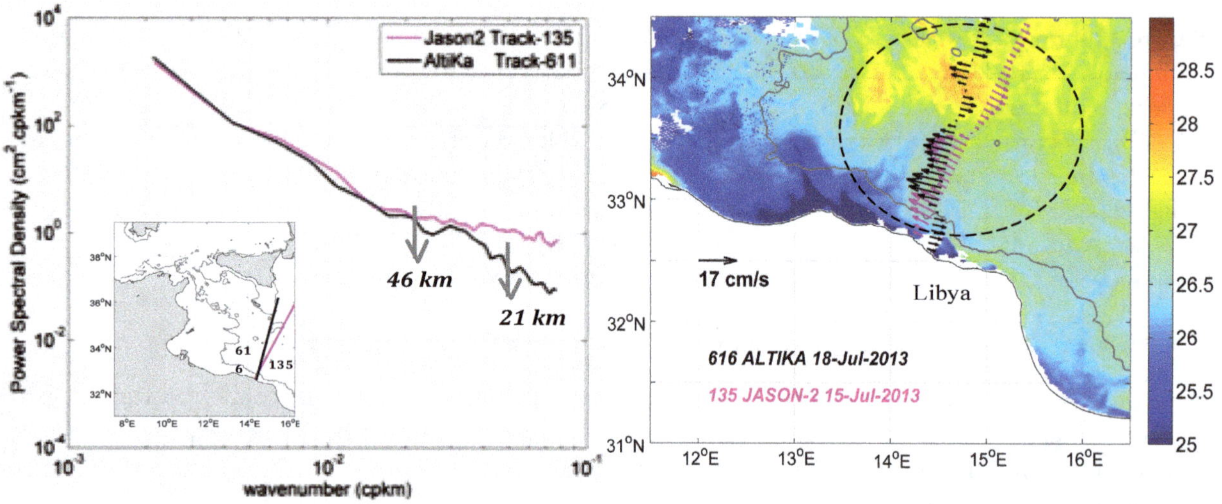

Figure 4. (**Left**) Wavenumber spectra from SSH observed in the Central Mediterranean Sea along the track 616 of SARAL/AltiKa (black) and along the track 135 of Jason-2 (pink); and (**Right**) comparisons of SST (in °C) with SARAL/AltiKa track 616 (black) and Jason-2 track 135 (pink) crosstrack velocities over the Tunisia-Lybian shelf for 18 July 2013. The SST pattern is derived from CMEMS products with 1/16° resolution. The 200 m isobath (in grey) is derived from the ETOPO2v1 database. The black dashed circle indicates the location of the Sidra Gyre.

To illustrate the impact of the improved observational capability of SARAL/AltiKa in terms of ocean circulation captured, the crosstrack absolute geostrophic velocities are computed from both the filtered SARAL/AltiKa and Jason-2 sea level data, adding the mean dynamic topography provided by AVISO to the SLA. They have then been compared to the information provided by high resolution Level 3 sea surface temperature (SST) 2D fields. Figure 4 (right) illustrates an example of this qualitative analysis. This zoom on the Tunisian-Libyan shelf for 18 July 2013 shows the presence of warmer waters north of the Lybian continental margin. From [10,11], it corresponds to the anticyclonic Sidra Gyre (black dashed circle), permanent feature of the regional circulation which advects warm and salty Ionian water through a surface current flowing northwestward offshore Libya. This feature is observed in both SARAL/AltiKa and Jason-2 current patterns but finer scales are derived from SARAL/AltiKa data and the undulations in the corresponding higher resolution surface currents perfectly correspond to the variations in the SST field.

Near the coast, on the Lybian shelf, the information provided by Jason-2 and SARAL/AltiKa diverge. In Jason-2, the surface currents still flows northwestward, while a southeastward current vein is observed in SARAL/AltiKa. For this date, SST data indicate the presence of colder waters flowing along the Libyan coast. They correspond to the AtlanticWaters advected Southeastward by the narrow Atlantic Lybian Current along the Tunisian-Libyan continental slope (see [10,11]). This very coastal current vein is captured by SARAL/AltiKa observations but not by Jason-2. This example shows how the better mesoscale resolution of SARAL/AltiKa allow to observe new signals, even near the coast, and then help to understand the fine-scale ocean dynamics.

2.2.2. SARAL-AltiKa Capabilities to Detect Coastal Currents: Comparisons With Jason-2 and HF Radar Data

In a second example, we refer to measurements of the surface velocity from radar. Among the various alternative sensors available, coastal High Frequency Radar is unique as it provides synoptic, high frequency and high resolution data at the boundary between the ocean and the atmosphere [12]. Here, we build up on a previous study [13] aimed at evaluating SARAL/AltiKa data in the coastal ocean (7–60 km from the shore). In that study, the altimeter performance was evaluated in the Ibiza Channel (Western Mediterranean) by estimating geostrophic currents and comparing them to high frequency (HF) radar surface velocities. Results showed that SARAL/AltiKa data were retrieved at a distance of only 7 km from the coast. The derived velocities revealed coherent mesoscale features with high temporal variability among the different cycles and with general reasonable agreement with HF radar fields (correlations of 0.54, which are significant at the 95% confidence level). In the present example, we also perform the comparison with Jason-2, whose track 187 crosses SARAL/AltiKa track 16 in the vicinity of Ibiza Island (Figure 5). The period of study is 18 April 2013 to 8 May 2014. The same processing as in [13] is applied for HF radar and as well as for both SARAL/AltiKa and Jason-2 missions. Surface geostrophic velocities from altimetry are computed by finite differences applied to the Absolute Dynamic Topography (ADT), which is obtained by adding SLA and the SMDT-MED2014 Mean Dynamic Topography [14]).

We then compare the mean across-track velocities for every satellite cycle with the equivalent interpolated HF radar vectors (3 days averages). As expected (Figure 5c,d) and due to the different repetitive cycle of each mission, the number of cycles available during the study period for SARAL/AltiKa is a factor of 3.5 less than for Jason-2. SARAL and HF radar mean velocities reveal a consistent pattern raising to positive significant correlations of 0.74 (Table 1). On the contrary, in the section sampled by Jason-2 (which is less than 10 km apart from SARAL/AltiKa), the time series are much noisier (both for radar and satellite) without a clear seasonal pattern. This translates in an insignificant correlation between Jason-2 and HF radar observations. However, RMS differences between satellite and HF radar velocities are of the same order (9 cm/s) for both missions. A more in depth investigation is needed to interpret and understand these results. Further works should span over a longer time period and interpret and extend the spatial coverage to an array of HF radars, taking advantage of the recent initiative to establish a pan-European HF radar network [12].

Table 1. Statistics of the comparisons between HF radar and altimetric across-track geostrophic velocities over the period comprised between 18 April 2013 and 8 May 2014. Twelve measurements are used for SARAL/AltiKa (repeat cycle of 35 days) and 36 measurements are considered for Jason-2 (repeat cycle of 9.9156 days; note that several cycles were missing for Jason-2).

	SARAL-AltiKa	Jason-2
Correlation	0.7382	-0.2481
Std (HF radar) (m/s)	0.0695	0.0518
Std (satellite) (m/s)	0.1055	0.0622
Rms Diff (HF radar – satellite)(m/s)	0.0942	0.0969

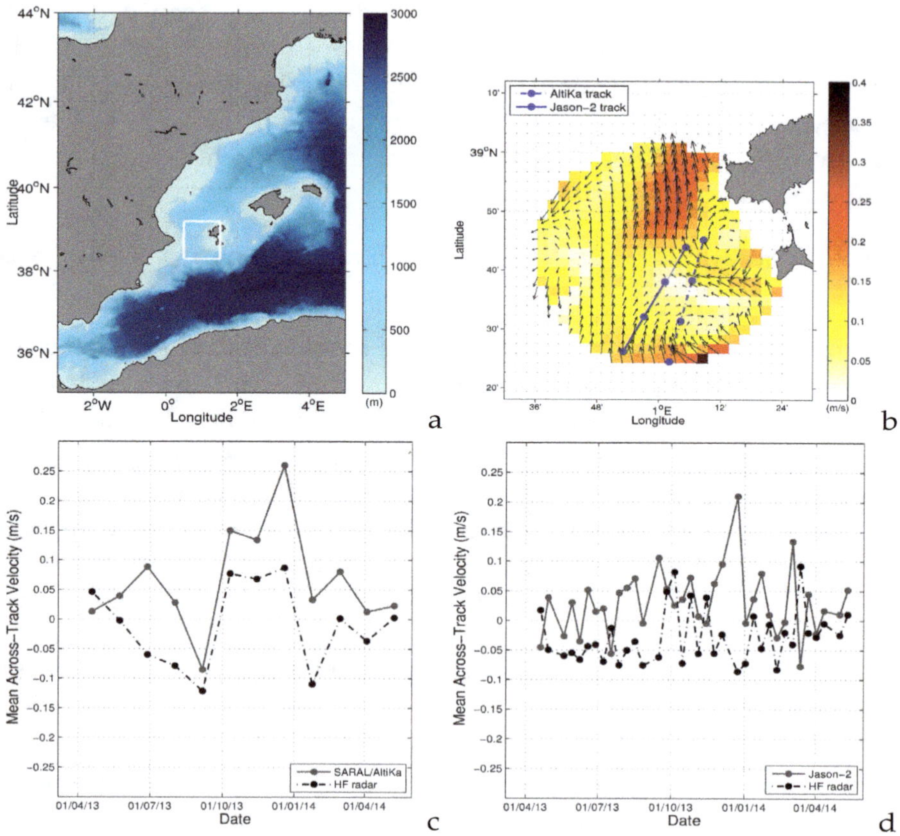

Figure 5. (**a**) Area of study in the Western Mediterranean. Colour contours represent bottom bathymetry and the white box corresponds to the area covered by the Ibiza HF radar shown in (**b**). (**b**) Coverage of SOCIB HF radar in the Ibiza Channel. The vectors correspond to surface velocities derived from HF radar (17 April 2013) field (three-day average). The colours are currents magnitude (in m/s). SARAL/AltiKa track 16 and Jason-2 track 187 are overlaid. The dots correspond to the position of the geostrophic velocity vectors derived from the altimeter missions. Note that the along-track delayed-time data provided at CMEMS is filtered with a 42 km filter and sub-sampled every 14 km (see details in [13]). (**c**) Comparison of the temporal variability of SARAL/AltiKa and HF radar mean cross-track velocity. X-axis is the number of cycles. (**d**) The same as (**c**) but for Jason-2.

2.3. SARAL/AltiK a Data Into Operational Systems

Altimetry is one of the most important data sources for assimilation into operational models. It provides global, real time, all-weather SSH measurements with high space and time resolution. Sea level is directly related to ocean circulation through the geostrophic approximation but sea level is also an integral of the ocean interior and is a strong constraint for inferring the 4D ocean circulation through data assimilation. Altimeters also measure SWH, which is essential for operational wave forecasting. High resolution from multiple altimeters is required to adequately represent ocean eddies and associated currents in models. Only altimetry can constrain the 4D mesoscale circulation in ocean models which is required for most operational oceanography applications.

There are very strong links between satellite oceanography and operational oceanography. The development of operational oceanography has been mainly driven by the development of satellite oceanography capabilities. The ability to observe the global ocean in near real time at high space and time resolution is indeed a prerequisite to the development of global operational oceanography and its applications. The first ocean parameter to be globally monitored from space was the SST on board meteorological satellites in the late 1970s. It is, however, the advent of satellite altimetry in the late 1980s that led the development of ocean data assimilation and global operational oceanography.

The satellite altimetry community was also keen to develop further the use of altimetry and this required an integrated approach merging satellite and in-situ observations with models.

2.3.1. SARAL/AltiK a Major Contributor to DUACS/CMEMS Multi-Mission Seal Level Products

For more than four years, SARAL/Altika products over ocean have been an important source of information for operational oceanography systems both in terms of assimilation and validation. Thanks to the efficiency of CNES and ISRO, the DUACS Sea Level system [15] was able to ingest SARAL/AltiKa data from 1 July 2013 onward in NRT, and merged them with the other flying satellites Jason-2, CryoSat-2 and HY-2A, to produce homogeneous set of Level 3 (along-track cross-calibrated SLA) and Level 4 (L4) products (multiple sensors merged as maps or time series).

Figure 6 shows the relative contribution of each mission in the multi-altimeter L4 products. This contribution (in %) is derived from a degrees of freedom of signal analysis described by [8]. During the first month, SARAL/AltiKa strongly contributes up to 40%, allowing us to mitigate the loss of Jason-1 few months earlier. Its contribution decreases by ~10% after the integration of HY-2A but rose again after HY-2A is removed of the system due to quality issues. During its whole repetitive period, SARAL/AltiKa was the second contributor, just after Jason-2 reference mission of the DUACS system because of an unexpected availability over the data over ocean as well as a very good timeliness of the OGDR and IGDR. With such a timeliness, and taking advantage of the "on the fly" DUACS system improvement implemented in February 2013, operational oceanography centres have been able since the beginning of the mission to access to homogeneous and cross calibrated DUACS SARAL/AltiKa products within a few hours. Another reason of this high contribution is the low noise level of SARAL/AltiKa enabled by its new Ka-band technology associated with its small footprint. Indeed, the energy removed from SARAL/AltiKa data by the DUACS filtering process is lower compared to what is removed from Jason-2 measurements by up to 20%. The good operational performances and data quality, routinely assessed by Calval teams, has allowed us to use SARAL/AltiKa in the fully operational Copernicus Marine Environment and Monitoring Service (CMEMS) from May 2015 onwards. The L3 and L4 Sea Level products are available on the CMEMS portal for the whole SARAL/AltiKa period (2013–2017).

Regional CMEMS products (Mediterranean Sea, Arctic, European Shelves, etc.) particularly benefit from SARAL/AltiKa inputs. The availability of data above 82° latitude allowed us to increase the density of observation in the Arctic area, in complement to CryoSat-2. More measures have also been available near the coast. It is highlighted in Figure 7, showing the increase of measurement use to compute an Envisat/SARAL/AltiKa mean profile compared to an ERS-2/Envisat mean profile, both profiles having been estimated using the same period length, 14 years. One can notice the data gain in the coastal areas such as the Danish straits, where the availability of the previous missions was poor. The coming reprocessed data DUACS DT2018 (CMEMS V4) will benefit from this new mean profile, improving not only the products during the SARAL/AltiKa period, but also the full ERS-1/ERS-2/Envisat period through the use of this new mean profile allowing a notably better description of the coastal mesoscale.

After the end of the repetitive mission in July 2016, the degradation of the data quality due to the lack of a precise MSS was strongly mitigated by the use of MSS_CNESCLS2015 [9], and the contribution of SARAL/AltiKa in the DUACS system remain significant, though now lower than the new Jason-3 and Sentinel-3A missions. However, the data now produced with this new orbit become even still more interesting for the MSS studies as the sea level is now measured over uncharted tracks. Next versions of MSS will benefits from these dataset, notably at 10–50 km wavelength all over the ocean and on the coastal areas. This will eventually lower the error budget of the Sentinel-3-A and B mission relying on these MSS, still reinforcing the crucial contribution of SARAL/AltiKa to the Sea Level products.

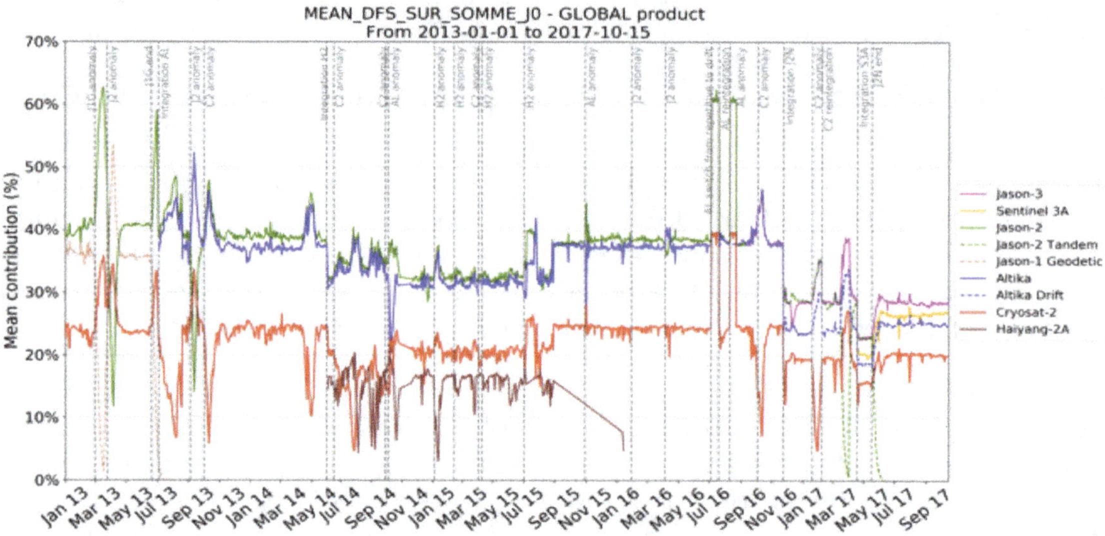

Figure 6. Relative contribution in percentage of each altimeter in the multi-mission NRT DUACS maps during the 2013–2017 period and link with altimeter events. The contribution is derived from the degrees of freedom of signal. The SARAL/AltiKa contribution is in blue, the plain curve corresponding to the repetitive period and the dotted curve to the drifting period.

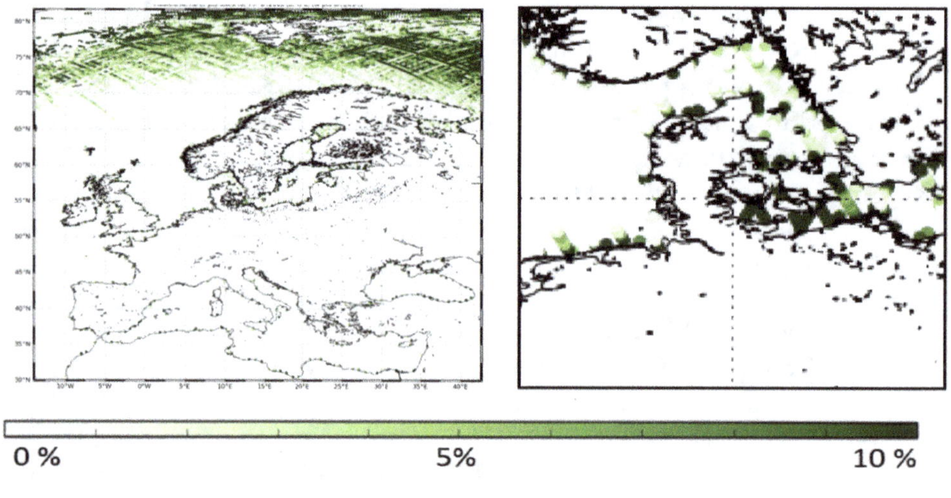

Figure 7. Relative increase in percentage of the number of point used to compute an Envisat/SARAL/AltiKa mean profile relatively to the previous a ERS-2/Envisat mean profile over: European Seas (**left**); and zooming over the Danish Straits (**right**).

2.3.2. Assimilation of SLA Data in the Mercator Ocean System

SARAL/AltiKa 35-day repeat orbit altimeter is well suited for the monitoring of mesoscale variability and to complement Jason-series observations (e.g., [16]). Its low noise level due to the use of Ka-band also allows a better resolution of along-track mesoscale signals [5]. SARAL/AltiKa thus provides essential altimeter observations for global ocean analysis and forecasting. The along track SARAL/AltiKa SLA observations has been successfully assimilated in the Mercator Ocean analysis and forecasting systems since 31 July 2013, only five months after the satellite launch. We analyse here its specific contribution in addition to other altimeters for global ocean analysis and forecasts. The new Copernicus global ocean system at 1/12° benefits from important improvements that lead to significant reduction of the analysis and forecast SLA errors. The observation errors for SST and SLA are now dynamically adjusted using an approach proposed by [17]. In 2015, SARAL/Altika diagnosed SLA observation errors have a global mean RMS of 3.6 cm, comparable to the Jason-2 error but lower than the one for HY-2A and CryoSat-2. This error estimate includes the mean dynamic topography, model representativity and processing error. The lower values are around 2 cm in the centre of the gyres but can reach more than 10 cm in high energetic regions, such as western boundary currents and the Antarctic circumpolar current.

Observing System Evaluation (OSE) experiments, covering the year 2015, have been carried out with this new system to assess the impact of an increase number of altimeters. Figure 8 shows the relative reduction of the SSH forecast Mean Square Error (MSE) due to the assimilation of SARAL/AltiKa ; Cryosat-2, Jason-2 and HY-2A being already assimilated. The impact is significant and highly regionally dependent. Innovation (observation minus model forecast) statistics are computed with Cryosat-2, Jason-2 and HY-2A data. The normalized SLA error difference between the experiment without and with SARAL/AltiKa SLA assimilated reaches 40% at mid-latitudes, whereas, in the same OSE but with the global 1/4° system, it reaches only 25% (not shown). This could be interpreted as a need of a higher density of observation to constrain the smaller scales of the 1/12° ocean model compared to the 1/4° system. The assimilation of SARAL/AltiKa SLA allows a gain of 0.8 day to get the same level of forecast error but without SARAL/AltiKa. The comparison of the estimated surface velocity fields in the different OSEs with subsurface drifters also shows an improvement with the addition of SARAL/AltiKa, even if three altimeter data sets have already been assimilated.

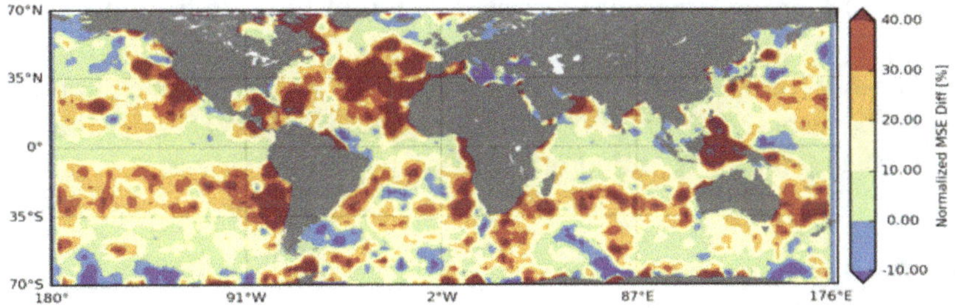

Figure 8. Normalized Mean Square forecast Error difference in percentage between the experiments assimilating three and four altimeter data sets under CryoSat-2, Jason-2 and HY-2A tracks with the 1/12° global system: (innov 3 alt)^2 − (innov 4 alt)^2 / (innov 3 alt)^2 − SARAL/AltiKa being the fourth altimeter.

2.3.3. Assimilation of SWH Data Into the Météo-France Operational Weather Forecast System

SARAL/AltiKa provides a very accurate SWH to improve the operational wave forecasting of Météo-France. During the last three years of operational use of SARAL/AltiKa, it has been demonstrated that the operational wave forecast is significantly improved for the global and regional scales. In addition, SARAL/AltiKa plays an important role in the correction of sea state in coastal zones. Figure 9 shows a snapshot of SWH from the operational Météo-France wave model (MFWAM) of a resolution of 10 km and dedicated to European seas during the storm happened in March 2016. The MFWAM model is routinely using SARAL/AltiKa, Jason-2 and CryoSat-2 altimeters wave data. The Jason-2 track overlaid on the snapshot indicates the accurate forecast at the peak of the event on 27 March 2016 at 21:00 UTC. The SARAL/AltiKa wave data were also used efficiently in the validation of high resolution coastal wave model of 200 m of resolution implemented at Météo-France. Figure 10 indicates a snapshot of SWH during the the storm of the 8 February 2016 which generated high waves in the English channel near the French coasts at Cherbourg. The SARAL/AltiKa track passing over the English channel and close to the French coast reveals the accurate forecast of the coastal model. Moreover with SARAL/AltiKa the assimilation of small wave heights (less than 1 m) in regional models MFWAM has been significantly improved for closed seas and lakes.

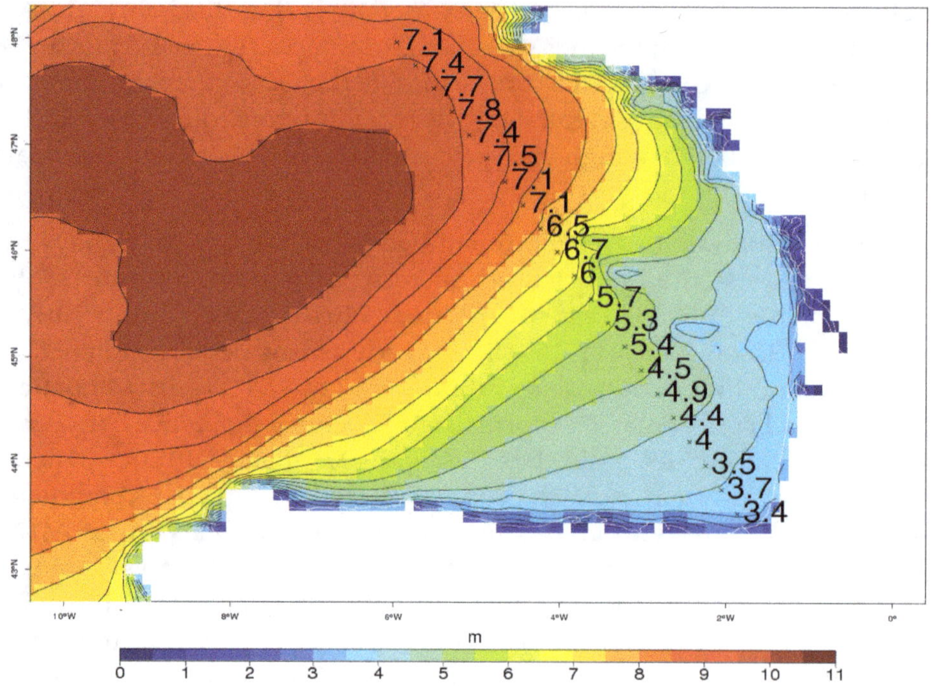

Figure 9. SWH (in m) from the regional wave model MFWAM dedicated to European seas on 27 March 2016 at 21:00 (UTC). The model MFWAM is driven by wind forcing from the atmospheric model of Météo-France ARPEGE (Action de Recherche pour Petite Echelle Grande Echelle) with a resolution of 10 km. The Jason-2 track in the gulf of Biscay indicates SWH (in m) during the peak of the event. SWH of more than 9 m was recorded off shore of the Brittany coast.

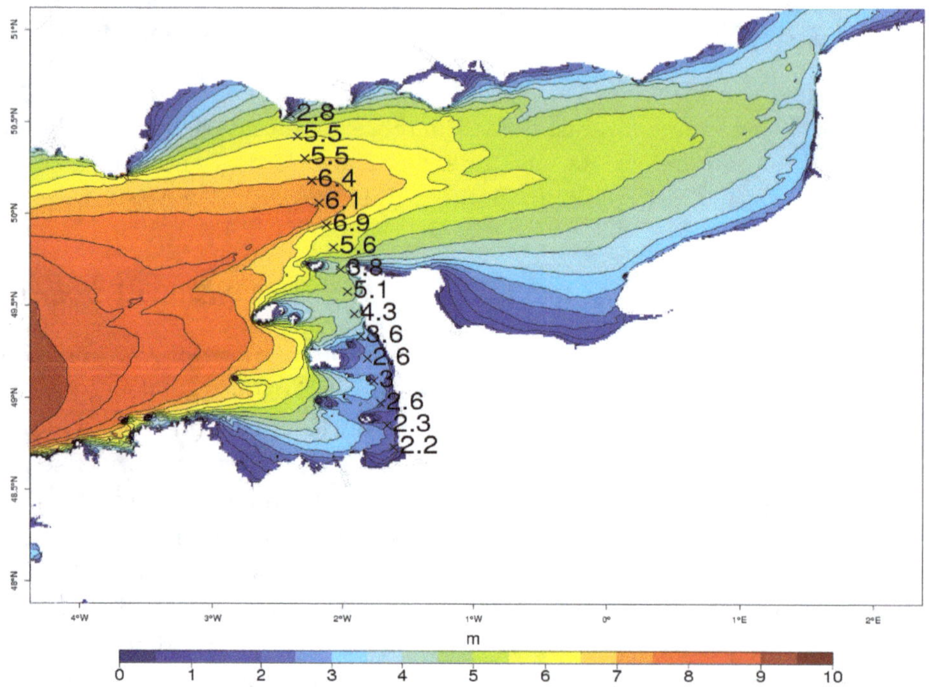

Figure 10. SWH (in m) in the English channel from the operational coastal model WaveWatch3 (WW3) with unstructured grid of a resolution of 200 m during the event of 8 February 2016 at 06:00 UTC. The model is driven by the wind forcing from the model ARPEGE. The SARAL/AltiKa track is overlaid and indicates SWH (in m) at the peak of the event near the coast.

2.4. *Accuracy of SWH Data*

The Comparative Analysis of Ku and Ka-Band Altimeters for Wave Observations

The measurements from both Ku- and Ka-band altimeters have been available concurrently. This case presented here aims to evaluate the suitability of both frequencies for wave and inland water studies. For this study the SARAL/AltiKa and Jason-2 data for year 2013 has been used. NOAA National Data Buoy Center (NDBC) buoy data have been used extensively to validate both SARAL/AltiKa and Jason-2 data individually for 2013. The SWH from both altimeters are collocated with NDBC buoy data at spatial and temporal intervals of 25 km and 30 min, respectively. Analysis has been carried out for various sea states and for the mid-latitude and tropical belts. The collocated points where difference of altimeter and observation is larger than three times the standard deviation of the difference are excluded from the analysis. The wave data for Cycle 4 from SARAL/AltiKa and Cycle 185 of Jason-2 have been analyzed over the Hudson's Bay area of Canada. This is the second largest bay in the world with average depth below 100 m and maximum depth of 270 m. The bathymetry of this region is shown in Figure 11a . Most of the Ku band altimeters does not show data availability at this shallow depth. Jason-2, for example, does not have data available over this region.

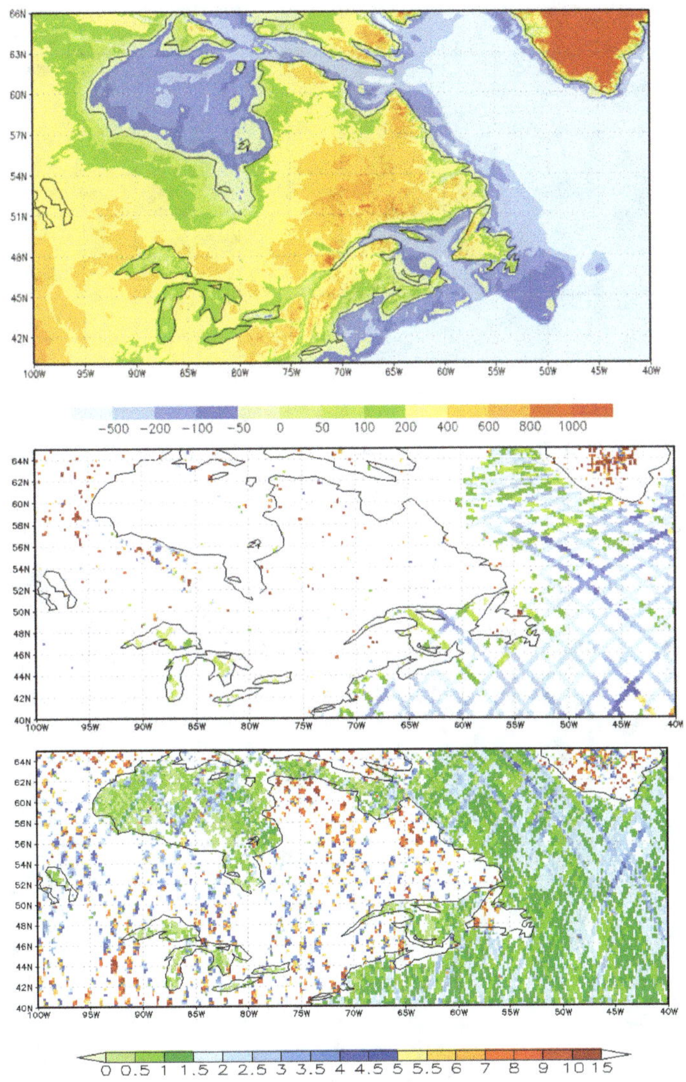

(a) (b) (c)

Figure 11. Availability of SWH data over very shallow depth from SARAL/Altika as compared to Jason-2. The example is Hudson Bay, with average depth of 100 m and maximum depth of 270 m: (**a**) bathymetry of Hudson Bay; (**b**) SWH from Jason-2; and (**c**) SWH from SARAL/AltiKa.

However, with emergence of SARAL/AltiKa, scientific fraternity is fortunate to get the valuable data from altimeter over this area. This opens the door to several applications for shallow waters including dams and reservoirs using Ka-band altimetry. Apart from being excellent in shallow water, Ka-band altimetry is known for its better approachability towards the coast. For example, over Indian subcontinent, the SWH data across a transect along 18°N has been shown in Figure 12 for Cycle 4 of SARAL/AltiKa and Cycle 185 of Jason-2. Very clearly, for the coast at about 83.5°E (marked by the yellow line), the SARAL/AltiKa has much better approachability, up to 84.2°E, as compared to Jason-2 (85.8°E). The comparison of SARAL/AltiKa and Jason-2 with NDBC buoys clearly shows that in calm and moderate sea conditions when SWH is between 0–2.5 m the performance of the SARAL/AltiKa is better than Jason-2 (Figure 13). The comparison of SWH with buoy observations for mid-latitudes and tropics clearly indicates that SARAL/AltiKa has excellent capabilities of observing waves for both the belts (Figure 14).

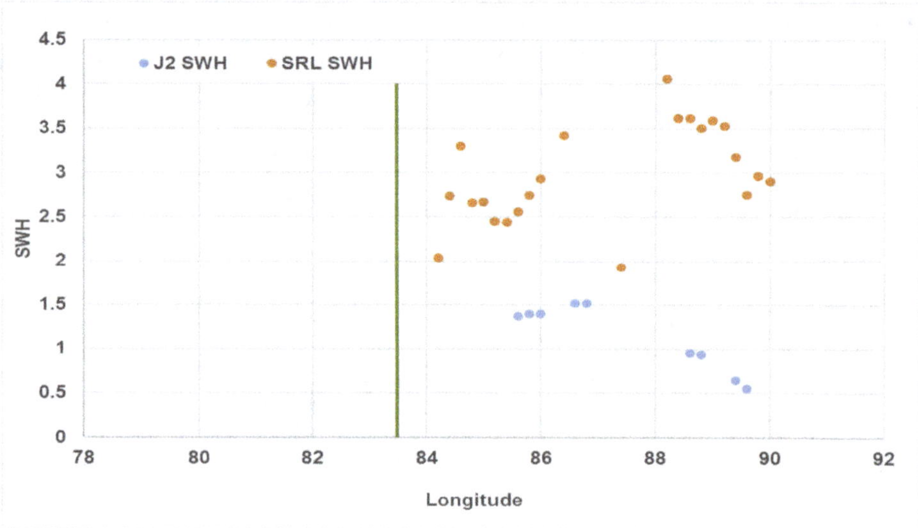

Figure 12. Availability of valid SWH data from SARAL/Altika and Jason-2 for transect along 18°N showing better available data near coast marked by green line.

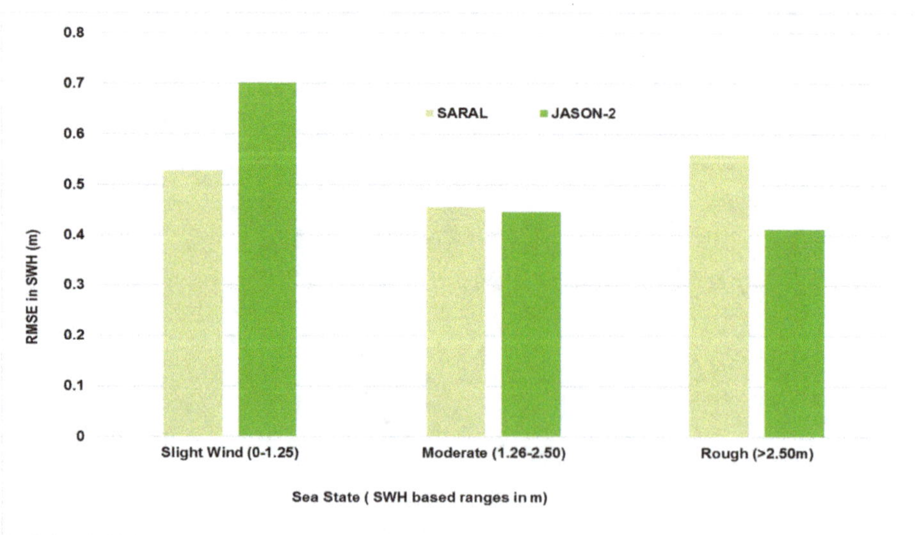

Figure 13. RMSE of SARAL/AltiKa and Jason-2 SWH with respect to buoy observed wave in various sea state categorized on basis of SWH.

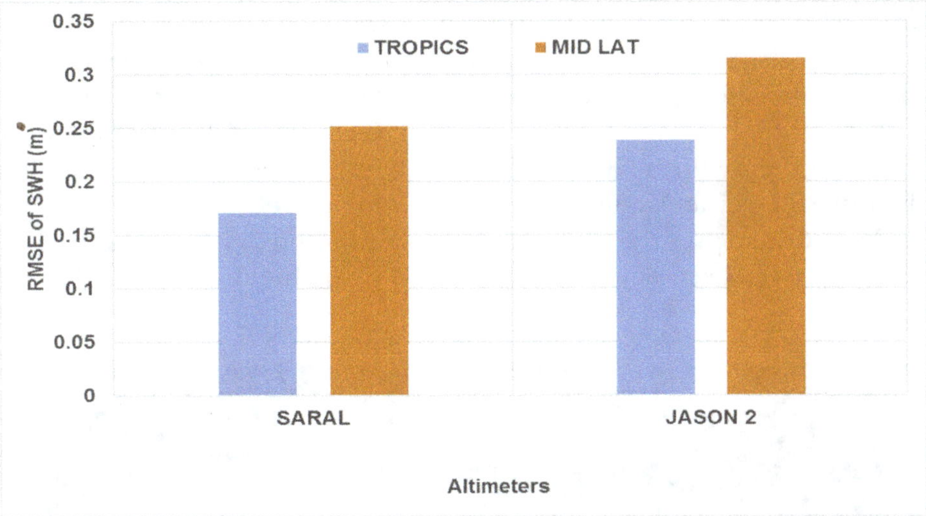

Figure 14. RMSE of SARAL/Jason-2 SWH with respect to buoy observed wave in mid-latitude (30–60° N) and Tropics (0–30° N).

3. Inland Waters

Worldwide, the number of operational ground stations for lake and river water levels and discharge monitoring has continuously decreased since the 1970s. Meanwhile, satellite capabilities for inland waters studies have emerged. In particular, many published studies have demonstrated that satellite altimetry allows measuring accurately water level along rivers and lakes [18,19]. For rivers, satellite altimetry has evolved from investigation of water height retrieval to the monitoring and discharge estimations. For lakes, it has evolved to the monitoring of water storage change [20]. Recent research focuses on integrating altimetry with other remote sensing techniques and hydraulic models to deal with key inland water resources issues such as flood (discharge, water spread and volume), water level monitoring (even in highly braided/fragmented river systems), rating curve generation for remote locations, reservoir operations, calibration of river/lake models, etc. Before it moves to a drifting orbit, SARAL/AltiKa contributes to meeting the growing demand of altimetry data for hydrological studies. Three examples of SARAL/Altika applications regarding rivers and lakes are provided in the following.

3.1. Rivers

3.1.1. Comparative Analysis of Water Level Retrieval Using Ku and Ka-Band Altimeters Over Brahmaputra River

Over inland water bodies, presence of sand bars, islands and vegetation contaminate radar return echo. Hence, with narrow footprint in Ka-band, altimeters enhance the possibilities of better data acquisition over the inland water bodies with multiple features. The decrease in time decorrelation between successive echoes has increased the pulse repetition frequency to 4000 Hz in Ka-band as compared to 2000 Hz with Ku-band altimeters. The limitation with higher frequency is that it has higher rainfall attenuation with increasing rainfall intensity, although total data loss due to this is less than 5%. In this study, river water levels were retrieved using Jason-2 and SARAL/AltiKa waveform datasets over Brahmaputra river for 2013. Three passes of Jason-2 (track number 242, 53 and 166) over the Brahmaputra river and nearby tracks of SARAL/AltiKa (track number 537, 810 and 438) were selected (see Figure 15). Time of pass of SARAL/AltiKa tracks and nearby pass data of Jason-2 were chosen. Beta parameter algorithm has been applied using the 20-Hz (Jason-2) and 40-Hz (AltiKa) waveform datasets to retrieve the river water levels with respect to mean sea level. A comparative analysis was carried out to assess the Ku-band vs. Ka-band altimetry performance for the river water level retrievals. Individual tracks were compared; it was found that SARAL/AltiKa performs slightly better for the upstream tracks, i.e., 537 vs. 242 and 810 vs. 53. For the downstream track where braiding pattern is high (track number 166, Jason-2 and 438, SARAL/AltiKa), Jason-2 performance decreased sharply (RMSE 1.19 m, Jason-2 vs. 0.41, SARAL/AltiKa) as compared to SARAL/AltiKa. This may be attributed due to higher band width (500 MHz) in Ka-band altimetry which provides shorter pulse duration; hence increase the vertical range resolution (0.3 m) as well as spatial resolution. Figure 16a,b presents the overall performance of the SARAL/AltiKa and Jason-2, respectively, taking into account all three tracks. The improved accuracy over the braided rivers highlight the high frequency Ka-band altimeter performance.

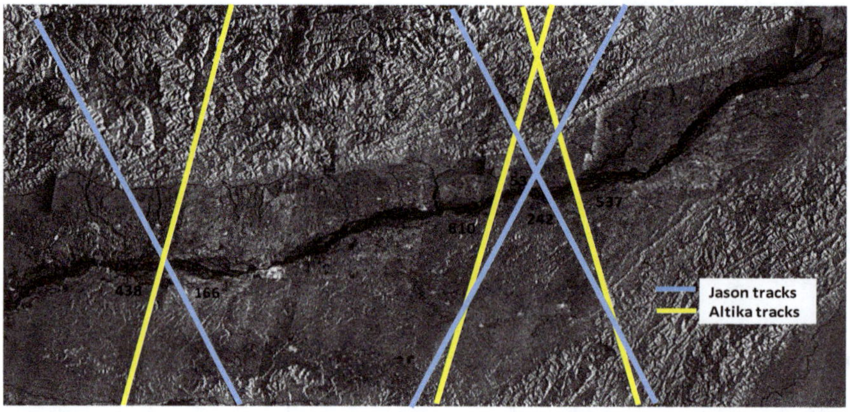

Figure 15. Brahmaputra river along with Jason-2 and SARAL/AltiKa tracks taken for the comparative analysis.

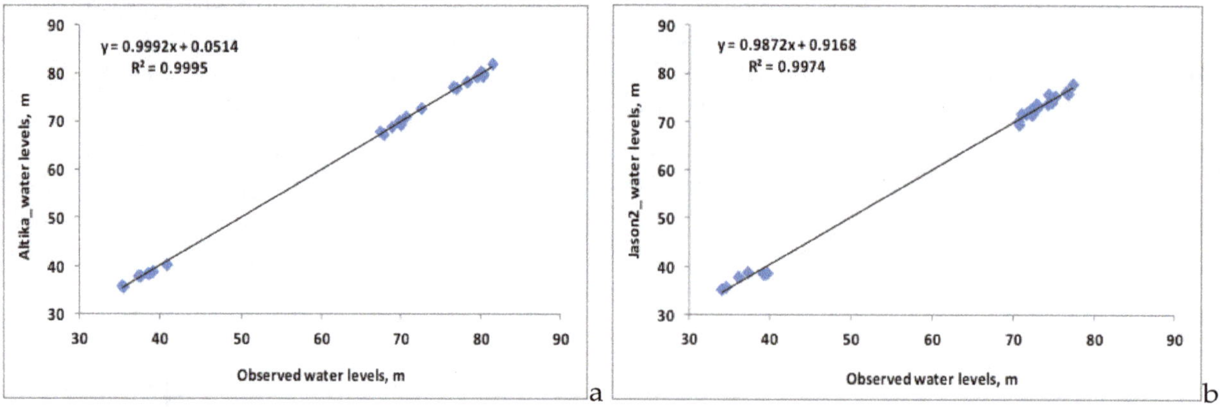

Figure 16. Overall relationship between observed versus retrieved water levels taking into account all study tracks: (**a**) SARAL/AltiKa ; and Jason-2 (**b**).

3.1.2. Discharge Estimates in the Congo Basin From Validated SARAL/AltiKa Measurements

The Congo basin, with an average discharge of 40,600 m^3/s and undergoing a 10% inter-annual variability [21,22], is the second largest hydrological basin on Earth. However, it is also one of the least monitored. Today, only a few tens of gauges still monitor the water level changes in the basin. In such a context, satellite altimetry provides a complement of information of highest interest [23]. A limited amount of works has been conducted to date to assess the quality of the SARAL/AltiKa data over rivers [24–26]. In the present study, we use the ranges issued at 40 Hz by the ICE1 retracking algorithm and applied the corrections for propagation in the atmosphere provided by global models such as ECMWF and Global Ionosphere Maps (GIM). The segment of height profiles at the crossing between the orbit ground track and a river reach (so called Virtual Stations) were subsequently processed with the median value of manually selected measurements being finally computed as the water height.

This procedure follows the results presented by [27,28] and has been widely used since then (see reviews in [18,20]). Two methods are usually used to assess the quality of the altimetry-derived series of water level: external comparison with in-situ gauge reading and internal comparison at cross-overs.

In the present study, we focus on the second method since the Congo basin is almost devoid of gauges. In addition, fortunately, the SARAL orbit forms crossovers at 1.5 day time difference over several rivers in the basin. Comparing water level measurements collected 1.5 days apart is not fully satisfactory but the positive side is that the difference is a conservative (over)estimate of the true measurement error since it includes the natural level variations. We selected 25 locations where the ascending and descending passes cross a river reach at less than 20 km, the maximum distance that we considered as relevant for a cross-over analysis. Within this dataset, the five cases correspond to reaches less than 250 m in width and 15 to reaches wider than 500 m (Figure 17). The colour code in Figure 17 stands for the standard deviation between the time series at the ascending and descending passes. The cumulative distribution of the cycle-by-cycle difference between all the measurement pairs (inset in Figure 17) shows that 90% of the measurement pairs present a difference in height less than 25 cm. The present result is consistent with the 10–30 cm errors found by [24,25] in the Ganga-Bramapoutra basin and by [26] in the Amazon basin. Noteworthy, there is no straightforward relationship with the width of the reach.

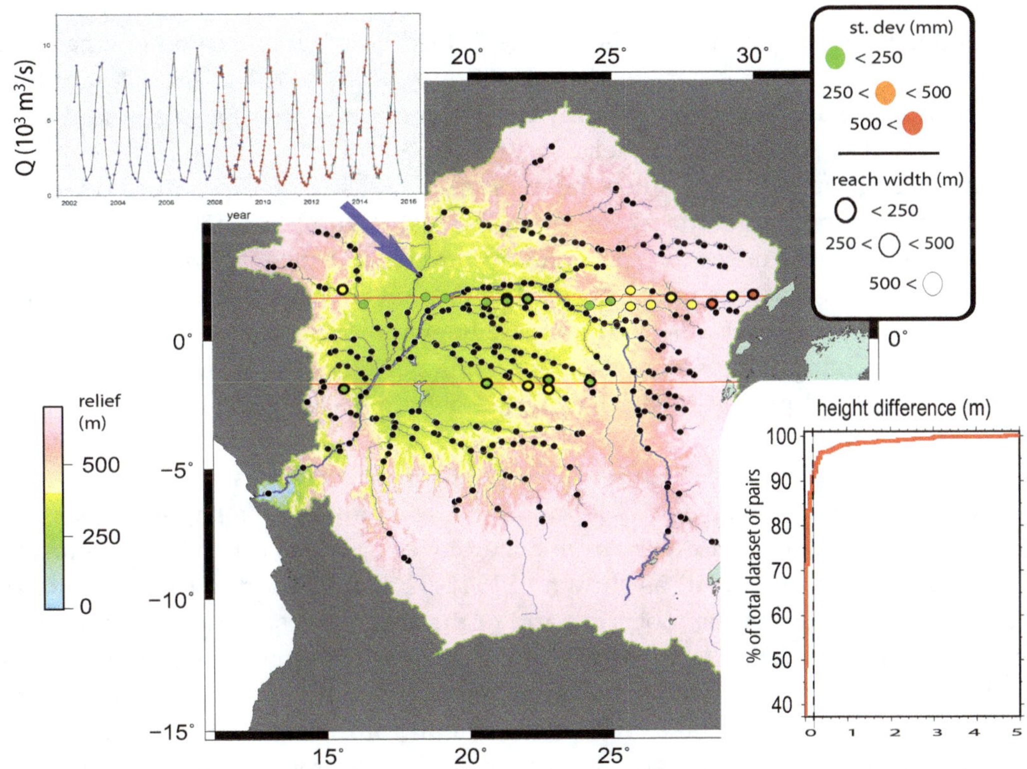

Figure 17. The Congo basin. The locations of the 350 SARAL/AltiKa Virtual Stations are shown with black dots. The +1.595° and −1.677° parallels in red are where the SARAL/AltiKa orbit forms crossover at 1.5 day temporal difference. The Virtual Stations used for the accuracy assessment at cross-overs are coded according the standard deviation between the two SARAL/AltiKa series of water level and the reach width, as recalled in the upper right inset. The cumulative histogram of the differences between measurement pairs is shown in the lower right inset. The dotted bar outlines the value of 20 cm in difference achieved by 90% of the population of height pairs. An example of Envisat + Jason-2 + SARAL/AltiKa discharge series on the Ubangui River derived from the rating curve established combining SARAL/AltiKa levels and MGB discharge is presented in the left hand inset.

One of the most important applications of water level measurements is their conversion into discharge estimates. Discharge estimates were obtained through the following steps: (1) same as first proposed by [29], compute distributed discharge between 2013 and mid-2014 by means of the rain-discharge model of the Institute of Hydraulic Research of Brazil (MGB-IPH) fed with rain fields from the JAXA/NASA Tropical Rainfall Measuring Mission (TRMM) mission; (2) compute height–discharge power laws compliant with the Manning equation; and (3) apply these rating curves to the SARAL/AltiKa series to derive discharge series. An example of such a discharge series is presented in the left hand inset of Figure 17. At this location on the Ubangui River, the Envisat/SARAL orbit and the Jason-2 orbit form a crossover right over the Ubangui River. Thus, a long continuous discharge series could be derived by applying the SARAL/AltiKa rating curve altogether to the Envisat, Jason-2 and SARAL/AltiKa times series of water level.

Overall, SARAL/AltiKa altimetry shows to be a powerful tool to estimate water levels with errors around 20 cm at more than 300 locations throughout the Congo basin, hence discharge in such a basin almost devoid of ground monitoring.

3.2. Lakes

A Case Study of the Tibetan Lakes

The Tibetan plateau is known by climatologists as the "Third Pole". It is located northeast of the Himalayan mountain range and has an average altitude of more than 4000 m. The Tibetan Plateau covers an area of 2.5 million square kilometres mainly made up of permafrost and covered by more than 1000 lakes of sizes larger than 1 km^2. Furthermore, the Tibetan Plateau is one of the regions of the world most sensitive to climate change since the average increase in temperature over the last few decades has been more than double the average for the rest of the Earth [30]. In such context, it has been demonstrated in many studies that lakes level changes in time are essential variables to track climate change impacts on water resources over the Tibetan plateau ([31], and many others). For example, an increase in temperature leads to an increase in evaporation but also accelerates glacier and permafrost thawing [32]. Changes in precipitation over long period of time also have direct impact on water contained in lakes [33]. Moreover it has been also shown in several studies that the lake level response time to climate change may be observable only over decades [20]. This has naturally placed the satellite altimetry as a useful tool for studying these links between lakes and climate changes. Due to its orbit the SARAL/AltiKa mission allows to monitor level changes on a large number of lakes over the Tibetan Plateau. It has moreover a very high accuracy over lakes, much better than that of Envisat or ERS-2 (placed on the same orbit) as shown in [34]. This is likely due to the choice of Ka band allowing a significant reduction of the footprint, and of the small ionospheric effect on the range measurement. The interest of this satellite is therefore significant since it allows extending the existing time series on the lakes over the Tibetan Plateau and may exhibit long term changes in an improved accuracy. Combining SARAL/Altika over lakes with other satellite altimeters is easily performed as a result of overlapping time of observations as shown in Figure 18. Inter-satellite instrumental biases are removed then to produce decadal level changes of lakes and thus observe and link the lake level with climate change (Figures 19–21). From these results we may draw two main conclusions:

- The responses of lakes to climate changes are not the same from one lake to another one and needs long term observations to be highlighted. For example, the Ziling lake which has grew up during more than 15 years seems over the last years to reach an equilibrium state which has been seen only with the SARAL/AltiKa measurements. It may be different for other lakes, as seen in Figures 20 and 21. It has been shown in [20] that long term changes of lakes over the Tibetan Plateau are highly variable and depend on regional climate change as well as the lake's bathymetry.

- Only SARAL/AltiKa allows extracting short term variability of water level of these lakes, since the other missions (ERS-2, Envisat, CryoSat-2, TOPEX/Poseidon, Jason-1 and Jason-2) were not precise enough to show these seasonal variabilities.

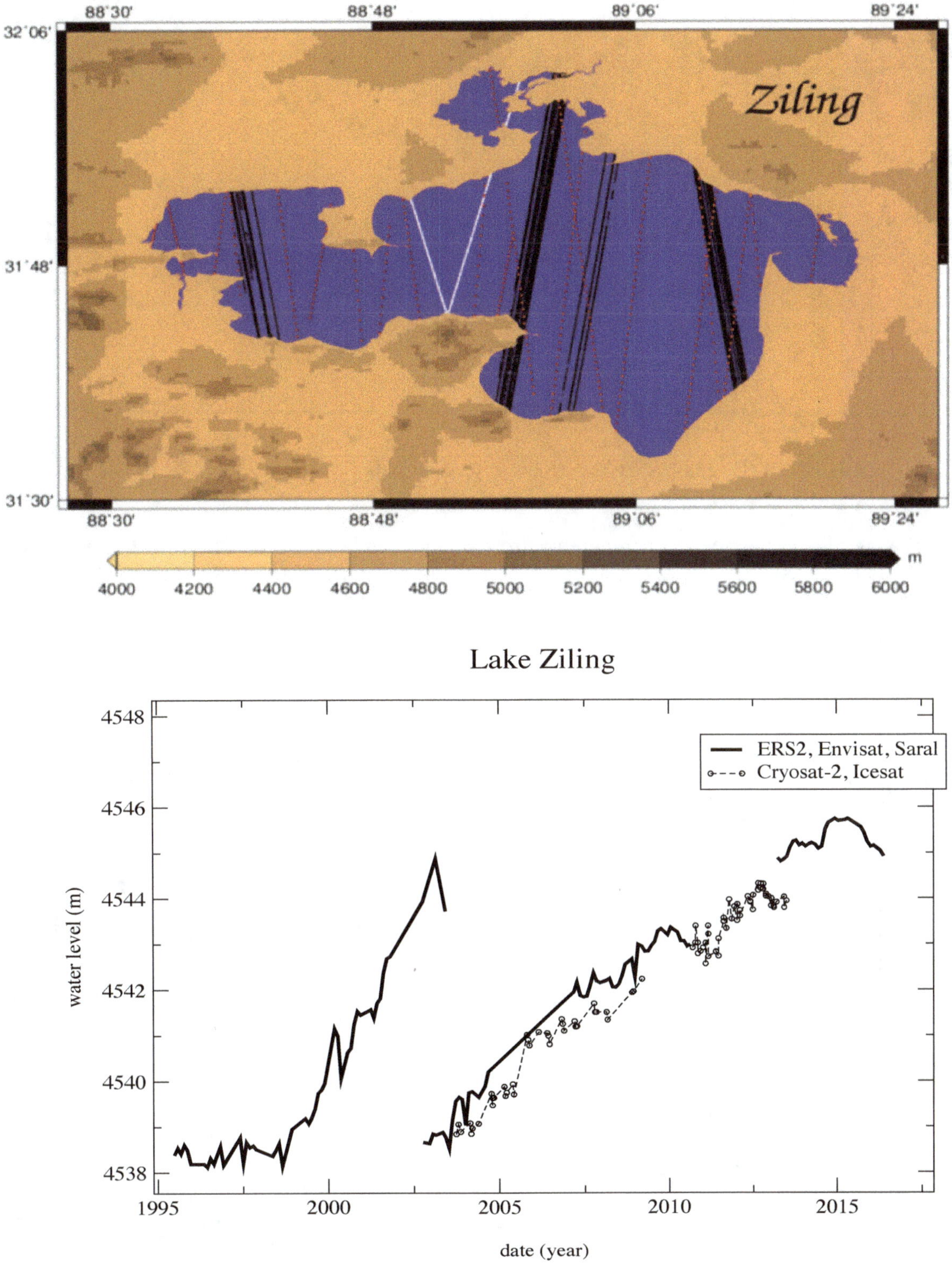

Figure 18. Tracks of satellite altimeters (black: ICESat; red: CryoSat-2; and white: ERS-2, Envisat and SARAL/AltiKa) over the Ziling Lake, one of the largest of the Tibetan Plateau (**top**); and water level changes from each satellite before correcting from inter-satellite tracks from 1995 to 2016 (**bottom**).

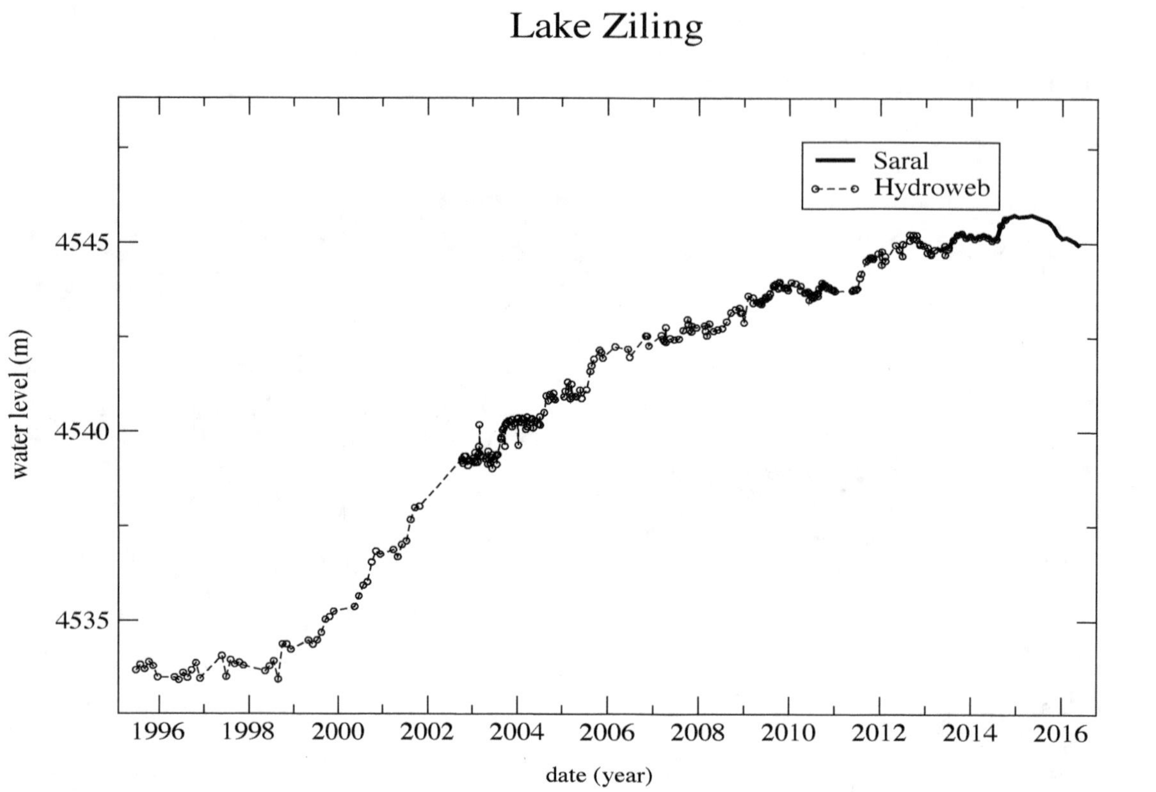

Figure 19. Water level change of the Ziling Lake from a multi-satellite data processing after removing the inter-satellite biases. After a long period of increasing starting in 1999, the lake level has stabilized in 2013/2014.

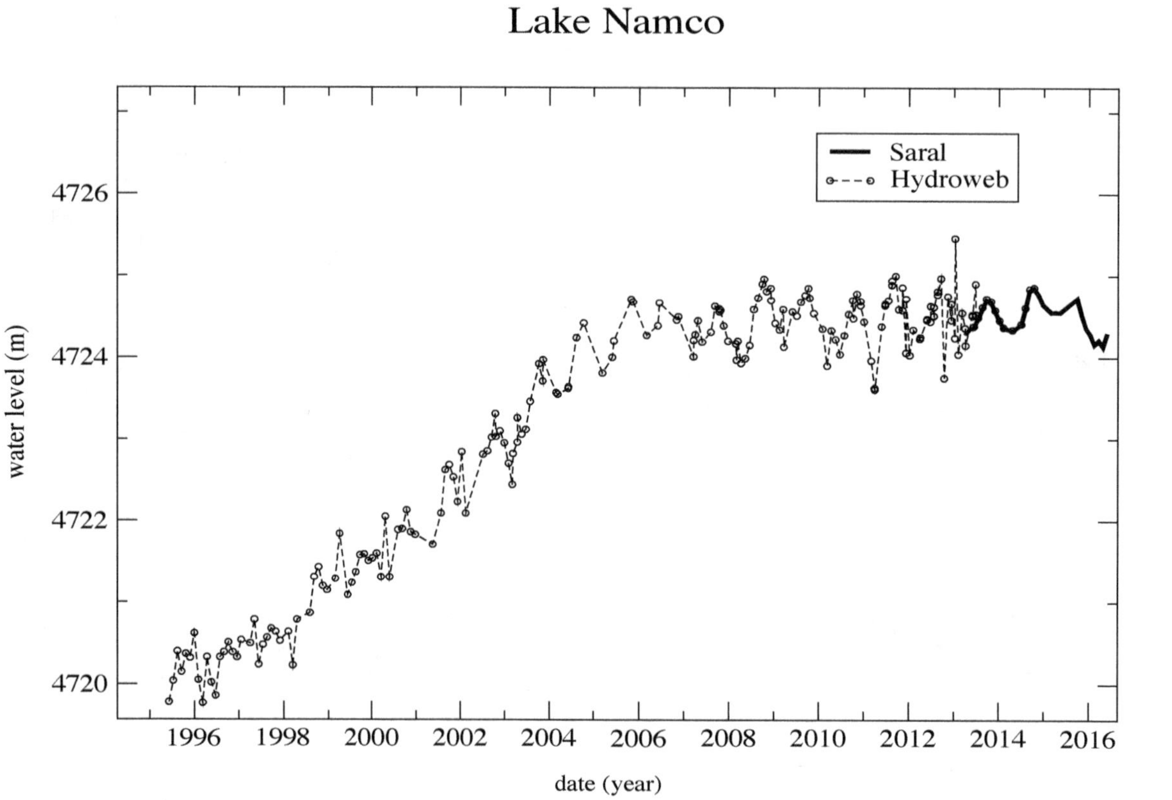

Figure 20. Water level change of the Namco lake from a multi-satellite data processing. After a long period of increasing, the lake level has stabilized in 2006/2007.

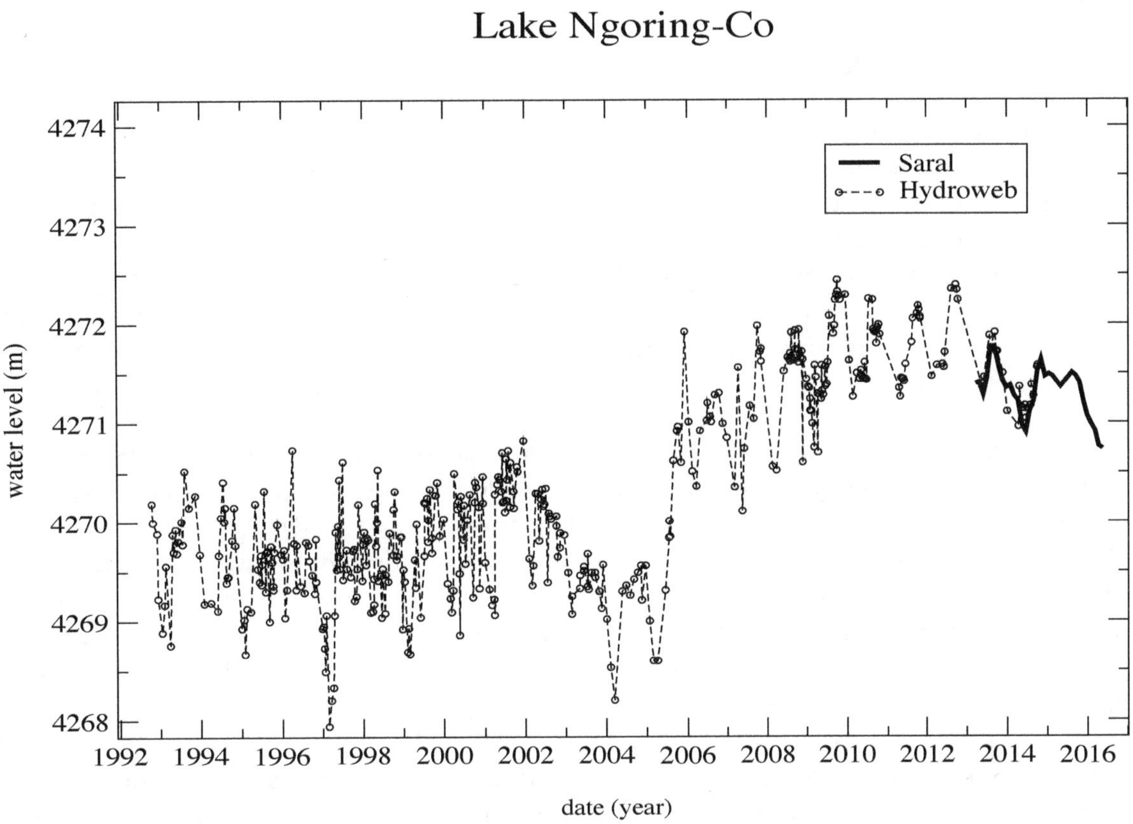

Figure 21. Water level change of Ngoring-Co Lake from a multi-satellite data processing. The lake was at equilibrium between 1992 and 2005 and has suddenly grown since 2006 to reach in a short time a new equilibrium, which slightly decreased after 2013.

4. Ice

For the last few decades, terrestrial ice has undergone large change and its survey is of a prime importance for climate study. Altimetry is one of the most powerful tools for ice sheet mass balance, sea ice thickness and iceberg detection studies. An altimeter operating in Ka- instead of Ku-band is far from being insignificant for these studies. The decrease of penetration depth within the snow, from 2 to 12 m for the Ku-band to less than 1 m for the Ka-band ensures the measurement of the snow surface. Moreover, the higher waveform rate and narrowed footprint of SARAL/AltiKa allows to better sample the small scale features and to better track the elevation transition. It is suitable to map the surface undulations or large crevasses above ice sheet or to detect leads on sea ice or small icebergs. Moreover, SARAL/AltiKa being on the same historical 35-day orbit than ERS-1-2 and Envisat allows to extend the survey for more than 20 years [35].

For sea ice, this phenomenon also seems to play a role but the major SARAL/AltiKa contribution yields from the reduced antenna aperture and in higher spatial resolution than previous altimeters. Indeed, the first echo is more likely to be at nadir decreasing influence of the off-nadir signal that often affects the retrieval measurement. Some studies in this issue confirm the ability of SARAL/AltiKa to detect leads and coastal polynyas as well as to represent spatial and temporal dynamics of water openings. Moreover, a meaningful contribution is the high sensitivity of the Ka-band backscatter to the snow and ice state, due to the altimeter wavelength. A few dB change may occur in a few weeks for any kind of snow or ice surface, even in the central part of the Antarctica ice sheet. It is observed in [36] a significant decrease of up to 15 dB at the end of ice season over Lake Baikal. The field measurements acquired along the SARAL/AltiKa tracks in Spring 2013 and Spring 2014 allow to suggest an important role of ice metamorphism and re-crystallisation.

4.1. *Ice Sheet*

Monitoring of the Antarctic Ice Sheet

Altimetry is probably one of the most powerful tools for ice sheet observation. Our vision of the Antarctic ice sheet has been deeply transformed since the launch of the ERS-1 satellite in 1991. With the launch of ERS-2, Envisat and SARAL/AltiKa, the series of altimetric observations now provides more than 20 years of continuous and homogeneous observations that allow monitoring of the shape and volume of ice sheets. The topography deduced from altimetry is one of the relevant parameters revealing the processes acting on ice sheet. Moreover, altimeter also provides other parameters such as backscatter and waveform shape [4] that give information on the surface roughness or snow pack characteristics.

The most surprising features of Ka-band observations above the Antarctica ice sheet is the very large space and time variability of the backscatter. While the Ku-band variability was about 0.5 dB over few cycles, it can reach 3 or 4 dB in Ka-band. This unexplained observation probably comes from the small wavelength (0.8 cm instead of 2.3 cm) that is very sensitive to snow metamorphism such as surface hoar. These structures are related to snow drift and sublimation and is one of the most unknown of the polar snow. Thus, one of the major limitation of radar altimeter above ice sheet is the penetration of the radar wave within the dry snowpack. For the classical Ku-band used by altimeters, the penetration reaches a few m. The slightest changes of surface echo or of volume echo affect the height retrieval. At the seasonal scale, the impact on height can be of few tens of centimetres. Moreover, this penetration may induce a long term artificial change in height caused by either change in the surface or snowpack properties. On the other hand, volume echo contains few interesting information on snow stratification and on ice grain size, the two volume components of Ku band. [37] show that the penetration depth in Ka-band is indeed between 50 cm and 1 m (see Figure 22). Moreover, they show that the stratification of the snow pack does not play a role in the volume component, the scattering by ice grain being the unique physics. The comparison between Ku and Ka-band may then correct from the penetration error and help in retrieve some snowpack parameter.

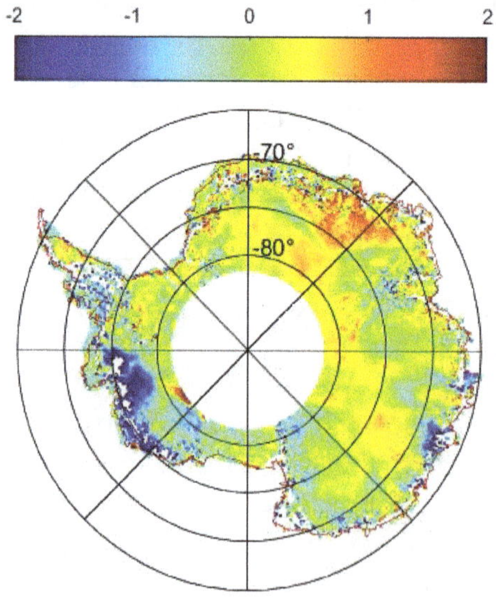

Figure 22. Difference (in m) between Ka and Ku-band height from the Envisat and SARAL/AltiKa altimeter over Antarctica. Due to the difference of penetration, in the central part, the Ku-band elevation is indeed below the Ka one. Near the coast, the smaller antenna aperture of Ka-band yields to a smaller slope error. In the West Antarctica sector, the negative pattern is due to the decrease of surface topography between the time of Envisat and of SARAL/AltiKa .

The characteristics of SARAL/AltiKa ensures a temporal stability of height retrieval that allows to obtain elevation trends with a good confidence. We well map the impressive loss of outlet glaciers of the West part of the continent. We are able to describe the large digging of Pine Island glacier from ERS-1 to Envisat and now its large enlargement during SARAL/AltiKa period [2].

4.2. Icebergs

Altimeters are powerful tools to study icebergs, either small (<3 km in length) or large (length > 16 km). Indeed, [38] demonstrated that small icebergs can produce a detectable signature in high resolution altimeter waveforms, at least in sea ice free ocean. Their method enables to detect, and to estimate the iceberg area from these signatures. The constraints on the method limit the detection to icebergs between 0.1 km^2 and 8 km^2. Nine satellite altimetry missions have already been processed to produce a 1992–present database of small iceberg location, area, volume and mean backscatter ([39,40]). Monthly mean probability of presence, area and volume of ice over a regular polar (100×100 km^2) or geographical ($1° \times 2°$) grid are also available and are distributed on the Centre ERS d'Archivage et de Traitement (CERSAT) website. The archive of SARAL/Altika has of course been processed and is part of the database. Comparison of the detection results from classical Ku band altimeters (such as Jason-series) showed that the SARAL/AltiKa higher waveform frequency of 40 Hz as well as the smaller footprint significantly improves the detection of the smallest icebergs [40] and that SARAL/AltiKa could be used as a reference to inter-calibrate the volume of ice estimates from the different sensors.

Altimeter can also be used to measure the freeboard elevation profile of large icebergs [41,42]. Combining iceberg tracks from the National Ice Center (NIC) and the archives of three Ku band altimeters, Jason-1, Jason-2 and Envisat, a database of the daily position, free-board profile, length, width, area and volume of large icebergs covering the 2002–2012 period is presented in [42]. Compared to Ku-band altimeters, the SARAL/AltiKa smaller footprint and higher waveform rate allows a better estimation of the elevation and backscatter profiles over icebergs. Figures 23 and 24 compare Jason-2 and SARAL/AltiKa data over a large iceberg (C16) detected in the Weddell Sea. They present MODIS images of the iceberg (Figures 23 and 24a) as well as the raw altimeter waveforms (Figures 23 and 24b) and the waveforms repositioned using the altimeter tracker position to better show the iceberg signature (Figures 23 and 24c). The iceberg elevation profile is estimated using a simple tracker based on the detection for each waveform of the first occurrence (bin) of a power gradient larger than a given threshold [42]. As the altimeter approaches the iceberg from the north near 65.3°S, the tracker starts to move up (Figures 23 and 24c,d) mitigating the sea and iceberg surfaces elevation as it can be seen. As the tracker is not locked on the iceberg surface, the strong echo from the iceberg starts to appear in the first gate of the waveforms, and then moves toward the nominal track point (0), while the echo from the sea surface moves away from zero (Figures 23 and 24b). The length over which the altimeter footprint contains both ocean and iceberg depends on the angle between the ground track and the iceberg, and on the footprint size. SARAL/AltiKa smaller footprint (5.6 km vs. 8.6 km for the last range bin) significantly reduces the transition length (see Figures 23 and 24d) by almost a factor 2 compared to Jason-2, especially for slanted angles. The 40 Hz waveform rate also combined with the smaller footprint lead to a more rapid detection by the tracker of the elevation transition between ocean and iceberg and iceberg and ocean. The detection of the iceberg edges although not perfect for slanted angles is significantly better than that of Ku band altimeters. Furthermore, as SARAL/AltiKa footprint is smaller, the probability that a waveform is not contaminated by ocean, i.e., that the elevation and backscatter measurements are more precise, is higher than that for Ku band altimeters. This can be clearly seen in Figures 23 and 24; whereas the SARAL/AltiKa backscatter clearly shows a decrease associated to the iceberg slope (Figure 23d) and sharp jumps between ocean and icebergs, it is difficult to identify the transition between iceberg and ocean for Jason-2.

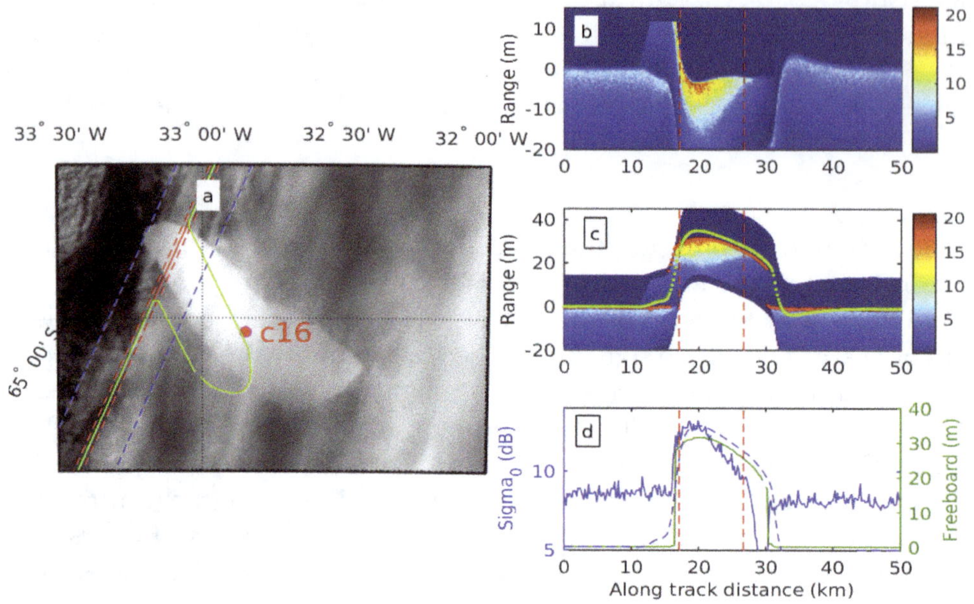

Figure 23. (**a**) MODIS image on 17 March 2013 17:34 UT and SARAL/AltiKa descending pass 102 Cycle 1 (17 March 2013 18:19UT). The red and blue dashed lines indicate the width of the first and last range bins, respectively, and the green line the elevation profile. (**b**) SARAL/AltiKa waveforms over C16. The dashed red lines indicate the iceberg limits from MODIS image. (**c**) Re-tracked waveforms using the tracker position, the green line represents the tracker position and the red line the elevation profile using a dedicated re-tracker. (**d**) Re-tracked elevation profile (green line), tracker position (dashed green line) and measured backscatter (blue line).

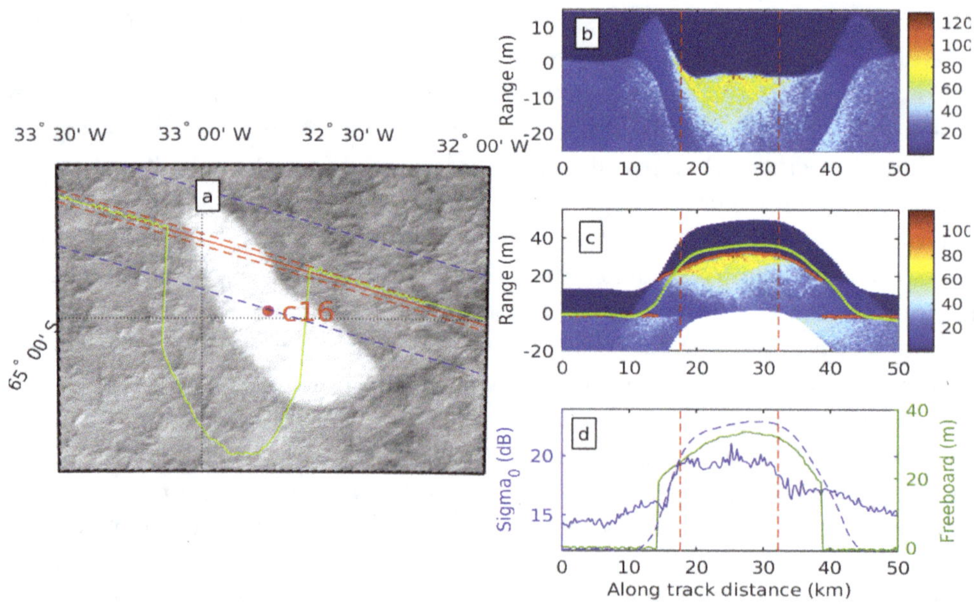

Figure 24. (**a**) MODIS image on 20 March 2013 16:45 UT and Jason-2 descending pass 180 Cycle 173 (20 March 2013 12:48UT). The red and blue dashed lines indicate the width of the fist range and last bins, respectively, and the green line the elevation profile. (**b**) Jason-2 waveforms over C16. The dashed red lines indicate the iceberg limits from MODIS image. (**c**) Re-tracked waveforms using the tracker position, the green line represents the tracker position and the red line the elevation profile using a dedicated re-tracker. (**d**) Re-tracked elevation profile (green line), tracker position (dashed green line) and measured backscatter (blue line).

4.3. Sea Ice

Regarding sea ice surfaces, SARAL/AltiKa offers mainly two advantages relatively to conventional Ku-band radar altimeters:

- First, its smaller footprint, higher vertical resolution (~30 cm) and higher horizontal sampling (~180 m) allow a better discrimination between ice floes and open sea ice fractures (generally referred to as leads), which strongly improves the measurement of sea level and freeboard height. These improvements are somehow counter-balanced by the technical issue of waveform saturation while the altimeter overflight surfaces with highly variable scattering. This effect tends to degrade the range estimation during the approach phase of specular surfaces like leads. However, the reactivity of the Attenuation Gain Controller can be easily corrected in future altimeter version.
- The second advantage of SARAL/AltiKa lies on its higher radar frequency: Unlike Ku-band radar altimeters, the Ka-band radar signal of AltiKa penetrates only part of the snowpack and possibly less than 3 cm [43]. The latter study uses this difference of penetration depth between Ku- and Ka-band altimeters to estimate a proxy of snow depth at the top of sea ice by combining SARAL/AltiKa (Ka-band) and CryoSat-2/SIRAL (Ku-band) (Figure 25). As the uncertainty related to the impact of snow depth on the freeboard-to-thickness conversion can be up to 100 %, this opportunity of measuring snow depth from Ka- and Ku-band radar altimetry represents a real breakthrough. In addition to the freeboard-to-thickness conversion, the snow depth measurement could be of hight interest for the quantification of energy, transfer between the atmosphere and the ocean as well as for the estimation of freshwater fluxes in the ocean.

The potential of combining Ka and Ku-band measurements to estimate snow depth at the top of sea ice is currently further investigated. In particular, the CryoVex airborne campaign carried, for the first time in March 2017, a Ka-band altimeter (Karen) in addition to the usual Ku-band altimeter (ASIRAS). The analysis of these highly accurate and colocated measurements will allow to provide more evidences on the interest of combining Ka and Ku band radar altimetry and could be a basis for the development of future polar altimetric missions.

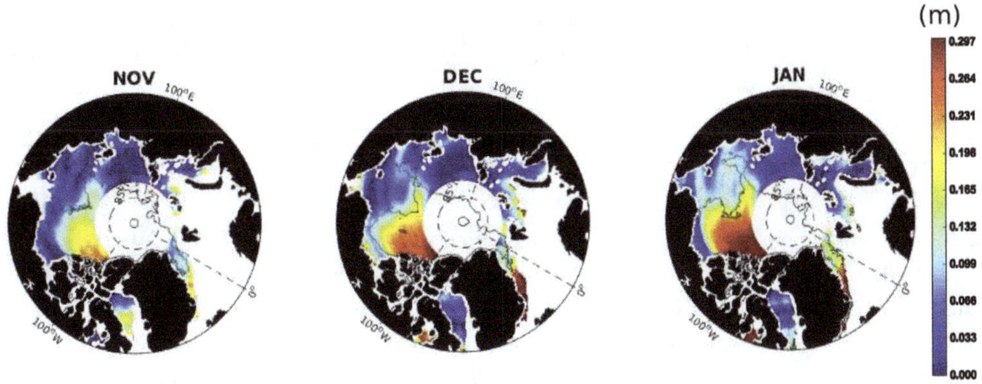

Figure 25. Maps of snow depth estimated from the combinaition of SARAL/AltiKa and CryoSat-2/SIRAL for November 2013, December 2013 and January 2014 [43].

5. Geodesy

Satellite altimetry has also become an effective remote sensing technique with important applications to geodesy [44]. Indeed, satellite altimetry has impact to geodesy by allowing mapping and monitoring a major part of the Earth surface and to make improvements on the measurements of the gravity field. The altimetric constellation with a number of successive and/or parallel missions with different orbit configuration and sampling characteristics are very supportive of geodetic applications.

The case of SARAL/AltiKa in its SARAL-DP configuration is of particular interest in this regard, as it is covering the earth with a very fine resolution well suited.

Ability to Find Uncharted Seamounts

Seamounts—undersea mountains of volcanic origin—are obstacles to flow, sites of enhanced mixing, habitat for fish, and possible hazards to submarines. Small seamounts are more common than large ones. Wessel et al. [45] found that all seamounts 2000 meters tall and taller appeared to have been found by the satellite altimeters of that time, but suggested that there might be as many as one hundred thousand seamounts between 1000 and 2000 m tall that had escaped detection. Since only about eight percent of the ocean floor has been sounded to one nautical mile resolution [46], the discovery of new seamounts will come mainly from new satellite altimetry. However, small seamounts remain difficult to find because they create only small anomalies in sea level (Figure 26). Smith [47] showed that the AltiKa altimeter on SARAL/AltiKa has much better precision in sea level measurement than previous Ku-band altimeters, and that this precision can be used to detect seamounts as small as 1 km tall. Marks [48] showed that a sea surface height profile made from the temporal median of SARAL/AltiKa repeat-track profiles can detect even smaller seamounts. It is now expected that SARAL's new geodetic mission will furnish data for finding many previously unknown seamounts. This will be a boon to scientists modelling ocean flow and mixing, to marine biologists and commercial fishermen, and perhaps also to navies that must navigate submarines.

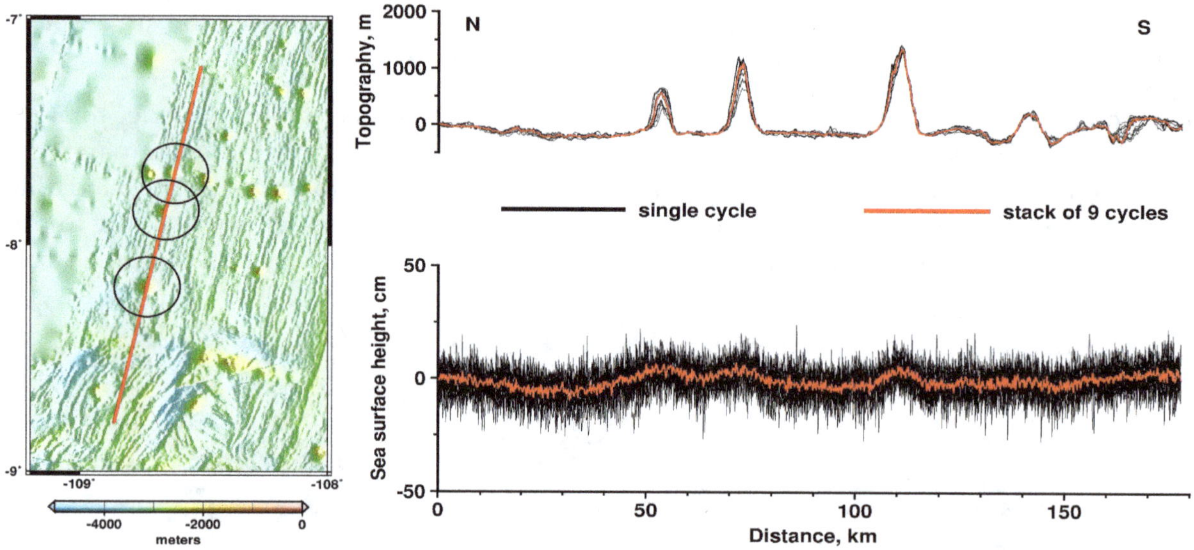

Figure 26. The red line shows the ground track of a SARAL/AltiKa repeat cycle (pass number 0396) over three small seamounts (in black circles). Seafloor topography is shown in the coloured image; the detailed portion is covered by a multi-beam echo sounder survey that confirms the position and height of the seamounts. Profiles along the red line in the previous image, showing: the sea floor topography under the satellite track (**top**); and the sea surface height anomaly (**bottom**). Black curves are individual results from each repeat cycle. The red curves are the median of all repeat cycles. The sea surface height anomaly produced by the seamounts is only 5 to 10 cm high, but is detectable, thanks to the better precision furnished by SARAL/AltiKa.

6. Conclusions

SARAL/AltiKa was launched on 25 February 2013. At the beginning, the SARAL/AltiKa mission was strongly motivated by a necessity by the research and operational oceanographic communities because of the need to fill the gap after the Envisat mission and before Jason-3 and Sentinel-3. Indeed, these objectives were completely fulfilled. However, the SARAL/AltiKa altimeter is also, and perhaps above all, the first oceanographic altimeter to use a high single frequency in Ka-band

(35.75 GHz). A little less than five years after launch, the satellite and the instruments onboard are performing well although the control of the reaction wheels appeared to be impossible at some stage, end of 2015 and beginning of 2016, leading the leave of the exact 35 repeat orbit and a shift to the Drifting Orbit Phase. Note this change had no impact of the quality and the availability of the SARAL/AltiKa data but only on the time/space sampling. As of today, all components of the altimetric system (i.e. AltiKa itself) are working properly.

Calibration and validation investigations have shown that the quality of the data meets the expectations and initial mission requirements and this is still correct [4]. Since the launch, the quality of all products appears to be continuously in line with the mission requirements. A large number of scientific investigations have been undertaken and SARAL/AltiKa has become a fully-fledged member of the altimetric constellation.

The advantages/drawbacks of the Ka-band summarize as follows:

- The higher frequency (35.75 GHz, to be compared to 13.5 GHz on Jason-2) leads to a smaller footprint (8 km diameter, to be compared to 20 km on Jason-2 and to 15 km for Envisat) and thus a better horizontal resolution.
- Ka-band allows to use a larger bandwidth (480 MHz to be compared to 320 MHz on Jason-2). This 480 MHz bandwidth provides a high vertical resolution (0.3 m) which is better with respect to other altimeters.
- The higher Pulse Repetition Frequency (4 kHz to be compared to 2 kHz on Jason-2) permits a decorrelation time of sea echoes at Ka-band shorter than at Ku-band, then allowing a better along-track sampling. This makes possible to increase significantly the number of independent echoes per second compared with Ku-band altimeters.
- Ka-band is much less affected by the ionosphere than one operating at Ku-band. This low ionospheric attenuation can even be considered as negligible, except for some exceptional ionospheric situations. It discards the need for a dual-frequency altimeter.
- Ka-band provides a better estimation of sea surface roughness than at Ku-band. The 8 mm wavelength in Ka-band is better suited to describing the slopes of small facets on the sea surface (capillary waves, etc.) and gives a more accurate measurement of the backscatter coefficient over calm or moderate seas, thus leading to a noise reduction of a factor of two compared to Jason-class altimeters for wave heights greater than 1 m. Moreover, the specificities of the Ka-band backscatter coefficient offer unique contributions in fields that where not foreseen (snow/ice morphology and its temporal variability (Section 4), soil moisture [4], etc.).
- With Ka-band, there is a lower radar penetration of snow and ice: penetration of snowpack is less than 3 cm for snow on sea ice and 1 m for continental ice, around ten times less than for Ku-band. The altimetric observation and height restitution thus correspond to a thin subsurface layer.
- A possible drawback of Ka-band was that the attenuation due to water or water vapour in the troposphere might affect Ka-band pulses in case of rain and increase significantlly the rate of missing data for strong rain rates. In fact, this was not found to be true in practice and rain had little influence on data availability and quality.

Regarding the oceans, SARAL/AltiKa provides an improved resolution of the SSH signal in particular of the mesoscales. Regarding the along-track SSH spatial resolution SARAL/AltiKa provides resolution down to 40–50 km wavelengths, whereas Jason and Envisat altimeters resolve down to 70–80 km wavelengths. Sentinel-3 SAR is still being analyzed, but it is suggested to also be resolved to 30–50 km wavelengths. **The coastal oceans,** directly benefit of the better discrimination in transition zones as well as this improved resolution of the measurements. Progress has been made in reducing the altimeter footprints and noise to approach closer to the coast, but also in processing algorithms, corrections, and products for coastal applications. The improved coastal altimeter data and their derived sea level and wind–wave data are being integrated into coastal observing systems. These data provide essential monitoring for both research and operational applications, in these coastal regions where in situ measurements are sparse.

Regarding inland waters, SARAL/AltiKa definitely improves the quality of hydrology products derived from satellite altimetry. Especially for water level time series over lakes and rivers it was shown that SARAL/AltiKa dramatically improved the quality due to its smaller footprint and its higher pulse repetition frequency. In particular, SARAL/AltiKa allows the access to smaller rivers and lakes that were not "visible" with standard altimeters. The impact of clouds or precipitation on the resulting water levels exists, but is not so strong that the data cannot be used for reliable and accurate estimation of water level time series. With regard to Envisat for example, the waveform disruption for SARAL/AltiKa for height estimation is better with SARAL/AltiKa. SARAL/AltiKa has the potential to provide better water level heights until 6 km near to the lake shore than Envisat. In addition, SARAL/AltiKa provides more homogenous water level heights over rivers than Envisat.

Regarding ice, the lesser radar penetration of snow is beneficial for various types of measurements. The difference between Ka and Ku-band height provides a good proxy of snow depth above sea ice that is a limitation to estimate ice thickness and is a climate indicator. Above ice sheet it ensures a mean to better constrain the waveform models and invert snowpack characteristics. Combined with a better spatial resolution, Ka-band in SAR mode for instance, would therefore allow closer monitoring of sea and continental ice. As for inland waters applications, ice applications gave evidence of the key importance of continuous observations along the same repeat orbit, as was possible with ERS-2, Envisat and SARAL/AltiKa (during the nominal phase).

Regarding geodesy, it appears that AltiKa is an excellent altimeter for resolving short-wavelength geoid anomalies. SARAL/AltiKa is now in its Drifting Phase, the dataset that would result would be a boon to marine geophysics, bathymetric estimation, and seamount mapping. Seamount size-frequency distribution models suggest that there may be as many as 105 seamounts between 1 and 2 km in height that are uncharted and were not detected by previous Ku-band altimeters. During its Drifting Phase, SARAL/AltiKa will surely find some of these.

A general observation is that the implementation of SARAL/AltiKa data into **operational systems** has been extremely easy both into operational oceanography centres for the SSH signal and into meteorological centres for the SWH in particular. It has also been the case for building databases of various levels of sophistication, for example for AVISO that integrates SARAL/AltiKa data as well as other altimeter data.

SARAL/AltiKa is in many respects a prototype of the altimetry of the future. SARAL/AltiKa represents the beginning of a new class of altimeters operating at Ka-band frequency with a small footprint and high pulse rate. The Ka-band is envisioned for several new altimetric satellite projects (i.e. AltiCryo from CNES, CryoSat-3 and SKIM from ESA) and the chosen band for the upcoming SWOT. It even opens more than previously the doors of interdisciplinarity. Indeed, the extended capabilities that are offered by the Ka-band allow to open even more widely some new frontiers of altimetry such as coastal oceanography, cryosphere, hydrology, beyond the traditional scope of the open ocean investigations. Ka-band altimetry, with SARAL/AltiKa as the today's most emblematic declination, can be seen also as a step towards improved resolution altimetry and a preparation for the NASA-CNES SWOT mission project and the KaRIN instrument (Ka-band Radar INterferometer).

Acknowledgments: The SARAL/AltiKa mission is an achievement of a very fruitful cooperation between ISRO and CNES. The contribution of EUMETSAT to data distribution is also strongly appreciated. We acknowledge the support of all the Investigators, PIs and Co-Is, of the SARAL/AltiKa Mission. Most of the studies performed by the French authors have been conducted and financed thanks to Centre National d'Etudes Spatiales (CNES), Centre National de la Recherche Scientifique (CNRS), and French Ministry of Research. The manuscript contents are solely the opinions of the authors and do not constitute a statement of policy, decision, or position on behalf of NOAA or the U. S. Government.

Author Contributions: Jacques Verron and Pascal Bonnefond coordinated the whole paper. For the ocean, Rosemary Morrow, Oscar Vergara, Marie-Isabelle Pujol and Gérald Dibarboure brought their contribution to Section 2.1; Florence Birol, Fatma Jebri and Ananda Pascual to Section 2.2; Yannice Faugère, Marie-Isabelle Pujol, Elisabeth Rémy, Mathieu Hamon and Lofti Aouf to Section 2.3; and Suchandra A. Bhowmick and Raj Kumar to Section 2.4. On inland waters, P. K. Gupta, A. K. Dubey, Taina Conchy and Stéphane Calmant brought their contribution to Section 3.1.1; and Jean-François Crétaux to Section 3.2. On ice, Frédérique Rémy brought her contribution to Section 4.1; Jean Tournadre to Section 4.2 ;and Kevin Guerreiro to Section 4.3. On geodesy, Walter H. F. Smith contributed to Section 5.

References

1. Vincent, P.; Steunou, N.; Caubet, E.; Phalippou, L.; Rey, L.; Thouvenot, E.; Verron, J. AltiKa: A Ka-band altimetry payload and system for operational altimetry during the GMES period. *Sensors* **2006**, 6, 208–234.
2. Verron, J.; Sengenes, P.; Lambin, J.; Noubel, J.; Steunou, N.; Guillot, A.; Picot, N.; Coutin-Faye, S.; Gairola, R.; Raghava Murthy, D.V.A.; et al. The SARAL/AltiKa altimetry satellite mission. *Mar. Geodesy* **2015**, *38*, 2–21.
3. The SARAL/AltiKa Satellite Altimetry Mission. *Mar. Geodesy* **2015**, *38*. Available online: http://www.tandfonline.com/toc/umgd20/38/sup1 (accessed on 12 January 2018)
4. Bonnefond, P.; Verron, J.; Aublanc, J.; Babu, K.N.; Bergé-Nguyen, M.; Cancet, M.; Chaudhary, A.; Crétaux, J.F.; Frappart, F.; Haines, B.J.; et al. The benefits of the Ka-band as evidenced from the SARAL/AltiKa altimetric mission: Quality assessment and specificities of AltiKa data. *Remote Sens.* **2018**, *10*, 83, doi:10.3390/rs10010083.
5. Dufau, C.; Orsztynowicz, M.; Dibarboure, G.; Morrow, R.; Le Traon, P.Y. Mesoscale Resolution Capability of altimetry: Present & future. *J. Geophys. Res. Oceans* **2016**, *121*, 4910–4927.
6. Ablain, M.; Raynal, M.; Lievin, M.; Thibaut, P.; Dibarboure, G.; Picot, N. Improving altimeter sea level calculation at small ocean scales. In Proceedings of the 2016 Ocean Surface Topography Science Team (OSTST) Meeting, La Rochelle, France, 31 October–1 November 2016.
7. Morrow, R.; Carret, A.; Birol, F.; Nino, F.; Valladeau, G.; Boy, F.; Bachelier, C.; Zakardjian, B. Observability of fine-scale ocean dynamics in the Northwest Mediterranean Sea. *Ocean Sci.* **2017**, *13*, 13–29.
8. Dibarboure, G.; Pujol. M.I.; Briol, F.; Le Traon, P.Y.; Larnicol, G.; Picot, N.; Mertz, F.; Ablain, M. Jason-2 in DUACS: Updated System Description, First Tandem Results and Impact on Processing and Products. *Mar. Geodesy* **2011**, 34, 214–241.
9. Pujol, M.-I.; Schaeffer, P.; Faugére, Y.; Raynal, M.; Dibarboure, G.; Picot, N. Gauging the improvement of recent mean sea surface models: A new approach for identifying and quantifying their errors. *J. Geophys. Res.* **2017**, submitted.
10. Jebri, F.; Birol, F.; Zakardjian, B.; Bouffard, J.; Sammari, C. Exploiting coastal altimetry to improve the surface circulation scheme over the central Mediterranean Sea. *J. Geophys. Res. Ocean.* **2016**, *121*, 4888–4909.
11. Sorgente, R.; Olita, A.; Oddo, P.; Fazioli, L.; Ribotti, A. Numerical simulation and decomposition of kinetic en e.g., in the Central Mediterranean: Insight on mesoscale circulation and en e.g., conversion. *Ocean Sci.* **2011**, *7*, 503–519.
12. Rubio, A.; Mader, J.; Corgnati, L.; Mantovani, C.; Griffa, A.; Novellino, A.; Quentin, C.; Wyatt, L.; Schulz-Stellenfleth, J.; Horstmann, J.; et al. HF Radar Activity in European Coastal Seas: Next Steps toward a Pan-European HF Radar Network. *Front. Mar. Sci.* **2017**, *4*, 8.
13. Pascual, A.; Lana, A.; Troupin, C.; Ruiz, S.; Faugère, Y.; Escudier, R. Assessing SARAL/AltiKa delayed-time data in the coastal zone: Comparisons with HFR observations. *Mar. Geodesy* **2015**, *38*, 260–276.
14. Rio, M.-H.; Pascual, A.; Poulain, P.-M.; Menna, M.; Barcelò-Llull, B.; Tintoré, J. Computation of a new mean dynamic topography for the Mediterranean Sea from model outputs, altimeter measurements and oceanographic in-situ data. *Ocean Sci.* **2014**, *10*, 731–744.
15. Pujol, M.I.; Faugère, Y.; Taburet, G.; Dupuy, S.; Pelloquin, C.; Ablain, M.; Picot, N. DUACS DT2014: The new multi-mission altimeter dataset reprocessed over 20 years. *Ocean Sci.* **2016**, *12*, 1067–1090.
16. Le Traon, P.Y.; Antoine, D.; Bentamy, A.; Bonekamp, H.; Breivik, L.A.; Chapron, B.; Corlett, G.; Dibarboure, G.; Digiacomo, P.; Donlon, C.; et al. Use of satellite observations for operational oceanography: Recent achievements and future prospects. *J. Oper. Ocean.* **2015**, *8* (Suppl. 1), 12–27.
17. Desroziers, G.; Berre, L.; Chapnik, B.; Poli, P. Diagnosis of observation, background and analysis-error statistics in observation space. *Quart. J. R. Meteorol. Soc.* **2005**, *131*, 3385–3396.
18. Calmant, S.; Seyler, F. Continental surface waters from satellite altimetry. *Géosciences* **2006**, *338*, 1113–1122.
19. Crétaux, J.F.; Birkett, C. Lake studies from satellite radar altimetry. *Comptes Rendus Geosci.* **2006**, *338*, 1098–1112.
20. Crétaux, J.-F.; Abarca Del Rio, R.; Berge-Nguyen, M.; Arsen, A.; Drolon, V.; Clos, G.; Maisongrande, P. Lake volume monitoring from Space. *Surv. Geophys.* **2016**, *37*, 269–305.

21. Alsdorf, D.; Beighley, E.; Laraque, A.; Lee, H.; Tshimanga, R.; O'Loughlin, F.; Dinga, B.; Moukandi, G.; Spencer, R.G. Opportunities for hydrologica research in the Congo basin. *Rev. Geophys.* **2016**, *54*, 378–409.
22. Laraque, A.; Bellanger, M.; Adèle, G.; Guebanda, S.; Gulemvuga, G.; Pandi, A.; Paturel, J.E.; Robert, A.; Tathy, J.P.; Yambele, A. Recent evolution of Congo, Oubangui and Sangha Rivers flows. *Acad. R. Sci. Belgique Geogr. Ecol. Trop.* **2013**, *37*, 93–100.
23. Silva, J.; Calmant, S.; Seyler, F.; Medeiros Moreira, D.; Oliveira, D.; Monteiro, A. Radar Altimetry aids Managing gauge networks. *Water Resour. Manag.* **2014**, 28, 587–603.
24. Dubey, A.K.; Gupta, P.; Dutta, S.; Pratap Sing, R. Water Level retrieval Using SARAL/AltiKa Observations in the braided Brahmapoutra River, Eastern India. *Mar. Geodesy* **2015**, 38, 549–567.
25. Frappart, F.; Papa, F.; Marieu, V.; Malbeteau, Y.; Jordy, F.; Calmant, S.; Durand, F.; Bala, S. Preliminary assessment of SARAL/AltiKa observations over the Ganges-Brahmaputra and Irrawaddy Rivers. *Mar. Geodesy* **2015**, *38*, 568–580.
26. Schwatke, C.; Dettmering, D.; Borgens, E.; Bosch, W. Potential of SARAL/Altika for Inland Water Application. *Mar. Geodesy* **2015**, 38, 626–643.
27. Frappart, F.; Calmant, S.; Cauhopé, M.; Seyler, F.; Cazenave, A. Results of Envisat RA-2 Derived levels Validation over the Amazon basin. *Remote Sens. Environ.* **2006**, *100*, 252–264.
28. Silva, J.; Calmant, S.; Rotuono Filho, O.; Seyler, F.; Cochonneau, G.; Roux, E.; Mansour, J.W. Water Levels in the Amazon basin derived from the ERS-2 and Envisat Radar Altimetry Missions. *Remote Sens. Environ.* **2010**, *114*, 2160–2181.
29. Paris, A.; Santos da Silva, J.; Dias de Paiva, R.; Medeiros Moreira, D.; Calmant, S.; Collischonn, W.; Bonnet, M.-P.; Seyler, F. Global determination of rating curves in the Amazon basin. *Water Resour. Res.* **2016**, *28*, 3787–3814.
30. Wang, B.; Bao, Q.; Hoskins, B.; Wu, G.; Liu, Y. Tibetan Plateau warming and precipitation change in East Asia. *Geophys. Res. Lett.* **2008**, *35*, L14702.
31. Liu, J.; Wang, S.; Yu, S.; Yang, D.; Zhang, L. Climate warming and growth of high-elevation inland lakes on the Tibetan Plateau. *Glob. Planet. Chang.* **2009**, *67*, 209–217.
32. Lei, Y.; Yang, K.; Wang, B.; Sheng, Y.; Bird, B.W.; Zhang, G.; Tian, L. Response of inland lake dynamics over the Tibetan Plateau to climate change. *Clim. Chang.* **2014**, *125*, 281–290.
33. Phan, V.H.; Lindenb, R.C.; Menenti, M. Geometric dependency of Tibetan lakes on glacial runoff. *Hydrol. Earth Syst. Sci. Discuss.* **2013**, *10*, 729–768.
34. Arsen, A.; Cretaux, J.-F.; Abarca-Del-Rio, R. Use of SARAL/AltiKa over mountainous lakes, intercomparison with Envisat mission. *J. Adv. Space Res.* **2015**, *38*, 534–548.
35. Remy, F.; Flament, T.; Michel, A.; Verron, J. Ice sheet survey over Antarctica with satellite altimetry: ERS-2, Envisat, SARAL/AltiKa, the key importance of continuous observations along the same repeat orbit. *Int. J. Remote Sens.* **2014**, *35*, 5497–5512.
36. Kouraev, A.; Zakharova, E.; Remy, F. Study of Lake Baikal ice cover from radar altimetry and in situ observations. *Mar. Geodesy* **2015**, *38*, 477–486
37. Rémy, F.; Flament, T.; Michel, A.; Blumstein, D. Envisat and SARAL/AltiKa observations of the Antarctic ice sheet: A comparison between the Ku-band and the Ka-band. *Mar. Geodesy* **2015**, *38*, 510–521.
38. Tournadre, J. Signature of Lighthouses, Ships, and Small Islands in Altimeter Waveforms. *J. Atmos. Ocean. Tech.* **2007**, *24*, 1143–1149.
39. Tournadre, J.; Girard-Ardhuin, F.; Legresy, B. Antarctic iceb e.g., distributions, 2002–2010. *J. Geophys. Res.* **2012**, 117, C05004.
40. Tournadre, J.; Bouhier, N.; Girard-Ardhuin, F.; Remy, F. Antarctic iceb e.g., distributions 1992–2014. *J. Geophys. Res.* **2016**, *121*, 327–349.
41. McIntyre, N.F.; Cudlip, W. Observation of a giant Antarctic tabular iceberg by satellite radar altimetry. *Polar Rec.* **1987**, *145*, 458–462
42. Tournadre, J.; Bouhier, N.; Girard-Ardhuin, F.; Remy, F. Large iceb e.g., characteristics from altimeter waveforms analysis. *J. Geophys. Res.* **2015**, 120, 1954–1974.
43. Guerreiro, K.; Fleury, S.; Zakharova, E.; Rémy, F.; Kouraev, A. Potential for estimation of snow depth on Arctic sea ice from CryoSat-2 and SARAL/AltiKa missions. *Remote Sens. Environ.* **2016**, *186*, 339–349.
44. Smith, W.H.F.; Barale, V.; Gower, J.; Alberotanza, L. The Marine Geoid and Satellite Altimetry. In *Oceanography from Space*; Springer: Dordrecht, The Netherlands, 2010.

45. Wessel, P. Global distribution of seamounts inferred from gridded Geosat/ERS-1 altimetry. *J. Geophys. Res.* **2001**, *109*, 19431–19441.
46. Wessel, P.; Chandler, M.T. The spatial and temporal distribution of marine geophysical surveys. *Acta Geophys.* **2011**, *59*, 55–71.
47. Smith, W.H.F. The Resolution of Seamount Geoid Anomalies Achieved by the SARAL/AltiKa and Envisat RA2 Satellite Radar Altimeters. *Mar. Geodesy* **2015**, *38*, 644–671.
48. Marks, K.M.; Smith, W.H.F. Detecting small seamounts in AltiKa repeat cycle data. *Mar. Geophys. Res.* **2016**, *37*, 349–359.

6

Monitoring Sea Level and Topography of Coastal Lagoons using Satellite Radar Altimetry: The Example of the Arcachon Bay in the Bay of Biscay

Edward Salameh [1,2,*], Frédéric Frappart [1,3], Vincent Marieu [4], Alexandra Spodar [4,5], Jean-Paul Parisot [4], Vincent Hanquiez [4], Imen Turki [2] and Benoit Laignel [2]

1 Laboratoire d'Etudes en Géophysique et Océanographie Spatiales (LEGOS), Université de Toulouse, IRD, CNES, CNRS, UPS, 31400 Toulouse, France; frederic.frappart@legos.obs-mip.fr
2 Normandie University, UNIROUEN, UNICAEN, CNRS, M2C, Morphodynamique Continentale et Côtière, 76000 Rouen, France; imen.turki@univ-rouen.fr (I.T.); benoit.laignel@univ-rouen.fr (B.L.)
3 Géosciences Environnement Toulouse (GET), Université de Toulouse, IRD, CNES, CNRS, UPS, 31400 Toulouse, France
4 Environnements et Paléoenvironnements Océaniques et Continentaux (EPOC), UMR 5805, allée Geoffroy St Hilaire, 33615 Pessac CEDEX, France; vincent.marieu@u-bordeaux.fr (V.M.); alexandra.spodar@univ-littoral.fr (A.S); Parisotjp@wanadoo.fr (J.-P.P.); vincent.hanquiez@u-bordeaux.fr (V.H.)
5 Laboratoire d'Océanologie et de Géosciences (LOG), UMR 8187, 59140 Dunkerque, France
* Correspondence: edward.salameh@legos.obs-mip.fr

Abstract: Radar altimetry was initially designed to measure the marine geoid. Thanks to the improvement in the orbit determination from the meter to the centimeter level, this technique has been providing accurate measurements of the sea surface topography over the open ocean since the launch of Topex/Poseidon in 1992. In spite of a decrease in the performance over land and coastal areas, it is now commonly used over these surfaces. This study presents a semi-automatic method that allows us to discriminate between acquisitions performed at high tides and low tides. The performances of four radar altimetry missions (ERS-2, ENVISAT, SARAL, and CryoSat-2) were analyzed for the retrieval of sea surface height and, for the very first time, of the intertidal zone topography in a coastal lagoon. The study area is the Arcachon Bay located in the Bay of Biscay. The sea level variability of the Arcachon Bay is characterized by a standard deviation of 1.05 m for the records used in this study (2001–2017). Sea surface heights are very well retrieved for SARAL ($R\sim0.99$ and $RMSE < 0.23$ m) and CryoSat-2 ($R > 0.93$ and $RMSE < 0.42$ m) missions but also for ENVISAT ($R > 0.82$ but with a higher $RMSE > 0.92$ m). For the topography of the intertidal zone, very good estimates were also obtained using SARAL ($R\sim0.71$) and CryoSat-2 ($R\sim0.79$) with RMSE lower than 0.44 m for both missions.

Keywords: radar altimetry; coastal altimetry; sea surface height; topography of the intertidal zone; ERS-2; ENVISAT; SARAL; CryoSat-2

1. Introduction

Coastal regions represent only 5% of Earth's land area, yet their societal and economical importance are larger than their surface area suggests [1]. The land area within 100 km from the coast accommodates about 39% of the global population according to the CIESIN (Center for International Earth Science Information Network) [2]. Coastal systems are experiencing high pressures due to population growth and the overexploitation of their resources. Anthropogenic pressures exacerbated by sea level rise and the increase of global temperature lead to a rapid and threatening environmental change of these systems, which requires effective long-term coastal management initiatives.

Common features of coastal systems are coastal lagoons, occupying 13% of coastal areas worldwide [3]. Coastal lagoons are defined as "inland water bodies, separated from the ocean by a barrier, connected to the ocean by one or more restricted inlets which remain open at least intermittently, and have water depth which seldom exceed a few meters" [4]. They are subject to forcings from rivers, wind stress, tides, precipitation to evaporation balance, and surface heat balance [4]. These ecosystems provide important services and societal benefits (e.g., food provision, recreational, water regulation, etc.); however their subsistence is threatened by global climate change [5]. Understanding the physical dynamics of these systems is of great importance in order to direct the planning and implementation of coastal management strategies in coastal lagoons.

In the need for a better understanding of lagoons' dynamics, satellite radar altimetry measuring the variation of the surface elevation could be a very useful tool providing key information, especially for non-monitored areas. However, using altimetry in coastal regions remains a great challenge due to numerous issues including land contamination in the footprint that impacts the radar echo (or waveform), but also the lower quality of the corrections applied to the distance between the satellite and the surface (or altimeter range) than over open ocean [6]. Despite these shortcomings, recent improvements in processing techniques (e.g., careful recovering of flagged data, applying specialized retracking, improving the correction terms) extended the capabilities of altimeters in coastal areas [7].

This study analyzes the performance of radar altimetry to monitor sea level and to provide, for the very first time, topography of the intertidal zone along the altimeter tracks in the Arcachon Bay, a coastal lagoon situated in the south-west of France. The choice of this lagoon as a case study was motivated by the coverage provided by ERS-2, ENVISAT, SARAL, and CryoSat-2 altimetry missions. Our goal is to assess the evolution of measurement accuracy at Ku-band (ERS-2, ENVISAT) and the benefits of the Ka-band (SARAL) in Low Resolution Mode (LRM). An assessment of instrumental performance was undertaken as well for observations made by the Ku-band satellite mission CryoSat-2, the first altimeter to operate in Synthetic Aperture Radar (SAR) mode.

The aim of this study is threefold: (i) to investigate the limitations and potential improvements of altimetry to monitor lagoons Sea Surface Height (SSH), (ii) to test the capability of altimetry to retrieve topographic variations and (iii) to optimize (time wise and accuracy wise) the processing of altimetry data in coastal lagoons environment. Tide gauge measurements and lidar topography datasets were used to evaluate SSHs and topography estimation made by satellite radar altimetry.

2. Study Area

The Arcachon Bay (44°40′N, 1°10′W) is a mesotidal shallow semi-confined lagoon, located in the southeast of the Bay of Biscay (Figure 1). The total lagoon surface (174 km^2) is composed of channels (57 km^2) that drain the intertidal area (117 km^2). The main channels have a maximum depth around 20 m and are extended by a complex network of secondary channels [8]. The tidal cycle is semi-diurnal with a weak diurnal inequality. The tide amplitudes vary from 0.8 to 4.6 m for neap and spring tides respectively. The Arcachon Bay connects to the Atlantic Ocean through two narrow passes of 1–1.5 km width and around 12 km long. The two passes are separated by the Arguin Bank. Important seawater exchanges, reaching up to 384.106 m^3 occur during each tidal cycle [9]. Freshwater inputs from small rivers and groundwater are coming mostly from the Eyre River and the Porges Canal, located south-east and north of the Bay respectively (see Figure 1). They represent more than 95% (73% and 24% respectively) of the total annual freshwater inflows [10]. The intertidal area is composed of a mix of muddy and sandy material [8]. A large zone of 70 km^2 of the mudflats in the inner lagoon is covered with Zostera noltii seagrass [11].

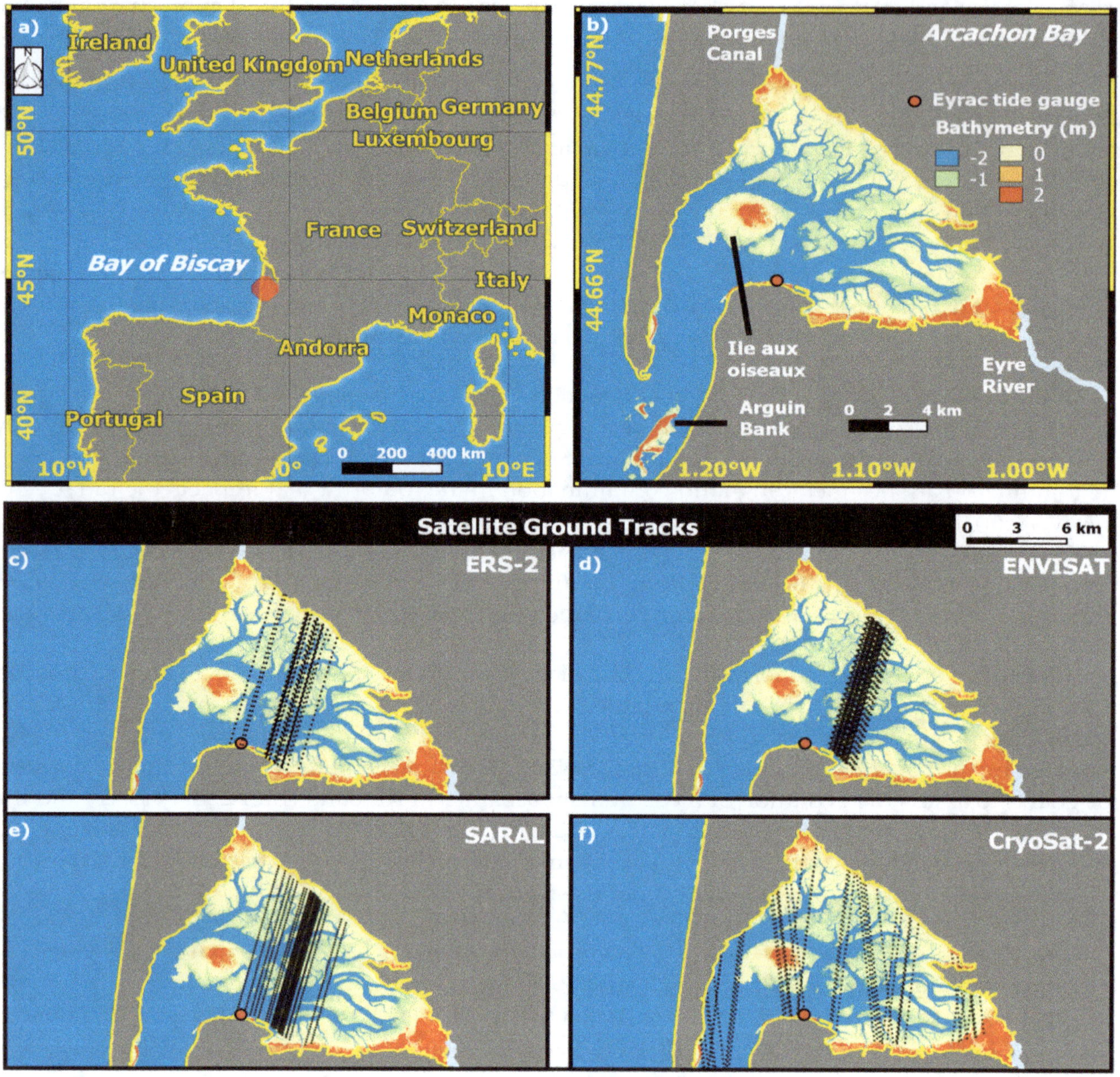

Figure 1. (**a**) The Arcachon lagoon is located in the Bay of Biscay along the south part of the French Atlantic coast. (**b**) The Arcachon lagoon is a mesotidal shallow semi-confined lagoon. Several altimetry missions' ground-tracks cover the lagoon: (**c**) ERS-2 (1993–2003, since 2003 ERS-2 has experienced a number of failures), (**d**) ENVISAT (2002–2010 on the nominal orbit), (**e**) SARAL (2013–2016 on the nominal orbit), and (**f**) CryoSat-2 (since 2010).

3. Datasets

3.1. Altimetry Data

Table 1 presents a summary of the main characteristics of the altimetry missions used in this study, which are described in more detail below.

Table 1. Main characteristics of the altimetry missions used in this study.

Mission	ERS-2	ENVISAT	SARAL	CryoSat-2
Agency	ESA	ESA	CNES/ISRO	ESA
Launch on	21/04/1995	01/03/2002	25/02/2013	08/04/2010
End date	06/07/2011	08/06/2012	Present	Present
Altimeter name	RA	RA-2	AltiKa	SIRAL
Radar frequency	Ku-band	Ku and S-bands	Ka-band	Ku-band
Altitude	785 km	790 km	790 km	717 km
Orbit inclination	98.52°	98.54°	98.54°	92°
Repetitivity	35 days	35 days	35 days	369 days
Ground-track spacing at the equator	85 km	85 km	85 km	7.5 km
Along track sampling	20 Hz (350 m)	18 Hz (~400 m)	40 Hz (175 m)	20 Hz (350 m)

3.1.1. ERS-2

The ERS-2 satellite (European Remote Sensing-2) was launched in 1995 by ESA (European Space Agency). Its payload is composed of several sensors, including a radar altimeter (RA), operating at Ku-band (13.8 GHz). It was sun-synchronously orbiting at an altitude of 785 km with an inclination of 98.52° with a 35-day repeat cycle. This orbit has a ground-track spacing about 85 km at the equator. ERS-2 provided observations of the topography of the Earth from 82.4° latitude north to 82.4° latitude south. ERS-2 data are available from 17 May 1995 to 9 August 2010 but with a limited coverage after 22 June 2003.

3.1.2. ENVISAT

ENVISAT (ENVIronmental SATellite) mission was launched on 1st March 2002 by ESA. It carries 10 instruments including the advanced radar altimeter (RA-2). RA-2 is a nadir-looking pulse-limited radar altimeter operating at two frequencies at Ku- (13.575 GHz) and S- (3.2 GHz) bands. ENVISAT orbits at an altitude of 790 km, with an inclination of 98.54°, on a sun-synchronous orbit with a 35-day repeat cycle, providing observations of the Earth surface (ocean and land) from 82.4° latitude North to 82.4° latitude South. This orbit was formerly used by ERS-1 and 2, with an equatorial ground-track spacing of about 85 km. ENVISAT remains on its nominal orbit until October 2010 [12]. From November 2010 to April 2012, ENVISAT was put into the extending phase consisting of a drifting on a 30-day orbit lowered by 17 km.

3.1.3. SARAL

SARAL (Satellite for Argos and ALtika) is a CNES-ISRO (Centre National d'Etudes Spatiales—Indian Space Research Organization) joint-mission that was launched on 25 February 2013. Its payload is composed of the AltiKa radar altimeter and bi-frequency radiometer, and a triple system for precise orbit determination: the real-time tracking system DIODE of DORIS instrument, a Laser Retroflector Array (LRA), and the Advanced Research and Global Observation Satellite (ARGOS-3). Its orbital characteristics are the same as ENVISAT (see above). The first four cycles of SARAL do not follow precisely the ENVISAT orbit. AltiKa radar altimeter is a solid-state mono-frequency altimeter that provides accurate range measurements. It is the first altimeter to operate at Ka-band (35.7 GHz). Its accuracy is expected to be about 1 cm over ocean. Over the coastal regions, it is expected to provide measurements significantly better than those from the previous Ku band missions. Improvements come from the reduced footprint of the Ka-band (about ten times smaller in surface than it is in Ku-band) and from the higher along-track sampling rate of 40 Hz (~175 km), twice that of ENVISAT [13].

3.1.4. CryoSat-2

CryoSat-2 mission was launched on 8 April 2010 by ESA. This mission is dedicated mainly to polar observations. However, its acquisitions can be useful for ocean and inland monitoring as it provides a global monitoring of the Earth's surface [14]. The mission's main payload consists of a radar altimeter, SIRAL (Synthetic Aperture Interferometric Radar Altimeter), operating at Ku-band

(13.575 GHz) in three different modes: Low Resolution Mode (LRM), Synthetic Aperture Radar mode (SAR), and Synthetic Aperture Interferometric mode (SARIn). CryoSat-2 orbits at an altitude of 717 km, with an inclination of 92°, on a non-sun-synchronous orbit with a 369-day repeat cycle. The equatorial ground-track spacing is about 7.5 km shifting every 30 days. The short inter-track distance increases the sampling over the study area during one cycle [15]. For this study region and period, CryoSat-2 operated in SAR mode.

All altimetry data used in this study come from the Geophysical Data Records (GDR) made available by the Centre of Topography of the Oceans and the Hydrosphere (CTOH—http://ctoh.legos.obs-mip.fr/). They are sampled along the altimeter track at 18 Hz for ENVISAT, 20 Hz for ERS-2 and CryoSat-2, and 40 Hz for SARAL (high-frequency mode commonly used over land and coastal areas where the surface properties are changing more rapidly than over the open ocean).

3.2. Ancillary Data

3.2.1. Arcachon-Eyrac Tide Gauge

The Arcachon-Eyrac tide gauge is managed by the French hydrographic service (Service Hydrographique et Océanographique de la Marine—SHOM) and the Gironde sea and land state office (Direction Départementale des Territoires et de la Mer—DDTM). It is operating since November 1967. The tide gauge (Figure 1) (1.163550021°W and 44.66500092°N) is a non-contact radar sensor providing sea level measurements at 1-min time intervals since June 2000. These data are made available by REFMAR. Altimetry and tide gauge data timeline is shown in Figure 2.

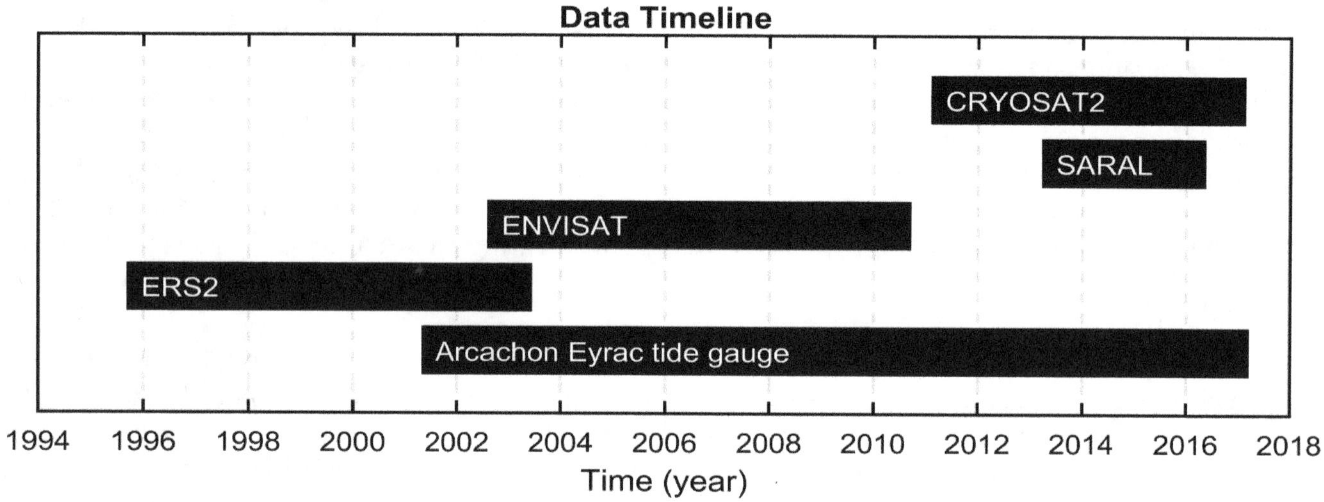

Figure 2. Temporal coverage of the in situ Arcachon Eyrac tide gauge and altimetry datasets used in this study.

3.2.2. Lidar-Derived Topography of the Intertidal Zone

The LIDAR data is extracted from RGE ALTI® product provided by the French national institute for geography and forest information (IGN). Arcachon Bay raw data have been acquired by airborne topographic LIDAR at low tide, on 25 June 2013, and interpolated on a regular 1 × 1 m grid for RGE ALTI® product. The controlled altimetric precision for this data is 0.2 m. For the convenience of the study, these data have been subsampled on a 10 × 10 m grid.

4. Methods

The following flowchart exhibits the different steps of the methodology (Figure 3). All steps are described in the corresponding sections.

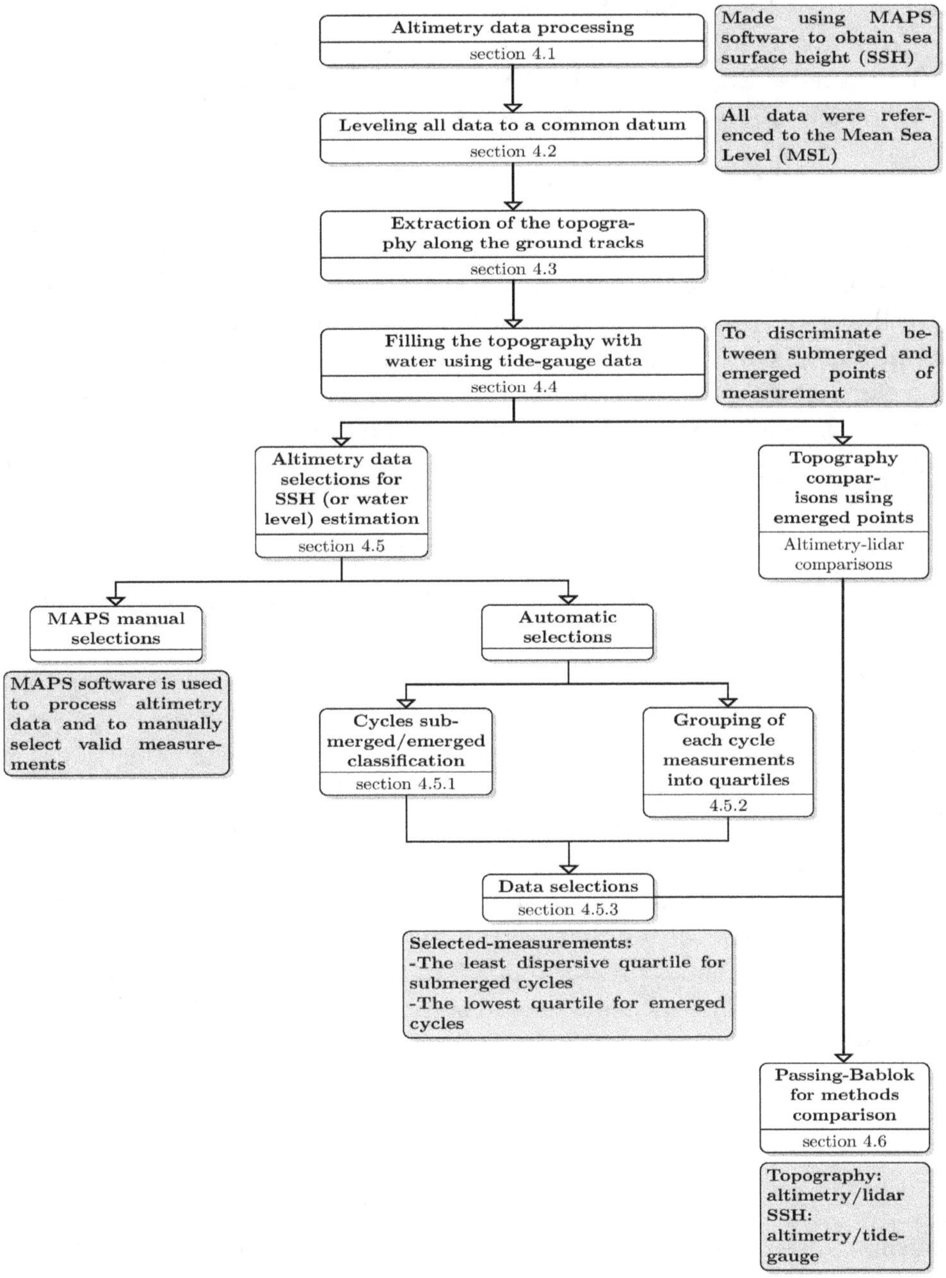

Figure 3. The different steps of the method presented in the form of a flow chart.

4.1. Altimetry Data Processing

The principle of radar altimetry is the following: the altimeter emits a radar pulse and measures the two-way travel-time from the satellite to the surface. The distance between the satellite and the Earth surface—the altimeter range (R)—is thus derived with a precision of a few centimeters. The satellite altitude (H) referred to an ellipsoid is also accurately known from orbitography modeling. Taking into account propagation delays due to interactions of electromagnetic wave in the atmosphere and geophysical corrections, the height of the reflecting surface (h) with reference to an ellipsoid or a geoid can be estimated as [16,17]:

$$h = H - \left(R + \sum \Delta R_{propagation} + \sum \Delta R_{geophysical}\right) \quad (1)$$

where H is the height of the center of mass of the satellite above the ellipsoid, R is the nadir altimeter range from the center of mass of the satellite to the sea surface taking into account instrumental corrections, $\Delta R_{propagation}$ and $\Delta R_{geophysical}$ are the sums of the environmental and geophysical corrections to apply to the range and respectively given by Equations (2) and (3).

$$\sum \Delta R_{propagation} = \Delta R_{ion} + \Delta R_{dry} + \Delta R_{wet} + \Delta R_{SSB} \quad (2)$$

ΔR_{ion} is the atmospheric refraction range delay due to the free electron content associated with the dielectric properties of the ionosphere, ΔR_{dry} is the atmospheric refraction range delay due to the dry gas component of the troposphere, ΔR_{wet} is the atmospheric refraction range delay due to the water vapor and the cloud liquid water content of the troposphere, ΔR_{SSB} is the range correction due to the interaction of the electromagnetic pulse emitted by the altimeter with the scatterometers within the footprint. It is known as Sea State Bias (SSB) and is the sum of the electromagnetic (EM), skewness, and tracker biases.

$$\sum \Delta R_{geophysical} = \Delta R_{ocean} + \Delta R_{solid\ Earth} + \Delta R_{pole} + \Delta R_{atm} \quad (3)$$

where ΔR_{ocean} is the value of the ocean tide, $\Delta R_{solid\ Earth}$ and ΔR_{pole} are the corrections respectively accounting for crustal vertical motions due to the solid Earth and polar tides, ΔR_{atm} is the dynamic atmosphere correction.

Altimetry-derived heights are automatically obtained from the GDR data using the Multi-mission Altimetry Processing Software (MAPS) that is commonly used for the selection of valid altimetry data and their processing over land and ocean [18–20]. More details on MAPS can be found in Frappart et al. [21]. As the ΔR_{SSB} values were most of the times flagged in the GDR, they were not applied for the consistency of the series of observations. ΔR_{SSB} corresponds to cm-level correction applied to the range as it varies typically from -1% to 4% of the Significant Wave Height (SWH) (e.g., Gaspar et al. [22]). In the Arcachon Bay, SWH is generally lower than 0.2 m but can exceptionally reach 0.5 to 0.7 m for wind intensities greater than 20 $m{\cdot}s^{-1}$ [23]. ΔR_{ocean} and ΔR_{atm} were not applied as they also affect the measurements from the Arcachon-Eyrac tide-gauge. Due to the width of the altimeters and radiometers footprints of several kilometers and several tenths of kilometers, radar echoes are a mix of returns from ocean, intertidal zone, and from the surrounding environment (forests, cities, roads, etc.) and microwave brightness temperatures are the combination of emissions from a very inhomogeneous environment. The same set of corrections used for land hydrology will be applied in this study.

The geophysical corrections applied to the range are derived from the Global Ionospheric Maps (GIM) and Era Interim model outputs from the European Centre Medium-Range Weather Forecasts (ECMWF) for the ionosphere and the dry and wet troposphere range delays respectively. The environmental corrections are obtained from the solid Earth tide [24] and polar tide [25] tables for SARAL, and using the IERS (International Earth Rotation and Reference Systems) convention for ENVISAT. Ranges used to derive altimeter heights were those processed with the Ice-1 retracking algorithm [26,27] (formulation in Appendix A) because they are present in the ERS-2, ENVISAT and SARAL GDRs and were shown to be more suitable for hydrological studies in terms of accuracy of water levels and availability of the data (e.g., Frappart et al. [28,29]). For CryoSat-2, altimeter heights were those processed by the Sea-Ice retracker [30].

4.2. Leveling to a Common Datum

The datasets used in this study are referenced to various datums. ENVISAT and CryoSat-2 altimetry data are referenced to WGS84 ellipsoid, SARAL altimetry data to the Topex/Poseidon (T/P) ellipsoid, the lidar-based topography of the intertidal zone to the French reference system NGF/IGN69, and the Arcachon-Eyrac tide-gauge to the French chart datum. A datum conversion from T/P ellipsoid to WGS84 is automatically performed for SARAL measurements using Equation (4) adapted from [31] in the new version of MAPS implemented for this study:

$$\Delta h = \frac{a'(1-e'^2)}{\sqrt{1-e'^2 sin^2\varphi}} - \frac{a(1-e^2)}{\sqrt{1-e^2 sin^2\varphi}} \quad (4)$$

where Δh is the variation of height at latitude φ due to the change of ellipsoid from T/P to WGS84 datum, a = 6,378,137 m and e = 0.081819190842621 are the semi-major axis and the eccentricity of the WGS84 datum, a' = 6,378,136.3 m and e' = 0.081819221456 are the semi-major axis and the eccentricity of the T/P datum.

For comparison purposes, they were all referenced to the Mean Sea Level (MSL). Datum conversion for all data (altimetry, topography, and tide gauge data) were made using the French maritime altimetry references [32] giving the vertical differences between the different datum (−44.8 m between WGS84 and the French chart, −1.98 m between the French chart and the French reference system, and −2.48 m between the French chart and the MSL).

4.3. Extraction of the Topography of the Intertidal Zone under the Altimeter Tracks

For comparison purposes, the topography of the intertidal zone was extracted along the altimeter ground tracks. For each altimetry measurement, the closest topography point was selected. The maximum distance obtained between an altimetry measurement and its corresponding topography point is 10 m, which is consistent with the spatial resolution of the lidar-derived topography (10 m). This extraction was performed for every ERS-2, ENVISAT, SARAL, and CryoSat-2 cycle as the orbit of the satellites is not exactly repetitive and generally varies within 1 km around the nominal track (Figure 1).

4.4. Manual Classifications of Altimetry Measurements and Cycles

Two types of classification based on ancillary data were made. The first one consists of separating altimetry measurements between submerged and emerged measurement points. This classification was made in order to use submerged points for SSH estimation, and emerged points for topography measurements. The resulting altimetry levels were compared to Arcachon Eyrac in-situ gauge measurements and lidar topography data respectively. To discriminate between land and water, the topography along the altimeter tracks is filled with the water levels measured at the Arcachon-Eyrac tide gauge. For each gridpoint of the Lidar-based topography, the gridpoint is considered submerged if the water level is greater than the topography, emerged if not (see Figures 6 and 7 in Section 5).

The second type is a classification by cycle and it was made to separate cycles with emerged land from completely submerged cycles. The cycle is considered as an emerged cycle if more than 20% of the altimeter ground track flew over land. This classification was performed to assess the performance of the automatic classification presented in Section 4.5.1.

4.5. Automatic Selections of Valid Altimetry Measurements

A method to automatically select the altimetry measurements (used for SSH estimation) over emerged and submerged areas was tested. It is composed of the following steps.

4.5.1. Classification of Cycles between Submerged and Emerged Cycles

This first step discriminates submerged from emerged cycles using intrinsic altimetry parameters in order to be completely independent of in-situ data. The rationale behind this step is the use of different selection criteria for submerged and emerged cycles. The classification was made using the unsupervised k-means clustering algorithm [33]. The number of clusters (k) was chosen to be two for submerged and emerged classes and the distance measure technique used is the cityblock (Manhattan) technique [34]. In contrast to the Euclidean distance (straight line distance between two points in Euclidean space), the cityblock distance is calculated as the distance in x plus the distance in y (in 2D). The backscattering coefficient and the peakiness (the average of these parameters per cycle) are the two parameters used to perform the clustering technique (peakiness formulation can be found in Appendix A). Before data clustering, the latter parameters were normalized and centered in order to unify their influence on distance. The use of these two parameters is based on a priori knowledge that they present significant descrepancies between water-dominated footprints and water-land-mixed footprints.

4.5.2. Grouping of Cycle's Measurements into Four Equal Parts

In this step, the altimetry measurements made during a given cycle were separated into four equally-sized groups. The four groups were separated by the following scores: Quartile 0 (the minimum), Quartile 1 (larger than 25% of the data points), Quartile 2 (the median), Quartile 3 (bigger than 75% of the data points), and Quartile 4 (the maximum). For each group, the standard deviation is computed to assess its dispersion.

4.5.3. Data Automatic Selections

For cycles classified as submerged cycles, the groups with the least standard deviation (least dispersion) are preserved. For cycles classified as emerged cycles, the first groups with the lowest 25% of values (values between Quartile 0 and Quartile 1) are preserved because these groups are the only groups able to reach the water when land emerges.

The results of the two types of selections are compared in Section 5.2.

4.6. Passing-Bablok Regression for Method Comparisons

Passing-Bablok is the regression method used to account for agreement and systematic bias between two methods (altimetry/tide-gauges or altimetry/lidar). We opted for Passing-Bablok method instead of ordinary linear regression because it is not sensitive to the outliers or the distribution of errors and because the independent variable (water level from tide gauge records or bathymetry from lidar) is not free of error. This robust, non-parametric method consists of fitting a line describing the relationship between the two variables (X and Y) and testing whether the slope is 1 and the intercept is 0. In-depth details on the method can be found in [35]. The results are presented as a scatter plot between X (method 1) and Y (method 2), a regression line, and a linear regression equation where the slope and the intercept represent proportional and constant systematic bias respectively. A statistical test of the assumption of linearity is performed using cumulative sum linearity test. Non-linear samples are not suitable for concluding on method agreement. Furthermore, a 95% Confidence Interval (CI) is also computed for the slope and the intercept to test the hypothesis that the slope is equal to 1 or the intercept is equal to 0. These hypotheses are accepted if 1 corresponds to the slope's 95% CI (if not, there is a proportional difference between the two methods) and if 0 corresponds to the intercept's 95% CI (if not, there is a constant difference or bias between the two methods).

4.7. Absolute Calibration of Altimetry Missions over the Intertidal Zone

Comparisons between altimetry-based and in situ SSH from tide gauge were performed. They require simultaneous in situ and altimetry measurements in the same terrestrial reference frame at the exact same location or comparison point (e.g., Cancet et al. [36]). The absolute altimeter bias ($Bias_{altimeter}$) is estimated as follows [37]:

$$Bias_{altimeter} = < h_{altimeter} - h_{in\ situ} > \quad (5)$$

where $h_{altimeter}$ and $h_{in\ situ}$ are the height of the reflecting surface estimated from altimeter and in situ measurements respectively. In this study, an absolute calibration is performed for estimating the bias of ERS-2, ENVISAT, SARAL and CryoSat-2 for sea level measurements and topography of the intertidal zone in the Arcachon Bay.

5. Results

Depending on their overflight time, radar altimetry missions acquired observations of the Arcachon Bay all over the tidal cycle. During low tide, they provide observations of the surface topography of the intertidal zone whereas, at high tides, they monitor the sea surface height. These changes in the nature of the reflecting surface, from wet sand and mud to sea water, modify the radar echo acquired by the altimeter. Lidar-based topography profiles were extracted along the altimetry ground tracks in the intertidal zone of the Arcachon Bay. They were filled with water using the record from Arcachon-Eyrac tide-gauge corresponding to the altimeter overflights. They were compared to the along track profiles of altimeter height, backscattering coefficients and waveform peakiness (not available for ERS-2 in the CTOH GDR and only available at 1 Hz for ENVISAT) estimated using the Ice-1 retracking algorithm for ERS-2, ENVISAT and SARAL and the SeaIce retracking algorithm for CryoSat-2 (see [27]) for details about the computation of these two latter parameters that are available in the altimeter GDRs). Examples for high and low tides using SARAL and CryoSat-2 data are respectively presented in Figures 4 and 5 (same type of figures is available in the Supplementary Material for ERS-2 (Figure S1) and ENVISAT (Figure S2)). SSH estimated using altimetry data was given by the median of submerged points' altimeter heights. For high and low tides, SARAL and CryoSat-2 missions show relatively good correspondence with in situ measurements while ERS-2 and ENVISAT show low accuracy, especially at low tides. ERS-2 is unable to accurately retrieve topography variations due to both, the coarse resolution of its footprint and changes in the range resolution between ocean (bandwidth of 300 MHz) and ice modes (bandwidth of 20 MHz). As for ENVISAT, despite the important bias obtained, altimetry measurements follow the topography variations and give an acceptable correlation coefficient as a result (see Sections 5.2 and 5.3). We observe a significant increase in the backscattering coefficients and waveform peakiness values between high and low tide (Figure 4c,d, and Figure 5c,d). This is due to the contribution of the emerging land to the received signal. A slight variation of these parameters is observed as well at the extreme sides for high tides where land and water fall together in the altimeter footprint.

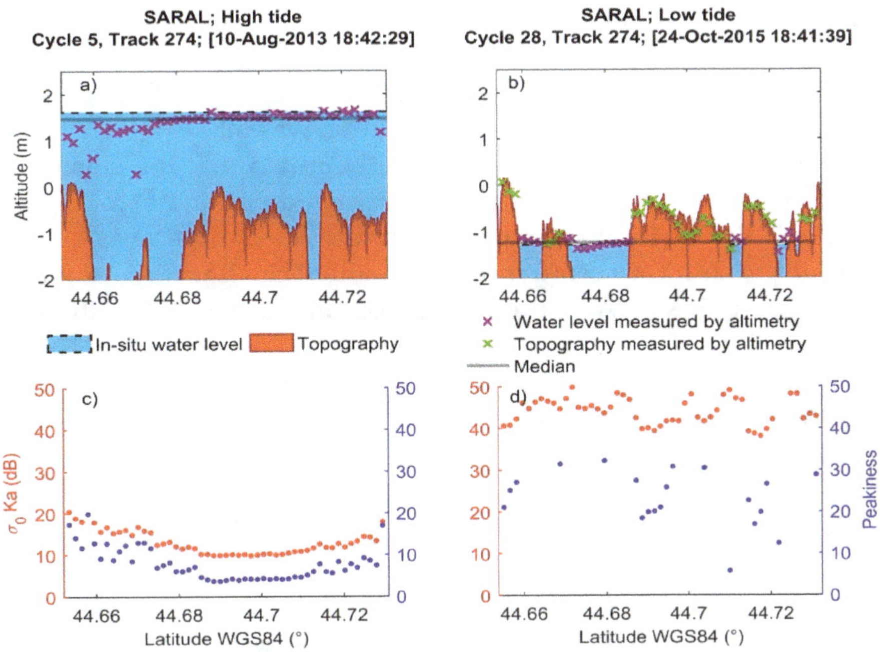

Figure 4. (**a**,**b**) Examples of SARAL along-track profiles of altimetry height over water (purple crosses) and land (green crosses) at high (**a**) and low (**b**) tides, the topography under the altimeter ground track is represented in brown and it is filled with water (in blue) using leveled tide-gauge records; (**c**,**d**) Variation of Ice-1 backscattering coefficients of Ka-band (red dots) and Peakiness (blue dots) at high (**c**) and low (**d**) tides.

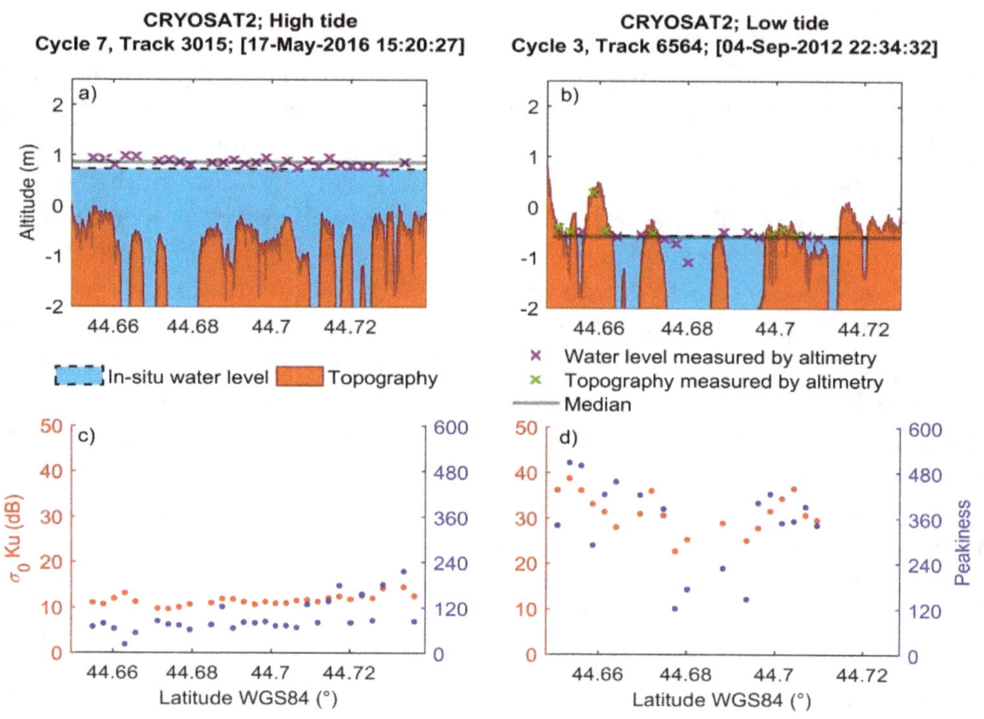

Figure 5. (**a**,**b**) Examples of CryoSat-2 along-track profiles of altimetry height over water (purple crosses) and land (green crosses) at high (**a**) and low (**b**) tides, the topography under the altimeter ground track is represented in brown and it is filled with water (in blue) using leveled tide-gauge records; (**c**,**d**) Variation of Retracker 1 backscattering coefficients of Ku-band (red dots) and Peakiness (blue dots) at high (**c**) and low (**d**) tides.

5.1. *Auto-Classification Using the k-Means Algorithm*

Using the k-means algorithm, the altimetry cycles were classified between emerged and submerged cycles. The evaluation of the auto-classification was made using the classification based on in-situ data as explained in Section 4.4. It should be noted that the clustering technique was made using two parameters (the backscattering coefficient and peakiness) for SARAL and CryoSat-2 (Figure 6), and using only the backscattering coefficient for ERS-2 and ENVISAT. SARAL showed the best classification results with 100% accordance with manual classification. A good consistency was shown also by CryoSat-2 and ENVISAT with ~80% accordance and ERS-2 with 82%.

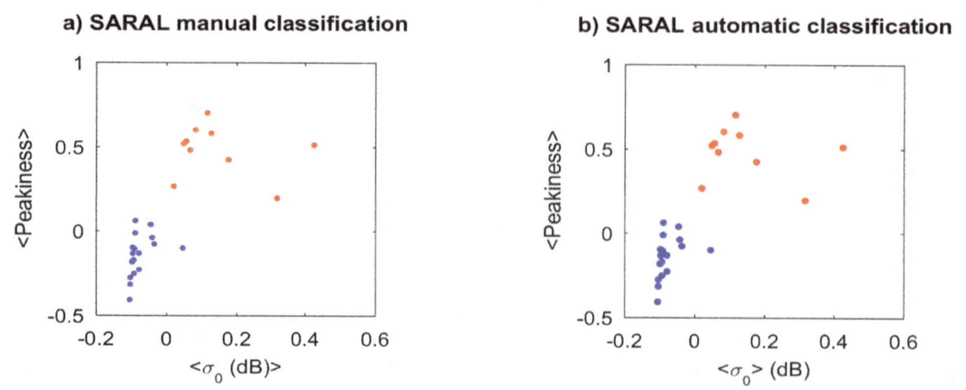

Figure 6. *Cont.*

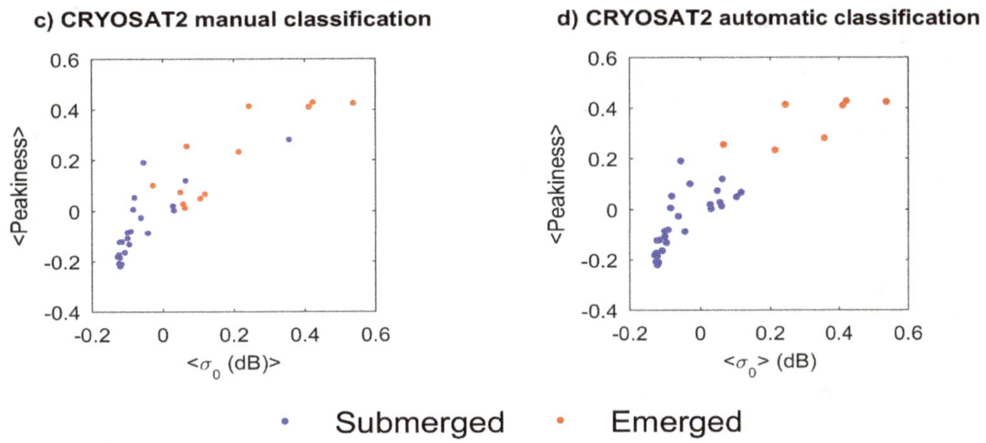

Figure 6. Comparisons between manual (left) and automatic (right) classifications (made using an unsupervised clustering technique) of cycles for SARAL (**a**,**b**) and CryoSat-2 (**c**,**d**).

5.2. Water Levels Comparison

For the selections made using MAPS, SSH estimation was given by the median of submerged altimetry measurements made during one crossing (cycle). An automatic selection of altimetry data was performed in addition to the manually refined data selections in MAPS. The automatic technique uses the less dispersive quartile of measurements in each cycle for cycles classified as submerged by the clustering technique, and the lowest quartiles for cycles classified as emerged. The SSH was given by the median of the chosen quartile without eliminating measurement points classified as emerged by the classification based on ancillary data. Figure 7 presents the comparisons made between SSHs acquired by altimetry and tide gauge measurements for MAPS manual selections (left) and for automatic selections (right) for ERS-2 (a and b), ENVISAT (c and d), SARAL (e and f), and CryoSat-2 (g and h).

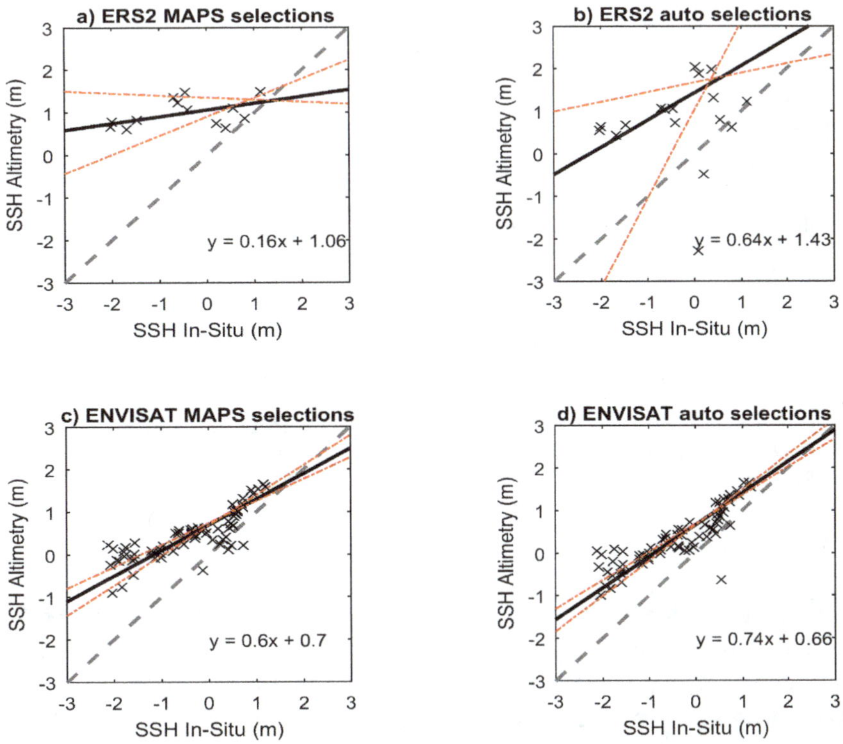

Figure 7. *Cont.*

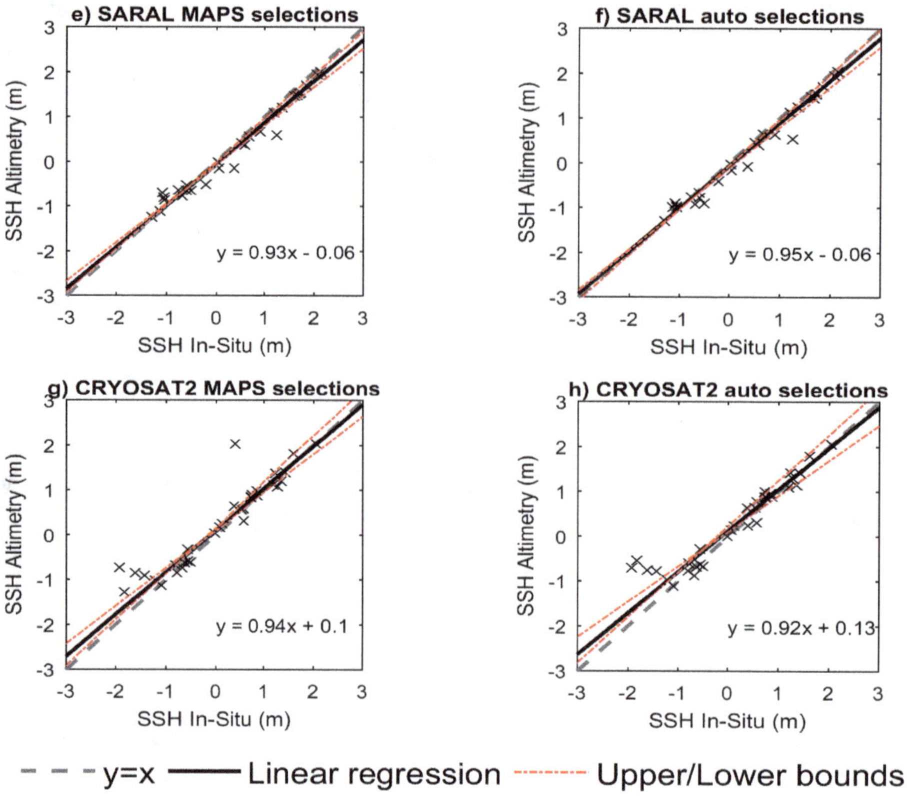

Figure 7. Comparisons between in-situ and altimetry-based SSH for ERS-2 (**a**,**b**), ENVISAT (**c**,**d**), SARAL (**e**,**f**), and CryoSat-2 (**g**,**h**) using two different approaches for selecting the valid measurements: MAPS manual selections (left) and automatic selections (right).

Linearity exists between altimetry and the tide gauge measurements, for ERS-2, SARAL, and CryoSat-2 data for the two types of selections and for the automatic selections made for ENVISAT (Table 2).

Table 2. Passing-Bablok regression model and statistical results for the comparisons made between in-situ and altimetry-based SSH (ERS-2, ENVISAT, SARAL, and CryoSat-2) for manual (MAPS) and automatic selections.

Mission	MAPS				Auto			
	ERS-2	ENVISAT	SARAL	CryoSat-2	ERS-2	ENVISAT	SARAL	CryoSat-2
Linearity [3]	1	0	1	1	1	1	1	1
Slope	0.16	0.60	0.93	0.94	0.64	0.74	0.95	0.92
Slope LB [1]	−0.05	0.51	0.87	0.85	0.23	0.67	0.91	0.79
Slope UB [2]	0.44	0.71	0.97	1.02	2.07	0.83	1.01	1.01
Intercept	1.06	0.70	−0.06	0.10	1.43	0.66	−0.06	0.13
Intercept LB	1.36	0.74	−0.09	0.10	1.68	0.69	−0.13	0.11
Intercept UB	0.91	0.69	−0.02	0.15	1.02	0.65	−0.05	0.22
R	0.39	0.82	0.99	0.93	0.11	0.86	0.99	0.95
RMSE (m)	1.75	1.04	0.23	0.42	1.70	0.92	0.22	0.39
Mean Bias (m)	1.47	0.88	0.18	0.23	1.50	0.79	0.17	0.24

[1] Lower Bound. [2] Upper Bound. [3] 1 for accepted linearity and 0 for refused linearity.

ERS-2 showed the most unsatisfactory results with slopes lower than 0.64 and intercepts higher than 1 (Table 2). The results obtained by the Passing-Bablok method could not be interpreted for ENVISAT MAPS selections since no linear relationship exists with the tide gauge measurements (a break is observed in the slope at 0 m (Figure 7c)). Low accuracy for MAPS selections made for ENVISAT could be inferred from the RMSE and the mean bias obtained (1.04 m and 0.88 m respectively). The automatic selections significantly improved the estimation of SSH obtained using ENVISAT data and rendered the linearity test positive. The slopes given by ENVISAT increased from 0.60 to 0.74 and were accompanied by an improvement of R (from 0.82 to 0.86) and RMSE that remains important (RMSE~0.92 m) (Table 2). As explained in Section 4.6, the method does not show proportional or constant biases if 1 belongs to the slope CI (slope(LB) < 1 < slope(UB)) and 0 belongs to the intercept CI (intercept (LB) < 0 < intercept (UB)) respectively. For SARAL, a better result is obtained using the automatic selection showing only constant bias with no proportional bias (the confidence interval of the slope contains the value 1). Slight proportional and constant biases are obtained for MAPS selections. MAPS selections present better results for CryoSat-2 that shows slight constant biases with no proportional biases for the two types of selections.

5.3. Topography Comparison

Figure 8 presents topography comparisons between lidar and altimetry measurements for ERS-2 (a), ENVISAT (b), SARAL (c), and CryoSat-2 (d). For topography comparisons all measurements from every cycle were used along with the corresponding lidar topography measurements. Table 3 lists all regression models obtained. Linearity was rejected only for ENVISAT data. Therefore, the slope and the intercept obtained for ENVISAT could not be interpreted. Good comparison results are obtained for SARAL giving a slope of 1.06 and an intercept at 0.01 with no proportional or constant bias observed for topography monitoring. CryoSat-2 provides as well acceptable results with slight systematic and proportional biases. Showing a positive linearity test, ERS-2 slope and intercept showed large differences from 1 and 0.

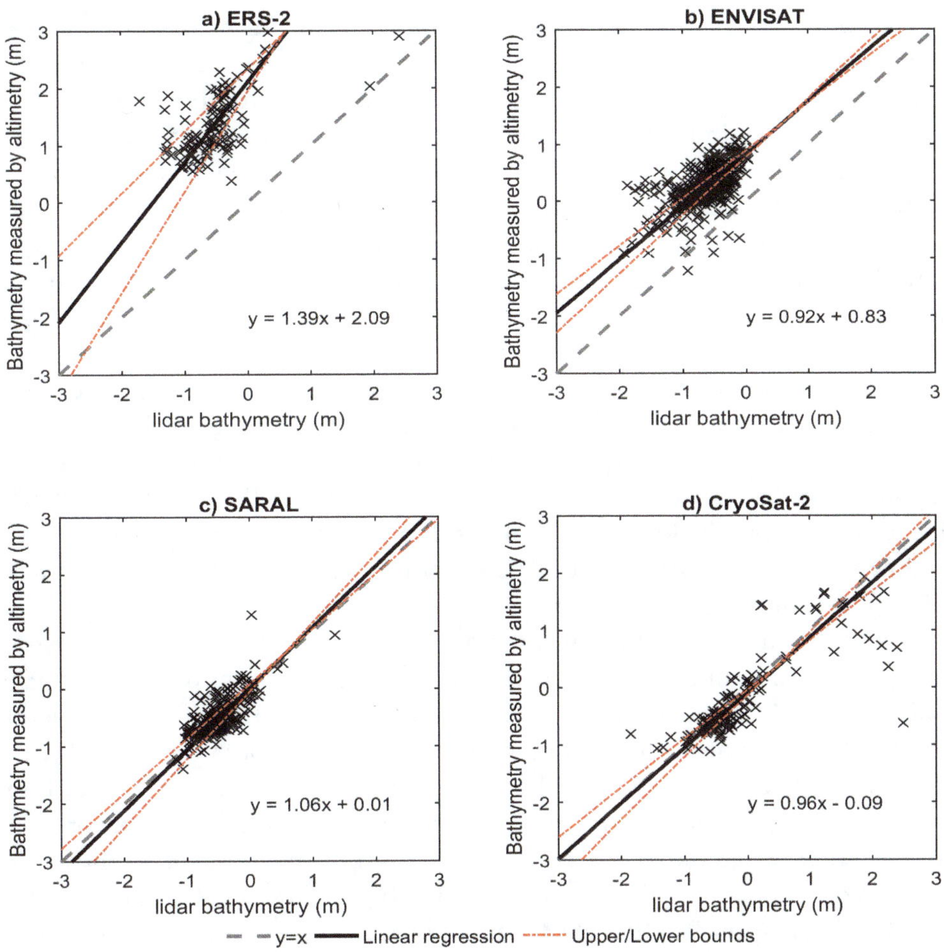

Figure 8. Comparisons between lidar and Altimetry-based topography measurements for: (**a**) ERS-2, (**b**) ENVISAT, (**c**) SARAL, and (**d**) CryoSat-2.

Table 3. Passing-Bablok regression model and statistical results for the comparisons made between lidar and altimetry-based topography estimation for ERS-2, ENVISAT, SARAL, and CryoSat-2.

Mission	ERS-2	ENVISAT	SARAL	CryoSat-2
Linearity Test [3]	1	0	1	1
Slope	1.39	0.92	1.06	0.96
Slope LB [1]	1.08	0.84	0.96	0.86
Slope UB [2]	1.75	1.02	1.19	1.09
Intercept	2.09	0.83	0.01	−0.09
Intercept LB	2.31	0.89	0.08	−0.04
Intercept UB	1.92	0.77	−0.04	−0.13
R	0.17	0.54	0.71	0.79
RMSE (m)	2.01	0.95	0.23	0.44
Mean Bias (m)	1.93	0.90	0.17	0.25

[1] Lower Bound. [2] Upper Bound. [3] 1 for accepted linearity and 0 for refused linearity.

6. Discussion

For each altimetry cycle of ERS2, ENVISAT, SARAL, and CryoSat-2, the water level of the Arcachon Bay was estimated by computing the median of submerged points, and compared to tide gauge measurements carried out at the same time (in a 5 min interval). Altimetry measurements above emerged points were considered as topography measurements and were compared to the nearest lidar bathymetry points.

Over coastal areas, the altimeter footprint is contaminated by signals from different surface types which reduce the accuracy of the measurements [6]. For water levels estimation, SARAL operating at Ka-band frequency shows significant improvement in the measurement accuracy (vertical resolution) comparing to the ERS-2 and ENVISAT Ku-band altimeters, allowing better observations in coastal areas like lagoons [38,39]. This improvement comes from the reduced footprint area (~6 km) of SARAL which reduces the impact of land on SARAL waveforms. It should be noted that the use of Ka-band is not the only factor that governs the size of the footprint, the larger bandwidth (~480 MHz) used by SARAL has an impact on the footprint size as well. According to [6], SARAL AltiKa's footprint is ~0.8 times smaller than a Ku-band altimeter for a given bandwidth. CryoSat-2 shows as well major improvements in the accuracy for water levels estimation comparing to ERS-2 and ENVISAT. Operating in the same band (Ku) as the latter missions, the better performance of CryoSat-2 comes from the use of the SAR mode of acquisition. The processing of CryoSat-2 SAR data is based on the delay Doppler technique that reduces the footprint size in the along-track direction which makes CryoSat-2 more suitable for coastal monitoring. The lower consistency with in-situ measurements was obtained using data from LRM missions operating at Ku-band (ERS-2 and ENVISAT) especially at low tides where its spatial resolution could not resolve the complex water-land mixture of the bay. Figure 7 shows that above mean sea level ENVISAT was capable of estimating the water level of the bay with a relatively low bias.

The results also show a great potential for radar altimetry to retrieve the topographic variation of the intertidal zone. SARAL and CryoSat-2 show a large consistency with the lidar measurements. ENVISAT was able to follow the variation of the intertidal zone bathymetry producing a correlation coefficient (R) of 0.57 but with an important bias (due to the large footprint of ENVISAT).

CryoSat-2 was the only mission that flew over the "ile aux oiseaux" sandbank (Figure 1b). Some altimetry measurements made by CryoSat-2 underestimated the elevation of topography in this area. It is very likely that the cause of the high discrepancy is due to the penetration of the electromagnetic wave in the dry soil (long time after high water). We calculated the maximum penetration depth that corresponds to the electromagnetic wave frequency and the dielectric properties of the sand (Appendix B). The differences obtained between the underestimated measurements made by altimetry and the corresponding lidar measurements are lower than the maximum penetration depth of 1.35 m which is consistent with our hypothesis. We eliminated the underestimated points from CryoSat-2 measurements and we observed an improvement in terms of slope (from 0.96 to 0.98), intercept (from −0.09 to −0.08), and R (from 0.79 to 0.89).

In this study, an automatic classification of cycles was performed to discriminate between submerged and emerged cycles. An automatic process was also envisaged to classify each measurement point, using the backscattering coefficient that tends to increase for land-dominated areas. However, an important limitation was encountered during the automating process. During low tides, some altimetry points above water showed as well high backscattering coefficients, which complicated the separation of emerged and submerged points using the backscattering coefficient parameter. The high values of the backscattering coefficient obtained for water at low tides is most likely due to smoother water surfaces in the channels than in the bay, caused by the reduced wave activity. This reduced roughness increases the specular reflection and thus the power received by the sensor (the backscattering coefficient). An example is shown in Figure 9 (left) for the cycle 20 of SARAL. The backscattering coefficient of water in large channels (zone 5 and zone 3) is lower than the backscattering coefficient of water in narrow channels (zone 4 and zone 2) showing values that match land-dominated areas (zone 1).

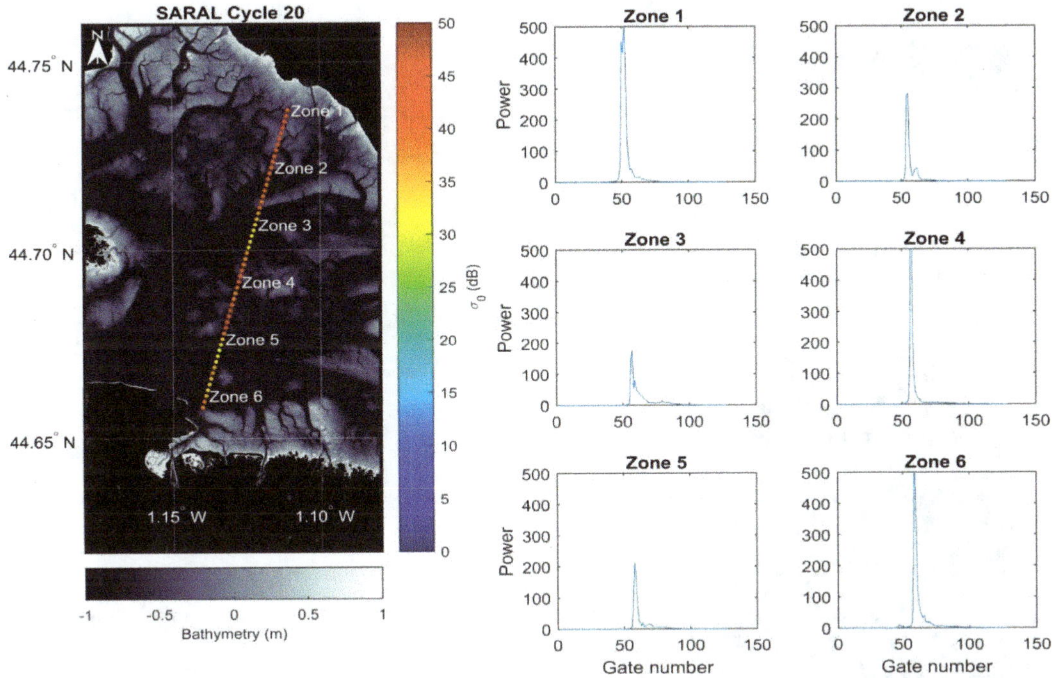

Figure 9. Spatial variation of the backscattering coefficient for SARAL (cycle 20) at low tide along with the corresponding waveforms of the indicated zones.

Figures 9–12 show the spatial variation of the backscattering coefficient for a given cycle and the waveforms corresponding to different points in the bay. Figures 10 and 12 correspond to cycles crossing the bay at high tides for SARAL and CryoSat-2 respectively (same type of figures is available in the supplementary material for ENVISAT (Figures S3 and S4)). For SARAL (Figure 10) and CryoSat-2 (Figure 12), we observe at the middle of the bay (zone 3 and zone 4) that the waveforms look like typical ocean waveforms. However, the waveforms at the edges of the bay (zones 1, 2, 5, and 6 for SARAL and zones 1 and 6 for CryoSat-2) show a sharp rise. It is not the case for ENVISAT at high tides (Figure S3). The waveforms of ENVISAT in the middle of the bay (zone 3 and zone 4) differ from the typical ocean waveforms due to the larger footprint of ENVISAT. It should be noted that the waveforms of CryoSat-2 at high tides maintain typical ocean waveforms closer to the coastline than SARAL due to the SAR mode of CryoSat-2 that increases the resolution in the along-track direction of the satellite. Another consequence of the SAR mode is the highly energetic waveforms.

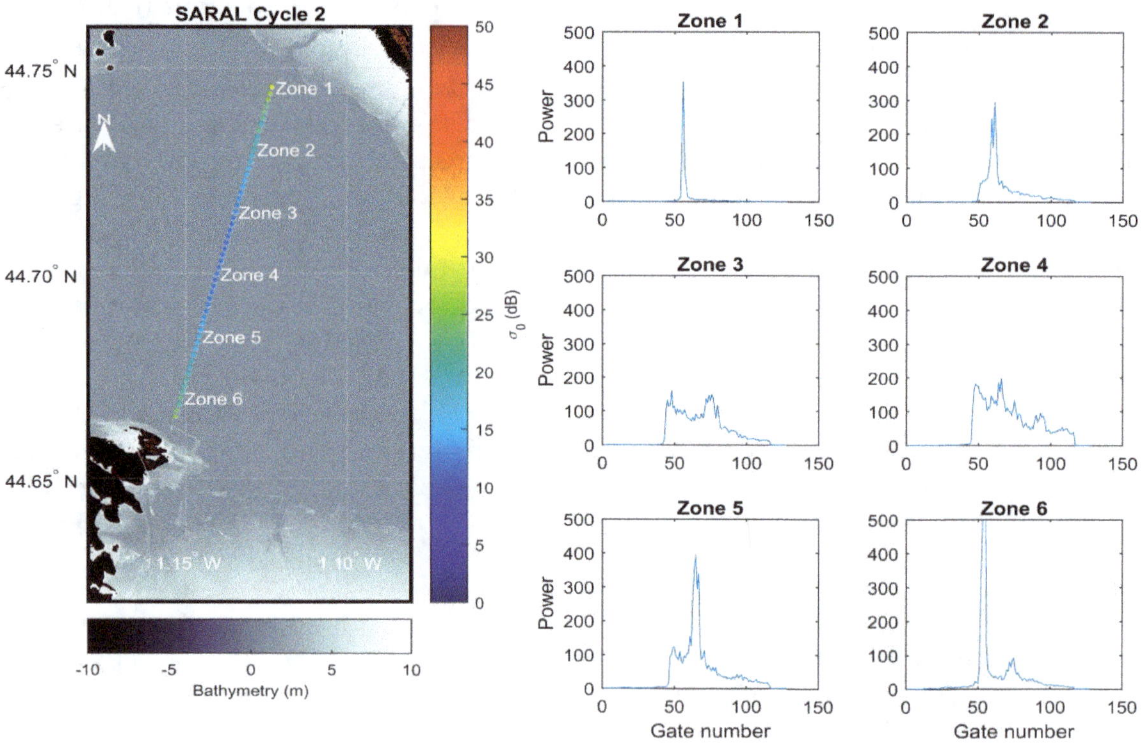

Figure 10. Spatial variation of the backscattering coefficient for SARAL (cycle 2) at high tide along with the corresponding waveforms of the indicated zones.

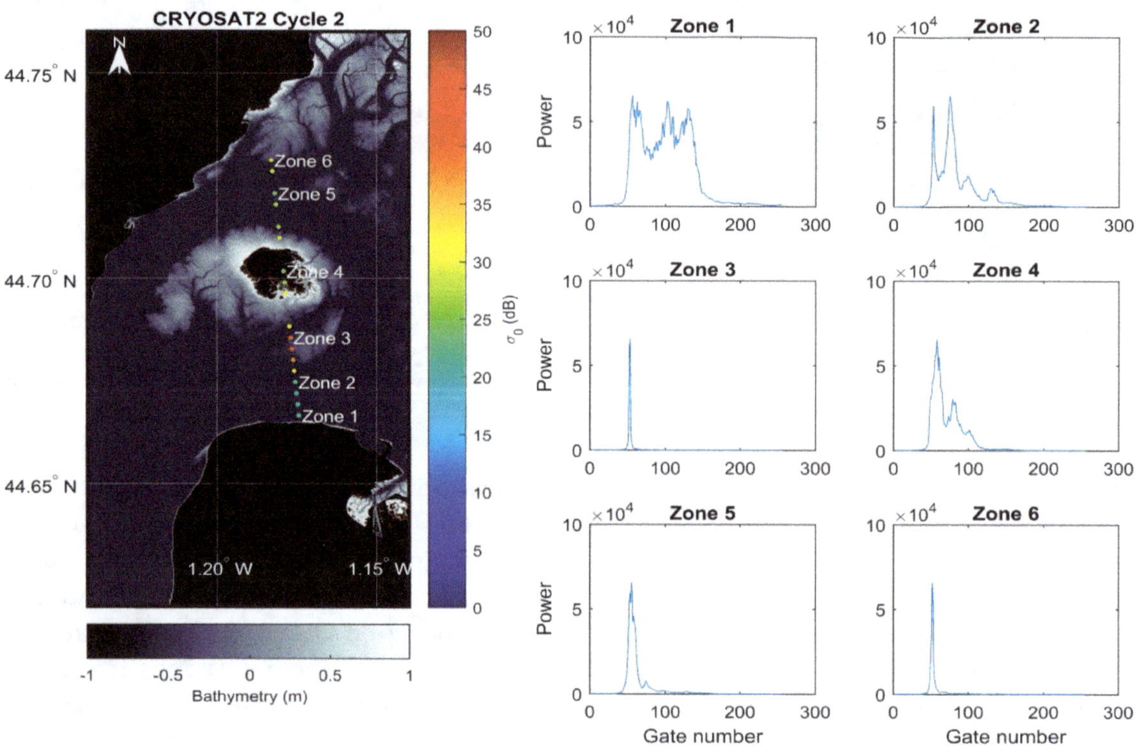

Figure 11. Spatial variation of the backscattering coefficient for CryoSat-2 (cycle 2) at low tide along with the corresponding waveforms of the indicated zones.

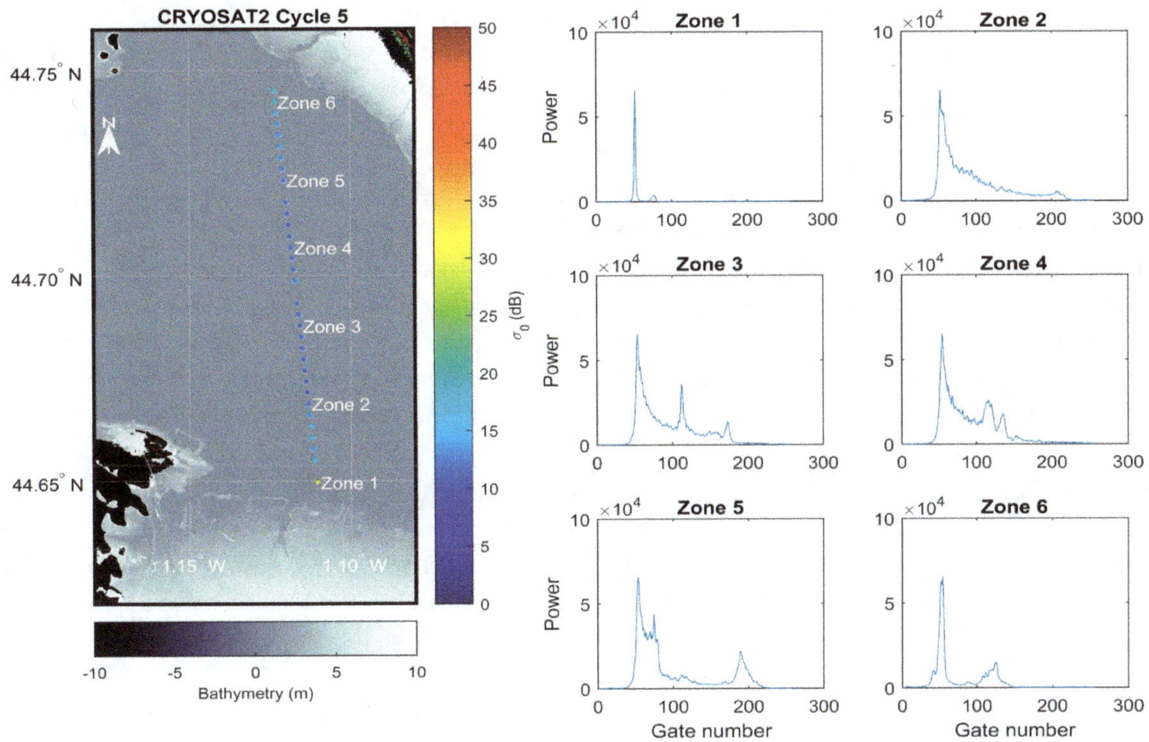

Figure 12. Spatial variation of the backscattering coefficient for CryoSat-2 (cycle 5) at high tide along with the corresponding waveforms of the indicated zones.

Figures 9 and 11 correspond to cycles crossing the bay at low tides for SARAL and CryoSat-2 respectively. At low tides, SARAL (Figure 9) shows peak shaped waveforms (relatively similar to ENVISAT waveforms at low tides (Figure S4)) while CryoSat-2 (Figure 11) shows mixed shaped waveforms. Typical ocean waveforms were not observed for low tides even in areas dominated by water.

The use of satellite radar altimetry to monitor SSH and topography has the potential to be applied in any tidal zone especially in remote areas where no tide gauge record or topography data are available. Radar altimetry parameters (SSH, σ_0 and peakiness) can be used to discriminate between high (parameter values almost constant for a given cycle) and low (larger variability) tides. Until now, only large submerged and emerged areas can be discriminated during low tides. Analyses of the radar echoes (waveforms) can provide valuable information to improve the characterization of the nature of the surface (emerged or submerged).

The new-era coastal altimetry satellites contribute to the advancement of coastal ocean observing systems [40]. The launch of the SWOT mission in 2021 will offer new opportunities for monitoring of coastal lagoons topography and changes in time thanks to (i) the acquisitions of elevation along two swaths of 60 km of width that will provide 2D maps, (ii) its better vertical accuracy, (iii) its shorter repeat period, (iv) the use of the Ka-band that has a smaller penetration depth.

7. Conclusions

This study is the first to demonstrate the capabilities of altimetry to retrieve the Sea Surface Height and the topography of the intertidal zone in a coastal lagoon. Data from four altimetry missions were analyzed over Arcachon's Bay coastal lagoon located in the South-East of the Bay of Biscay: (i) ERS-2, (ii) ENVISAT, (iii) SARAL, and (iv) CryoSat-2. The two first missions were operating at Ku-band in low resolution mode whereas SARAL is the first mission to operate in Ka-band in LRM and CryoSat-2 is the first mission to operate at Ku-band in SAR mode. Altimetry measurements were validated using tide gauge and lidar bathymetry datasets.

Considering data from the classical LRM missions operating at Ku-band (ERS-2 and ENVISAT), SSH and topography retrievals from ERS-2 are inaccurate whereas ENVISAT measurements provide better results for SSH (R > 0.82 and RMSE < 1.04 m) than for topography (R~0.54 and RMSE~0.95 m). However, SARAL and CryoSat-2 showed very satisfying correspondence with ancillary data with correlation coefficient (R) higher than 0.93 and RMSE lower than 0.42 m for SSH estimations and R higher than 0.71 for topography with RMSE lower than 0.44 m. It should be noted that the standard deviation of the whole sea level records is 1.05 m.

Furthermore, a successful and simple statistical method (based on a combination of a non-supervised clustering technique and grouping altimetry data into quartiles) was introduced in order to select valid altimetry data. The purpose of this method is to accelerate the processing phase of altimetry data and to optimize the accuracy of Sea Surface Height estimation. This selection method showed very satisfying results with similar accuracy performance as the manual selections made using MAPS software.

The most important limitation lies in the large footprint of altimeters. This coarse resolution prevented us from automatically discriminating land from water using intrinsic altimetry parameters like the backscattering coefficient or waveforms peakiness. However, we were able to discriminate acquisitions made during low tides from the ones made during high tides, because at the scale of the bay good contrast between low and high tides was obtained by the backscattering coefficient and the peakiness.

While altimetry was able to monitor the topography of the bay, caution must be taken when using altimetry data to retrieve topographic variations in similar environments. Dry sandy features are most likely problematic due to the important penetration depth of electromagnetic waves in such mediums. It is recommended to use this approach in areas exposed frequently to water.

Supplementary Materials
Figure S1: (a,b) Examples of ERS-2 along-track profiles of altimetry height over water (purple crosses) and land (green crosses) at high (a) and low (b) tides, the topography under the altimeter ground track is represented in brown and it is filled with water (in blue) using leveled tide-gauge records; (c,d) Variation of Ice-1 backscattering coefficients of Ku-band (black dots) at high (c) and low (d) tides., **Figure S2**: (a,b) Examples of ENVISAT along-track profiles of altimetry height over water (purple crosses) and land (green crosses) at high (a) and low (b) tides, the topography under the altimeter ground track is represented in brown and it is filled with water (in blue) using leveled tide-gauge records; (c,d) Variation of Ice-1 backscattering coefficients of Ku-band (black dots) at high (c) and low (d) tides. **Figure S3**: Spatial variation of the backscattering coefficient for ENVISAT (cycle 21) at high tide along with the corresponding waveforms of the indicated zones. **Figure S4**: Spatial variation of the backscattering coefficient for ENVISAT (cycle 9) at low tide along with the corresponding waveforms of the indicated zones.

Acknowledgments: This work was supported by the CNES. It is based on SWOT space mission and the SWOT COTEST project and was also supported by the Normandy Region.

Author Contributions: Edward Salameh and Frédéric Frappart conceived and designed the experiments; Edward Salameh, Frédéric Frappart, Alexandra Spodar, Vincent Marieu, and Jean-Paul Parisot performed the experiments and analyzed the data; Edward Salameh and Frédéric Frappart wrote the paper. All authors were involved in the analysis of the results and contributed to the discussion of results and the preparation of the manuscript.

Appendix A.

The Ice-1 (OCOG—Offset Centre of Gravity) retracking algorithm approach consists of replacing the waveform with a box that has the same center of gravity as the waveform. The box is defined using three measurements:

$$\text{centre of gravity} = \frac{\sum_{n=1+aln}^{n=N-aln} n y^2(n)}{\sum_{n=1+aln}^{n=N-aln} y^2(n)} \tag{A1}$$

$$\text{amplitude} = \sqrt{\frac{\sum_{n=1+aln}^{n=N-aln} y^4(n)}{\sum_{n=1+aln}^{n=N-aln} y^2(n)}} \tag{A2}$$

$$\text{width} = \frac{\left(\sum_{n=1+\text{aln}}^{n=N-\text{aln}} y^2(n)\right)^2}{\sum_{n=1+\text{aln}}^{n=N-\text{aln}} y^4(n)} \quad \text{(A3)}$$

where N is the total gate number, aln is the number of eliminated gate in the starting and ending of waveform and y(n) is the power of the nth gate. The middle of the leading edge point (LEP) is given by:

$$\text{LEP} = \text{centre of gravity} - 0.5 \times \text{width} \quad \text{(A4)}$$

Peakiness is a measure of how sharply peaked an echo is. It is the ratio of the maximum power (highest bin value) to the accumulated echo power above the retracking point. The peakiness in AltiKa GDR is given by:

$$\text{Peakiness}_{\text{AltiKa}} = \frac{P_{max} \times N_{right}}{\sum_{i=1}^{N_{WF}} P_i} \quad \text{(A5)}$$

where N_{WF} is the total number of the registering gates, and N_{right} is a number of points to the right from the tracking point (i.e., above the retracking point). For SIRAL:

$$\text{Peakiness}_{\text{SIRAL}} = \frac{P_{max} \times 30}{\sum_{i=1}^{N_{WF}} P_i} \quad \text{(A6)}$$

Appendix B.

The penetration depth (δp) of an electromagnetic wave in an homogeneous medium can be expressed as a function of its dielectric permittivity. According to [41], it can be written as follows:

$$\delta_p = \frac{\lambda\sqrt{\varepsilon\prime}}{2\pi\varepsilon''} \quad \text{(A7)}$$

where λ is the wavelength of the electromagnetic wave, $\varepsilon\prime$ and ε'' are the real and complex parts of the dielectric permittivity respectively. Empirical relationships between dielectric permittivity (ε) and soil nature and its water content were established. In this study, we used the empirical relationships from [42] that have the following form:

$$\varepsilon = a_0 + a_1S + a_2C + (b_0 + b_1S + b_2C)m_v + (c_0 + c_1S + c_2C)m_v \quad \text{(A8)}$$

where S and C are the sand and clay fraction of soil given in percent, m_v is the volumetric water content given in percent, a_0, a_1, a_2, b_0, b_1, b_2, c_0, c_1, c_2 are the constants of the empirical relationship given in Table 2 of (Hallikainen et al., 1985).

Maximum penetration depth are obtained for dry soils (i.e., $m_v = 0$). (A1) becomes:

$$\max(\delta_p) = \frac{\lambda\sqrt{a_0\prime + a_1\prime S + a_2\prime C}}{2\pi(a_0'' + a_1''S + a_2''C)} \quad \text{(A9)}$$

Over the sandbank known as "île aux oiseaux", we consider $S = 100$ and $C = 0$. This latter equation becomes:

$$\max(\delta_p) = \frac{\lambda\sqrt{a_0\prime + 100a_1\prime}}{2\pi(a_0'' + 100a_1'')} \quad \text{(A10)}$$

References

1. Agardy, T.; Alder, J. Coastal Systems. In *Ecosystems and Human Well-Being: Current State and Trends*; Hassan, R., Scholes, R., Ash, N., Eds.; Island Press: Washington, DC, USA, 2005; pp. 513–550.
2. CIESIN—Center for International Earth Science Information Network. Gridded Population of the World (GPW) Version 3 Beta. Available online: http://sedac.ciesin.columbia.edu (accessed on 20 September 2017).

3. Barnes, R.S.K. *Coastal Lagoons (Cambridge Studies in Modern Biology 1)*; Cambridge University Press: Cambridge, UK, 1980.
4. Kjerfve, B.; Magill, K.E.; Baruch, B.W. Geographic and hydrodynamic characteristics of shallow coastal lagoons. *Mar. Geol.* **1989**, *88*, 187–199. [CrossRef]
5. Chapman, P.M. Management of coastal lagoons under climate change. *Estuar. Coast. Shelf Sci.* **2012**, *110*, 32–35. [CrossRef]
6. Vignudelli, S.; Kostianoy, A.G.; Cipollini, P.; Benveniste, J. (Eds.) *Coastal Altimetry*; Springer: Berlin/Heidelberg, Germany, 2011.
7. Cipollini, P.; Calafat, F.M.; Jevrejeva, S.; Melet, A.; Prandi, P. Monitoring Sea Level in the Coastal Zone with Satellite Altimetry and Tide Gauges. *Surv. Geophys.* **2017**, *38*, 33–57. [CrossRef]
8. Deborde, J.; Anschutz, P.; Auby, I.; Glé, C.; Commarieu, M.-V.; Maurer, D.; Lecroart, P.; Abril, G. Role of tidal pumping on nutrient cycling in a temperate lagoon (Arcachon Bay, France). *Mar. Chem.* **2008**, *109*, 98–114. [CrossRef]
9. Plus, M.; Dumas, F.; Stanisiè Re, J.-Y.; Maurer, D. Hydrodynamic characterization of the Arcachon Bay, using model-derived descriptors. *Cont. Shelf Res.* **2009**, *29*, 1008–1013. [CrossRef]
10. Rimmelin, P.; Dumon, J.-C.; Maneux, E.; Gonçalves, A. Study of Annual and Seasonal Dissolved Inorganic Nitrogen Inputs into the Arcachon Lagoon, Atlantic Coast (France). *Estuar. Coast. Shelf Sci.* **1998**, *47*, 649–659. [CrossRef]
11. Blanchet, H.; De Montaudouin, X.; Chardy, P.; Bachelet, G. Structuring factors and recent changes in subtidal macrozoobenthic communities of a coastal lagoon, Arcachon Bay (France). *Estuar. Coast. Shelf Sci.* **2005**, *64*, 561–576. [CrossRef]
12. Envisat, C.Z. RA-2 Advanced Radar Altimeter: Instrument Design and Pre-Launch Performance Assessment Review. *Acta Astronaut.* **1999**, *44*, 323–333.
13. Verron, J.; Sengenes, P.; Lambin, J.; Noubel, J.; Steunou, N.; Guillot, A.; Picot, N.; Coutin-Faye, S.; Sharma, R.; Gairola, R.M.; et al. The SARAL/AltiKa Altimetry Satellite Mission. *Mar. Geodesy* **2015**, *38*, 2–21. [CrossRef]
14. Nielsen, K.; Stenseng, L.; Andersen, O.; Knudsen, P. The Performance and Potentials of the CryoSat-2 SAR and SARIn Modes for Lake Level Estimation. *Water* **2017**, *9*, 374. [CrossRef]
15. Wingham, D.J.; Francis, C.R.; Baker, S.; Bouzinac, C.; Brockley, D.; Cullen, R.; de Chateau-Thierry, P.; Laxon, S.W.; Mallow, U.; Mavrocordatos, C.; et al. CryoSat: A mission to determine the fluctuations in Earth's land and marine ice fields. *Adv. Space Res.* **2006**, *37*, 841–871. [CrossRef]
16. Frappart, F.; Blumstein, D.; Cazenave, A.; Ramillien, G.; Birol, F.; Morrow, R.; Rémy, F. Satellite Altimetry: Principles and Applications in Earth Sciences. In *Wiley Encyclopedia of Electrical and Electronics Engineering*; John Wiley & Sons, Inc.: Hoboken, NJ, USA, 2017; pp. 1–25.
17. Chelton, D.B.; Ries, J.C.; Haines, B.J.; Fu, L.L.; Callahan, P.S. Satellite Altimetry. In *Satellite Altimetry and the Earth Sciences: A Handbook of Techniques and Applications*; Fu, L.L., Cazenave, A., Eds.; Academic Press: San Diego, CA, USA, 2001; pp. 1–131.
18. Frappart, F.; Fatras, C.; Mougin, E.; Marieu, V.; Diepkilé, A.T.; Blarel, F.; Borderies, P. Radar altimetry backscattering signatures at Ka, Ku, C, and S bands over West Africa. *Phys. Chem. Earth* **2015**, *83–84*, 96–110. [CrossRef]
19. Frappart, F.; Roussel, N.; Biancale, R.; Martinez Benjamin, J.J.; Mercier, F.; Perosanz, F.; Garate Pasquin, J.; Martin Davila, J.; Perez Gomez, B.; Gracia Gomez, C.; et al. The 2013 Ibiza Calibration Campaign of Jason-2 and SARAL Altimeters. *Mar. Geodesy* **2015**, *38*, 219–232. [CrossRef]
20. Biancamaria, S.; Frappart, F.; Leleu, A.-S.; Marieu, V.; Blumstein, D.; Desjonquères, J.-D.; Boy, F.; Sottolichio, A.; Valle-Levinson, A. Satellite radar altimetry water elevations performance over a 200 m wide river: Evaluation over the Garonne River. *Adv. Space Res.* **2017**, *59*, 128–146. [CrossRef]
21. Frappart, F.; Papa, F.; Marieu, V.; Malbeteau, Y.; Jordy, F.; Calmant, S.; Durand, F.; Bala, S. Preliminary Assessment of SARAL/AltiKa Observations over the Ganges-Brahmaputra and Irrawaddy Rivers. *Mar. Geodesy* **2015**, *38*, 568–580. [CrossRef]
22. Gaspar, P.; Ogor, F.; Le Traon, P.-Y.; Zanife, O.-Z. Estimating the sea state bias of the TOPEX and POSEIDON altimeters from crossover differences. *J. Geophys. Res.* **1994**, *99*, 24981–24994. [CrossRef]
23. Parisot, J.-P.; Diet-Davancens, J.; Sottolichio, A.; Crosland, E.; Drillon, C.; Verney, R. Modélisation des agitations dans le bassin d'Arcachon. *Xèmes Journées Sophia Antip.* **2008**, 435–444. [CrossRef]

24. Cartwright, D.E.; Edden, A.C. Corrected Tables of Tidal Harmonics. *Geophys. J. R. Astron. Soc.* **1973**, *33*, 253–264. [CrossRef]
25. Wahr, J.M. Deformation induced by polar motion. *J. Geophys. Res. Solid Earth* **1985**, *90*, 9363–9368. [CrossRef]
26. Wingham, D.J.; Rapley, C.G.; Griffiths, H. New Techniques in Satellite Altimeter Tracking Systems. In Proceedings of the IGARSS 86 Symposium, Zurich, Switzerland, 8–11 September 1986; pp. 1339–1344.
27. Bamber, J.L. Ice sheet altimeter processing sheme. *Int. J. Remote Sens.* **1994**, *15*, 925–938. [CrossRef]
28. Frappart, F.; Minh, K.D.; L'hermitte, J.; Cazenave, A.; Ramillien, G.; Le Toan, T.; Mognard-Campbell, N. Water volume change in the lower Mekong from satellite altimetry and imagery data. *Geophys. J. Int.* **2006**, *167*, 570–584. [CrossRef]
29. Frappart, F.; Calmant, S.; Cauhopé, M.; Seyler, F.; Cazenave, A. Preliminary results of ENVISAT RA-2-derived water levels validation over the Amazon basin. *Remote Sens. Environ.* **2006**, *100*, 252–264. [CrossRef]
30. Laxon, S. Sea ice altimeter processing scheme at the EODC. *Int. J. Remote Sens.* **1994**, *15*, 915–924. [CrossRef]
31. Jekeli, C. *Geometric Reference System in Geodesy*; Division of Geodesy and Geospatial Science School of Earth Sciences, Ohio State University: Columbus, OH, USA, 2006.
32. SHOM. Références Altimétriques Maritimes (RAM). Available online: http://diffusion.shom.fr/ (accessed on 10 February 2018).
33. Hartigan, J.A.; Wong, M.A. A K-Means Clustering Algorithm. *J. R. Stat. Soc.* **1979**, *28*, 100–108.
34. Craw, S. Manhattan Distance. In *Encyclopedia of Machine Learning*; Sammut, C., Webb, G.I., Eds.; Springer Science & Business Media: Boston, MA, USA, 2011.
35. Passing, H.; Bablok, W. A new biometrical procedure for testing the equality of measurements from two different analytical methods. *J. Clin. Chem. Clin. Biochem.* **1983**, *21*, 709–720. [PubMed]
36. Cancet, M.; Bijac, S.; Chimot, J.; Bonnefond, P.; Jeansou, E.; Laurain, O.; Lyard, F.; Bronner, E.; Féménias, P. Regional in situ validation of satellite altimeters: Calibration and cross-calibration results at the Corsican sites. *Adv. Space Res.* **2013**, *51*, 1400–1417. [CrossRef]
37. Ménard, Y.; Jeansou, E.; Vincent, P. Calibration of the TOPEX/POSEIDON altimeters at Lampedusa: Additional results at Harvest. *J. Geophys. Res.* **1994**, *99*, 24487. [CrossRef]
38. Vu, P.; Frappart, F.; Darrozes, J.; Marieu, V.; Blarel, F.; Ramillien, G.; Bonnefond, P.; Birol, F.; Blarel, F. Multi-Satellite Altimeter Validation along the French Atlantic Coast in the Southern Bay of Biscay from ERS-2 to SARAL. *Remote Sens.* **2018**, *10*, 93. [CrossRef]
39. Bonnefond, P.; Verron, J.; Aublanc, J.; Babu, K.; Bergé-Nguyen, M.; Cancet, M.; Chaudhary, A.; Crétaux, J.-F.; Frappart, F.; Haines, B.; et al. The Benefits of the Ka-Band as Evidenced from the SARAL/AltiKa Altimetric Mission: Quality Assessment and Unique Characteristics of AltiKa Data. *Remote Sens.* **2018**, *10*, 83. [CrossRef]
40. Liu, Y.; Kerkering, H.; Weisberg, R.H. *Coastal Ocean Observing Systems*; Elsevier (Academic Press): London, UK, 2015.
41. Ulaby, F.; Moore, R.; Fung, A. *Microwave Remote Sensing: Active and Passive. Volume 2, Radar Remote Sensing and Surface Scattering and Emission Theory*; Addison-Wesley: Boston, MA, USA, 1982.
42. Hallikainen, M.; Ulaby, F.; Dobson, M.; El-rayes, M.; Wu, L. Microwave Dielectric Behavior of Wet Soil-Part 1: Empirical Models and Experimental Observations. *IEEE Trans. Geosci. Remote Sens.* **1985**, *GE-23*, 25–34. [CrossRef]

7

On the Desiccation of the South Aral Sea Observed from Spaceborne Missions

Alka Singh [1,*], Ali Behrangi [1,2], Joshua B. Fisher [1] and John T. Reager [1]

1 Jet Propulsion Laboratory, California Institute of Technology, Pasadena, CA 91109, USA; ali.behrangi@jpl.nasa.gov (A.B.); Joshua.B.Fisher@jpl.nasa.gov (J.B.F.); John.Reager@jpl.nasa.gov (J.T.R.); alka.singh@jpl.nasa.gov (A.S.)

2 Department of hydrology and atmospheric sciences, The University of Arizona, Tucson, AZ 85721, USA; behrangi@email.arizona.edu

* Correspondence: alka.singh@jpl.nasa.gov

Abstract: The South Aral Sea has been massively affected by the implementation of a mega-irrigation project in the region, but ground-based observations have monitored the Sea poorly. This study is a comprehensive analysis of the mass balance of the South Aral Sea and its basin, using multiple instruments from ground and space. We estimate lake volume, evaporation from the lake, and the Amu Darya streamflow into the lake using strengths offered by various remote-sensing data. We also diagnose the attribution behind the shrinking of the lake and its possible future fate. Terrestrial water storage (TWS) variations observed by the Gravity Recovery and Climate Experiment (GRACE) mission from the Aral Sea region can approximate water level of the East Aral Sea with good accuracy (1.8% normalized root mean square error (RMSE), and 0.9 correlation) against altimetry observations. Evaporation from the lake is back-calculated by integrating altimetry-based lake volume, in situ streamflow, and Global Precipitation Climatology Project (GPCP) precipitation. Different evapotranspiration (ET) products (Global Land Data Assimilation System (GLDAS), the Water Gap Hydrological Model (WGHM)), and Moderate-Resolution Imaging Spectroradiometer (MODIS) Global Evapotranspiration Project (MOD16) significantly underestimate the evaporation from the lake. However, another MODIS based Priestley-Taylor Jet Propulsion Laboratory (PT-JPL) ET estimate shows remarkably high consistency (0.76 correlation) with our estimate (based on the water-budget equation). Further, streamflow is approximated by integrating lake volume variation, PT-JPL ET, and GPCP datasets. In another approach, the deseasonalized GRACE signal from the Amu Darya basin was also found to approximate streamflow and predict extreme flow into the lake by one or two months. They can be used for water resource management in the Amu Darya delta. The spatiotemporal pattern in the Amu Darya basin shows that terrestrial water storage (TWS) in the central region (predominantly in the primary irrigation belt other than delta) has increased. This increase can be attributed to enhanced infiltration, as ET and vegetation index (i.e., normalized difference vegetation index (NDVI)) from the area has decreased. The additional infiltration might be an indication of worsening of the canal structures and leakage in the area. The study shows how altimetry, optical images, gravimetric and other ancillary observations can collectively help to study the desiccating Aral Sea and its basin. A similar method can be used to explore other desiccating lakes.

Keywords: lake level; lake volume; evaporation; streamflow; Gravity Recovery and Climate Experiment (GRACE); altimetry; Landsat; Aral Sea

---

1. Introduction

Lakes and reservoirs store 87% of the Earth's total fresh open water [1]. Unfortunately, many of them are gradually receding over the years due to climatic or/and anthropogenic forcings [2].

There are several feedbacks between the lake and its environment. Thus, quantifying the changes in the lake is critical to understanding the interactions among various components of the region better. For example, fluctuations in a lake/reservoir can be linked to different climatic changes at a regional scale, including desertification, dust storms, melting of glaciers, and changes in the vegetation and land types. For instance, lake volume loss reduces its heat capacity, and thus it can warm up and cool off faster than before.

One of the examples of massive lake volume reduction is the Aral Sea. In the 1960s, Soviet Russia started the world's second largest irrigation program, under which the Amu Darya and the Syr Darya rivers were diverted across the Karakum desert. The average annual combined pre-irrigation streamflow of the Amu Darya and the Syr Darya into the Aral Sea was 56 km^3 [3], which was reduced to less than 10 km^3 by 2002 [4]. The world's fourth-largest freshwater lake (until the 1960s) eventually separated into two parts, the vast South Aral Sea located in the south and the small North Aral Sea situated in the north. After the construction of Dike Kokaral dam between the north and south part of the Aral Sea, the Syr Darya streamflow has stabilized the North Aral Sea. However, in the past few decades, the perennial Amu Darya has transformed to an intermittent river as it runs through the desert and Khorezm oasis before merging into the South Aral Sea [5]. Consequently, the South Aral Sea continued its journey of desiccation and became a hypersaline and almost non-habitable lake. The Aral Sea is in the lowland climate zone [6], but studies have suggested that it may move towards a monsoon climate [4], which is characterized by seasonal climate change due to warming and cooling of the Aral Sea.

Hydrologic analysis of the Aral Sea and surrounding regions has been an active area of research in the last two decades [7–11]. The present study tries to advance previous studies by quantifying different hydrological parameters of the lake using various remote-sensing datasets that can complement each other. This is an essential step as in situ observations are limited and often not available. A comprehensive analysis of the dynamics of the South Aral Sea is performed using a mass balance equation as an example of how integrated multisensor data can be used to monitor different hydrological variables in an endorheic basin. This study demonstrates a framework of how a lake volume calculated from different remote-sensing data can act as a thermometer of the hydrological state of the basin. Lake volume variations are used to estimate evaporation loss from the lake, and to evaluate existing evaporation products, which are hard to validate otherwise. Furthermore, we estimated runoff from two different methods. Runoff is a product of hydrological processes acting in the watershed. To evaluate the cause of change in runoff, the saptio-temporal changes of the entire basin are analyzed.

2. Study Area

Here we study the South Aral Sea (Figure 1, blue polygon) and the Amu Darya basin (Figure 1, green polygon). The South Aral Sea is a remnant of the vast Aral Sea where the Amu Darya terminates. It has a shallow, broad east lobe and a deep, elongated west lobe, and a narrow channel connects them towards the north. The Amu Darya primarily runs through Turkmenistan and Uzbekistan covering an area of nearly 617,000 km^2. It receives water almost entirely from glaciers in the Pamir Mountains and the Hindu Kush and mainly originates from Tajikistan and northern Afghanistan, forming the border between the two countries. The high Pamir and Tian Shan ranges are significantly wet, creating massive glaciers and snowfields, and the Amu Darya streamflow brings the water to the severely arid southeast of the Aral Sea. The streamflow from these glaciers is heaviest during the spring thaw.

The Amu Darya basin mask is derived from the global river basin database obtained from http://www.wsag.unh.edu/Stn-30/stn-30.html [12,13]. The South Aral Sea land-water mask time series are generated from the multi-temporal Landsat dataset [14].

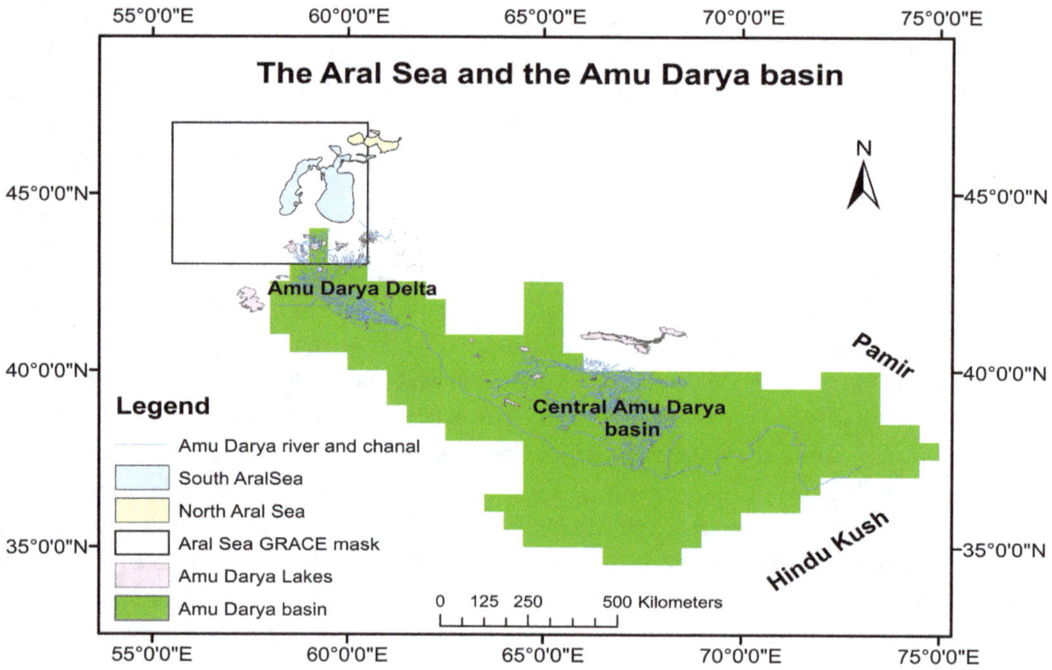

Figure 1. Study area: the Amu Darya originates from the glaciers of Hindukush and Pamir and terminates into the South Aral Sea (blue polygon).

3. Observation and Model Data

3.1. Lake Water Storage (LWS)

Remote-sensing-based lake-water-storage estimation can be done using altimetry and optical images. Since 1992, satellite altimetry is used successfully for monitoring the water levels of large rivers, lakes, and floodplains [15–17]. Satellite altimetry measures the water surface height along its pass based on the reflection from the ground of the emitted radiations from the satellite. Different missions have different pass location, repeat cycle, and footprints. However, there are significant inter-pass gaps, which lead to limited coverage of the terrestrial water bodies. For instance, the East Aral Sea could not have reliable altimetry observations for more than four years (2012–2016) as none of the altimetry mission's pass was over the dry lake. Singh et al. (2015) demonstrated the use of Landsat and bathymetry to obtain the water level, as an alternative to altimetry. In this study, a linear regression between altimetry water level and the Gravity Recovery and Climate Experiment (GRACE) terrestrial water storage (TWS) is examined as another alternative to estimate water level (discussed in Section 4.1).

Many studies intersected the digital elevation model (DEM) with the water level obtained from satellite altimetry to compute water storage changes of a lake/reservoir [18–20]. In this study also, the South Aral Sea volumetric variations are estimated by intersecting the Aral Sea bathymetry with the satellite altimetry water level time-series obtained from the Database for Hydrological Time Series of Inland Waters (DAHITI), developed by the Technische Universität München (http://dahiti.dgfi.tum.de/en/) [21]. Jason-1, Jason-2, and Jason-3, (10 days repeat pass) observe the East Aral Sea, while the West Aral Sea has additional observations from Envisat and Saral/Altika (35 days repeat pass).

Multiple Landsat missions' (Thematic Mapper (TM), Enhanced Thematic Mapper Plus (ETM+), and Operational Land Imager (OLI)) near-infrared (NIR) images are mosaiced and classified to generate land-water mask time series. Landsat bathymetry intersection is used to create water level time-series (Singh et al. 2015).

3.2. *Terrestrial Water Storage (TWS)*

Since 2002, variations in the terrestrial water storage within the Earth's system is routinely observed by GRACE [22–24]. The availability of monthly observations from the time variable GRACE mission has revolutionized the hydrological estimations by providing a possibility for the closure of water budget over a study area. In this study, 0.5°-gridded, JPL mascon solutions are downloaded from https://podaac.jpl.nasa.gov [25–27]. The mascon cell size varies with latitude, and for this region, it is 4° × 5°. Thus the GRACE signal includes mass changes occurring in the large extended area (4° × 5°) as shown in Figure 1 (black box). The study uses GRACE signals from the lake region (Figure 1, black box) to estimate its water level. However, the lake area is less than one-tenth of the area observed by GRACE, but lake volume is a major driver in the mass variation of the region. Additionally, the GRACE signal from the Amu Darya basin is also analyzed to examine the spatiotemporal pattern of the basin and to estimate long-term streamflow into the East Aral Sea. For the analysis of the Amu Darya basin, the GRACE mascon solution is scaled to sub mascon 0.5° resolution by multiplying it with 0.5° gain factor obtained from https://podaac.jpl.nasa.gov. These gain factors derived from the Community Land Model do not have a lake component. Therefore, we can not use the scaled mascon product for the lake analysis, and the mass conservation within a mascon unit is more reliable than in 0.5° sub-mascon level.

3.3. *Vegetation Index*

The global Moderate-Resolution Imaging Spectroradiometer (MODIS) based normalized difference vegetation index (NDVI) product (MYD13C2) is retrieved from the National Aeronautics and Space Administration (NASA) Land Processes Distributed Active Archive Center (LP DAAC) [28].

3.4. *Evapotranspiration (ET)*

Since 2002, the MODIS satellite has been providing a wide range of information about global dynamics at 250 m to 1 km spatial resolution. The spectral signatures captured by its 36 spectral bands are used for many applications such as optical imaging of the Earth, estimation of radiation budget, and calculation of vegetation indices. The study investigated two MODIS derived ET products: MOD16 and Priestley-Taylor Jet Propulsion Laboratory (PT-JPL).

Global monthly 0.5° MOD16 ET product is downloaded from http://ntsg.umt.edu/project/modis/mod16.php [29,30]. The MOD16 algorithm [31] is based on the Penman–Monteith (1965) equation for ET.

The PT-JPL actual evapotranspiration (AET) product [32] is based on the Priestley-Taylor potential evapotranspiration (PET) formulation. To reduce PET to AET, Fisher et al. introduced ecophysiological constraint functions based on atmospheric moisture (vapor pressure deficit and relative humidity) and vegetation indices (NDVI and soil-adjusted vegetation index). The driving equation in the model is the following:

$$AET = ET_s + ET_c + ET_i \tag{1}$$

where ET_s, ET_c, and ET_i are evaporation from the soil, canopy and intercepted water, respectively. Each is calculated explicitly based on relative surface wetness, green canopy fraction, plant temperature constraint, plant moisture constraint and soil moisture constraint. No calibration or site-specific parameters are required for this approach.

In addition to MODIS-derived ET, we have also evaluated ET from global hydrological models (GHMs). Monthly ET estimates from the Water Gap Hydrological Model (WGHM) obtained by personal contact and the National Oceanic, and Atmospheric Administration's (NOAA) Global Land Data Assimilation System (GLDAS) is retrieved from the [33,34].

3.5. *In Situ Data*

The Aral Sea was one of the well-sampled inland water bodies on the planet until the 1980s. However, following the collapse of the USSR, field research into its advanced stages of desiccation has reduced significantly [4]. The Amu Darya monthly streamflow (2003–2010) and the historical annual hydrological data of the lake (1780–2009) are obtained from the http://www.cawater-info.net/ database. The Aral Sea bathymetry (from the 1960s) at 1 m contour spacing is provided by Prof. Renard [35] by personal communication. The East Aral Sea section of the bathymetry is updated to 30 m spatial resolution by Singh et al. [36] by combining Landsat and altimetry, and it is publicly available. Therefore, in this study, the West Aral Sea analysis is done using 1960s bathymetry while the East Aral Sea analysis used the updated bathymetry of Singh et al.

3.6. *Precipitation*

Monthly precipitation estimates are obtained from the latest Global Precipitation Climatology Center (GPCC V6) [37], and Global Precipitation Climatology Project (GPCP V2.3).

All datasets (except streamflow, Landsat, and altimetry) are harmonized to monthly, mm level water height on 0.5° (180° meridian) grid. Landsat-based 30 m resolution masks of the East and West Aral Sea are resampled to a 0.5° grid. The land-water mask time series is interested in the respective monthly evaporation and precipitation data to estimate the volumetric variations in evaporation and precipitation only from the water body.

4. Methods and Results

In this section, we demonstrate how remote-sensing observations are used to quantify the water budget components of the Aral Sea and Amu Darya basin. The retrieval method, challenges, solutions, and results are discussed as follows:

4.1. *South Aral Sea Volume Dynamics*

For hundreds of years (historical records began in the 1780s) until the 1960s, the Aral Sea level fluctuation was less than 5 m (Figure 2a). However, predominantly due to the massive irrigation project in the 1960s, it receded in shape and size dramatically. By 1985, the small North Aral Sea separated from the vast South Aral Sea. The North Aral Sea went through some fluctuations as seen in Figure 2a (red plot) due to many failed dam-building attempts between the two parts of the Aral Sea. It eventually stabilized after the construction of dike Kokaral dam in 2005 (Figure 2a, magenta plot). However, the South Aral Sea continued to shrink and further separated into the shallow East Aral Sea and the deep West Aral Sea by 2003 [9].

The annual water level of the South Aral Sea was estimated by a set of altimetry sensors (i.e., Jason1, Jason2, Envisat and Saral/altika) and compared with the available in situ water level. The results show a good agreement (Figure 2a) with more than 0.99 correlation and ~60 cm RMSE (root mean square error) or 1.8% normalized RMSE (by the mean of the observed data). Figure 2b (red and blue plots) shows that East and West Aral Sea had similar water level until 2009 and they were disconnected for the first time in late 2009. However, the 2010 flood resumed the equipotential status between the two parts of the South Aral Sea for nearly a year. Afterward, they started their independent paths of desiccation. The reason behind their different progression since 2010 has been that the East Aral Sea reached approximately its lake bed later in 2009, so it needs to rise for about 2 m to get reconnected with the West Aral Sea. The West Aral gets water only from the East Aral overflow; there was a 3–4 months lag between the increase in water level in 2010 among the two parts. Meanwhile, the West Aral continuously undergoes evaporation loss, as its lakebed is 13 m below mean sea level (MSL), while the flat East Aral dries up at 27 m.

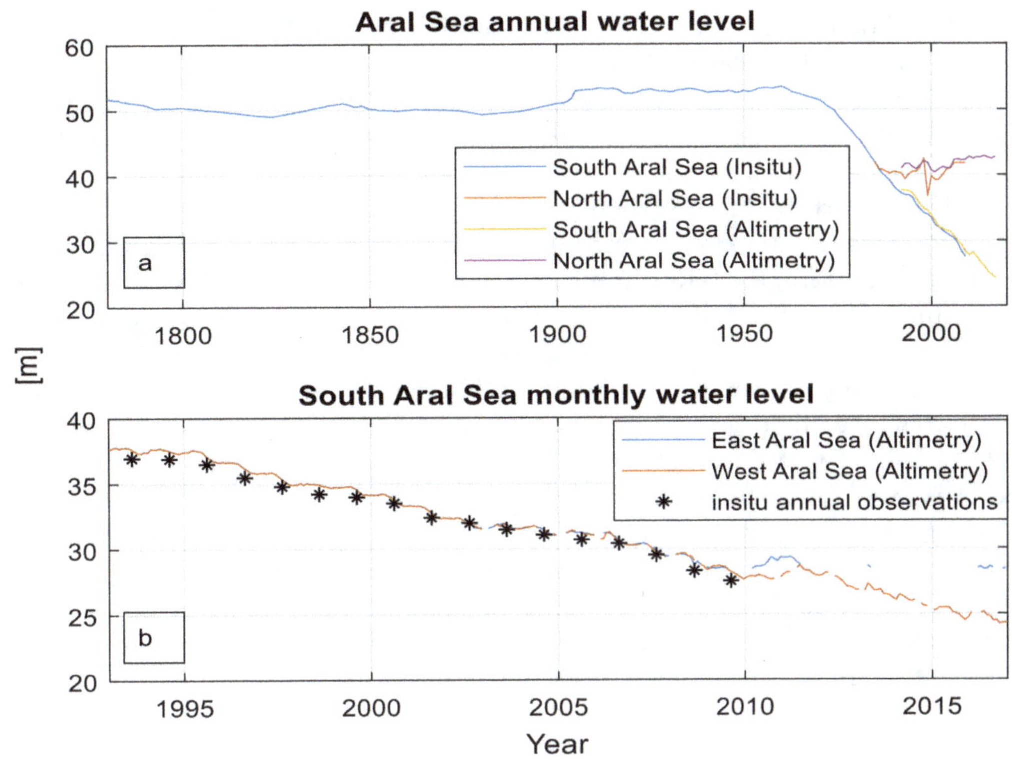

Figure 2. Water level above mean sea level (**a**) historical annual Aral Sea data (**b**) altimetry-based monthly South Aral Sea observations.

Satellite altimetry missions have continuously observed the West Aral Sea. However, the East Aral Sea lacks reliable observations after its water level went below 28.5 m (Figure 2b, red plot) because none of the altimetry missions has passed through the dry lake. Therefore, other remote-sensing observations have to be explored to explain the missing water level for nearly four years from 2011. Singh et al. (2015) demonstrated that Landsat together with bathymetry can provide 0.97 correlation with the altimetry water level and can act as a good alternative. However, optical images fail in bad weather (dust storm, and cloud). Consequently, observations from Landsat are not available for many months, especially during winters over the Aral Sea (Figure 3b). Therefore, we examined another potential way to fill the gaps by linear regression between GRACE-based terrestrial water storage (TWS) from the Aral Sea region (Figure 1). Study area black box) and the East Aral Sea water level from altimetry. First, seasonal components from the TWS and water level time series are removed. Then deseasonalized TWS and water level (Figure 3a) are linearly regressed using nine years (2003–2011) of monthly data. The best fit model ($r^2 = 0.82$, RMSE = 49 cm) is used for the estimation of the deseasonalized water level for the 2003–2017 time frame, and then the altimetry-based seasonal component is added. The derived GRACE-based water level and altimetry observations showed 0.9 correlation and ~54 cm RMSE or 1.8% normalized RMSE. Figure 3b shows that GRACE-based estimates agreed well with the altimetry observations except for the 2010 flood. The GRACE-based estimate observed an early peak in spring 2010, while altimetry saw it in summer 2010. It can be attributed to the integral nature of the GRACE signal, which observes mass changes not only from the surface water but also from the soil moisture and groundwater.

Although the TWS variability in this mascon also includes signals from the West Aral Sea, it obtains water only from the East Aral Sea overflow. Therefore, the empirical relation between the increase in the TWS and water level of the East Aral Sea remains valid. The East Aral acquires a similar water surface area as the West Aral as soon as it floods due to its flat topography and eventually has similar evaporation loss progression. However, the impact of consecutive evaporation loss from the West Aral on TWS may lead to the overestimation of the water-level drop, but it soon reaches the cutoff as the East Aral dries up.

Therefore, the GRACE-derived sea level for the East Aral Sea stops at 27 m, as in late 2014, and it resumes with an increase in the TWS of the lake. Whenever the two parts of the South Aral are connected, then they have mostly similar progression but different water levels since 2010.

However, due to the limited spatial resolution of GRACE, the GRACE mask (Figure 1, black box) extends beyond the lake. Spring 2010 was relatively wet, (which increased the soil moisture in the entire 4° × 5° box and the lake size was at its diminished level. It seems that whenever snow accumulation or soil moisture variation is more than lake mass change, TWS has less ability to estimate the water level. Furthermore, the GRACE mask also includes the Amu Darya delta, which absorbs a significant amount of the floodwater before it reaches the South Aral Sea (discussed in Section 4.2). Figure 3b also shows that in spring 2009, GRACE-based estimates were nearly 29.5 m, which cannot be correct because there is no altimetry observation for that period, suggesting that water level had to be below 28.5 m. In this case, the time series is interpolated to fill the data gaps.

Nevertheless, for rest of the time-series, TWS in the GRACE box is driven by the lake mass variation, and thus the GRACE-based water level shows good agreement with altimetry observations.

Figure 3b shows the complete disappearance of the East Aral Sea in 2014, consistent with the analysis of the Landsat data (Singh et al. 2016). The East Aral Sea bed is approximately 27 m above MSL, and when its water level reaches 28.5 m from MSL, the lake expands enough to be observed by altimetry missions. Therefore, to obtain a complete time series, the water level is estimated from a combination of altimetry (when the water level is above 28.5 m) and Landsat (when the water level is below 28.5 m). GRACE-based estimates are used when neither altimetry nor Landsat could is available.

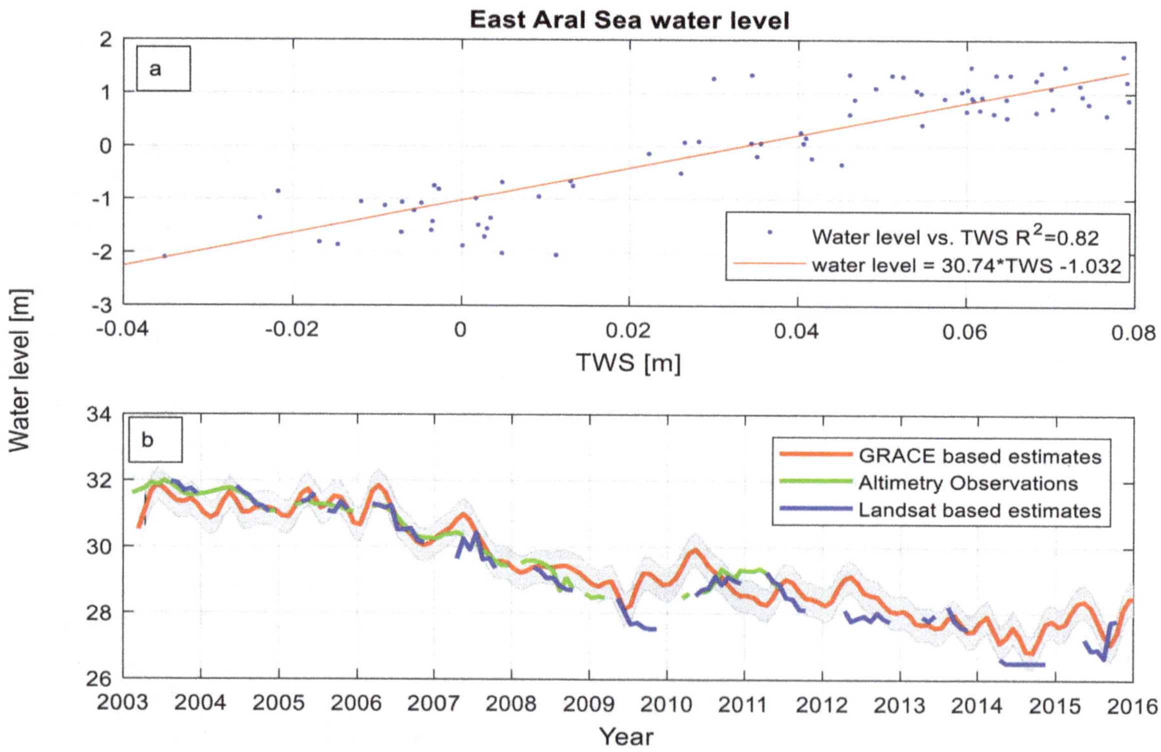

Figure 3. The East Aral Sea water level. (**a**) Best fit between de-seasonalized terrestrial water storage (TWS) and altimetry water level; and (**b**) altimetry water level observations compared with the derived water height from Landsat and Gravity Recovery and Climate Experiment (GRACE). The gray area shows the uncertainty range of the GRACE-based estimate, calculated by the ± root mean square error (RMSE).

In this study, we have three independently generated datasets of lake structure. Lake level is estimated by satellite altimetry, lake water-surface-area or contours are estimated by the Landsat imageries and bathymetry of the lake derived from 1960s field observations. A combination of any of the two can calculate the third, for example, the altimetry and Landsat/MODIS combination can generate the bathymetry of a lake [36,40] or estimate lake volume from [18,41]. In this work, the bathymetry of the East and West Aral Sea are intersected with their respective water levels to obtain water surface area time-series due to data gaps in the Landsat observations (Figure 3b). Volumetric variations of the lakes are calculated using a truncated pyramid model (Equation (2)) by integrating the change in water level and surface area [18].

$$VV(t) = \sum_{t=1}^{n} \frac{1}{3} \times (H_t - H_{t-1}) \times \left(A_t + A_{t-1} + \sqrt{(A_t \times A_{t-1})}\right) \quad (2)$$

where,

$VV(t)$ = Volumetric variations with respect to the initial state (t_0) at the nth month
A_t = Area of the water extent at month t
A_{t-1} = Area of the water extent at the previous month
H_t = Level of the water body at month t
H_{t-1} = Level of the water body at the previous month.
n = Number of months.

Between 1993 and 2017, the South Aral Sea lost approximately 195 km^3 water (Figure 4a). Since 2009, the East Aral Sea has been fluctuating on its almost flat lakebed. Eventually, most of the streamflow evaporates within a year. The West Aral Sea seasonally receives some water when the East Aral Sea water level exceeds 28.5 m above MSL. However, due to evaporation loss, the West Aral Sea was still losing water at approx. 2.7 km^3/annum rate and reduced below 25 m by late 2016 (Figure 2b). Considering the geometry of the West Aral Sea (Figure 4b), nearly 43 km^3 of its volume might still exist below the 25 m water level. If the current trend of water loss (almost 2.7 km^3/year) continues, then the West Aral Sea might disappear by nearly 2032. The seasonal expansion and shrinking of the East Aral Sea are neither enough to revive the east part nor maintain the West Aral Sea due to massive evaporation loss. The South Aral Sea dynamics are driven by the Amu Darya streamflow and evaporation loss from the lake. Therefore, we further explored these two parameters.

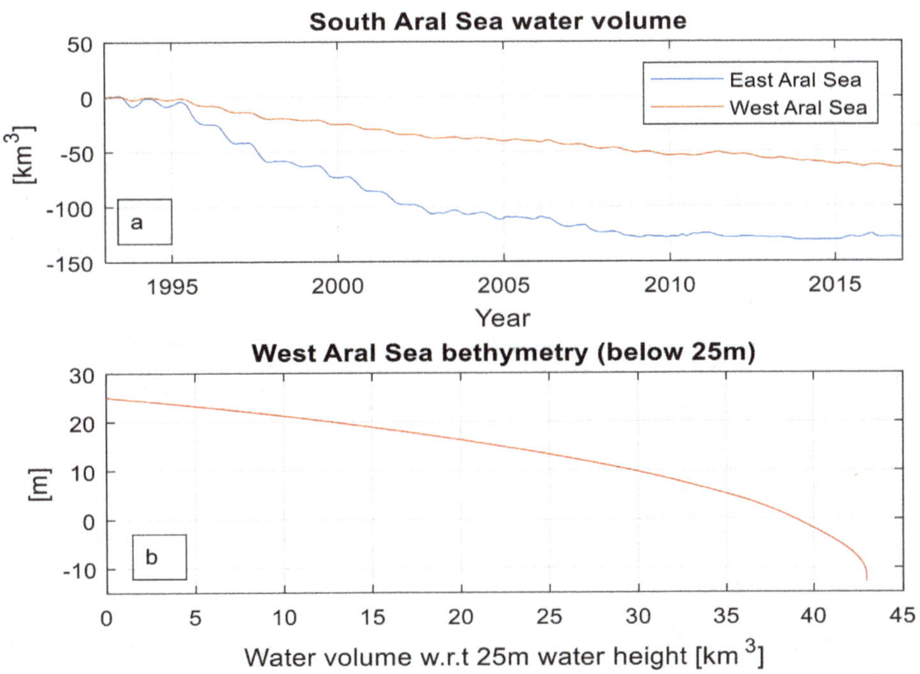

Figure 4. (**a**) the South Aral Sea volume variations and (**b**) the West Aral Sea bathymetry below 25 m MSL.

4.2. Evaporation from the South Aral Sea

The Aral Sea is a terminal inland water body, and evaporation is the only major water outlet. The annual evaporation rate of the lake was nearly 66 km^3 [9] before the 1960s. The historical vertical flux data from the Aral Sea shows a substantial difference between the total annual precipitation and evaporation to/from the waterbody (Figure 5a). However, the recent annual fluxes from the lake shown in Figure 5b demonstrate that MOD16 ET is almost equal to GPCP precipitation. It cannot be realistic, knowing the fact that the lake is shrinking and it has an additional Amu Darya streamflow into it.

Figure 5b shows that the difference between annual AET and PET has decreased due to accelerated AET with the decreasing lake size. The positive feedback between sea surface temperature (SST) and evaporation can explain the increase in the rate of AET from the Aral Sea. As the lake loses water, it not only reduces water surface area but also becomes shallower. Shallow water like the shrunken East Aral Sea (especially since 2009) heats up faster than deep water as it has less volume per square area. With decreasing lake area, specific humidity near the lake surface is decreased, and thus the rate of evaporation is increased. Another distinguishing factor of the East Aral Sea is its increasing salinity, which is above 130 ppm [10]. The salinization of the lake leads to vertical stratification, which is characterized by a rapid change in water temperature and salinity level at a given horizontal or vertical region. Consequently, the surface of the lake heats up faster as the salt concentration is not distributed evenly but stratified from lower salt concentration at the surface to the highest at the bottom of the lake [42].

Considering the significance of the evaporation for the water balance of the Aral Sea, we compared potential ET from PT-JPL, and actual ET estimates from WGHM, GLDAS, MOD16, and PT-JPL for the South Aral Sea.

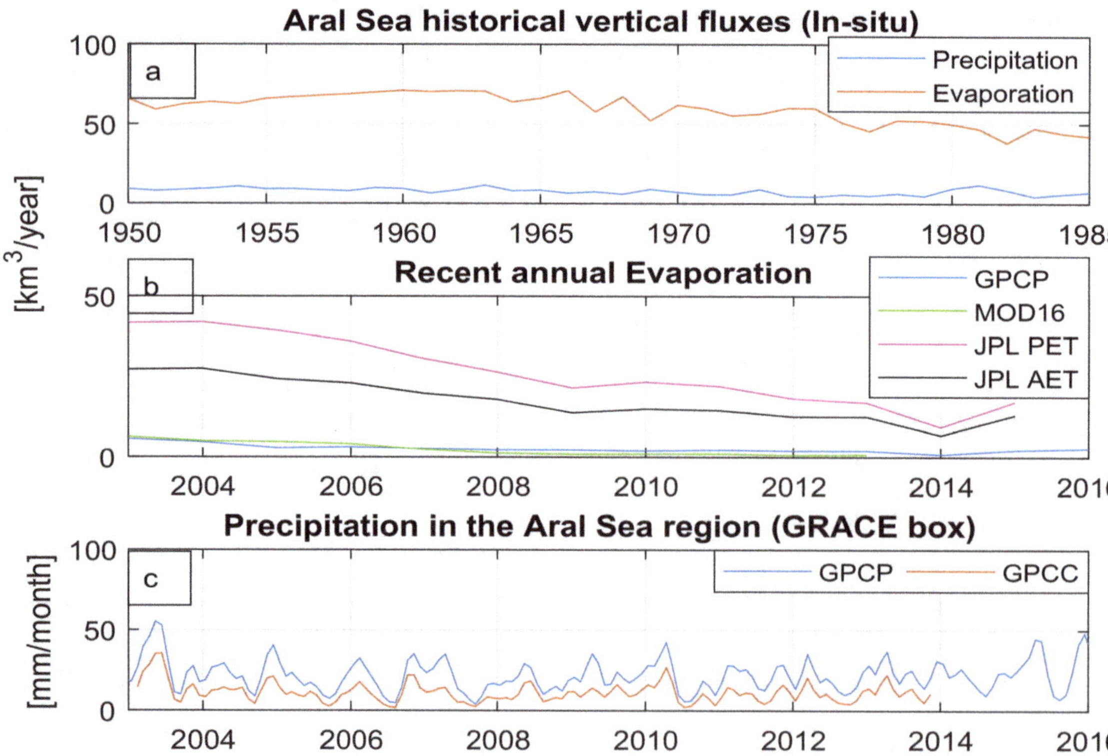

Figure 5. Time series of the South Aral Sea precipitation and evaporation; (**a**) historical annual in-situ observations; (**b**) recent annual evapotranspiration (ET) and Global Precipitation Climatology Project (GPCP) observations; and (**c**) monthly precipitation observations over the lake.

Evaporation from open water is considerably different from land. Terrestrial open water evaporation dynamics needs to be added in the current evaporation estimates taking into account the land-water conversion. The presence of a lake and its size have a significant impact on its surrounding climate. Total evaporation decreases with the reducing size of the lake, but most of the ET models do not use a dynamic water mask. Consequently, the impact of available water in the system can be miscalculated. In this study, the time-variable Landsat-based water-surface masks are intersected with all the ET estimates to calculate volumetric evaporation loss from the waterbody.

As an independent estimate of evaporation, we back-calculated evaporation from the lake using lake volume dynamics (Figure 4a) and based on the water balance equation (Equation (3)):

$$\mathrm{BCE} = \Delta(VV(t)) + R - P \tag{3}$$

where,

BCE = Back-calculated evaporation from the lake (magenta plot in Figure 6)
$\Delta(VV(t))$ = Diffrential of the lake volume (calculated by Equation (2)) with respect to its previous month
R = Amu Darya streamflow into the South Aral Sea
P = Precipitation

This alternative and independent approach to calculate the lake evaporation provides an additional tool for evaluating the evaporation products. However, groundwater infiltration into the lake has been ignored in this estimation due to non-availability of the data. The lake volume estimates have limited uncertainty compared to other hydrological parameters because of the well-established water level estimation methods from satellite altimetry with cm-level accuracy [21]. However, the bathymetry of the lake is more than half a century old, and it has experienced some restructuring due to floods [18]. This may introduce an unknown uncertainty (likely to be small) in the estimation of the lake volume. Nevertheless, the estimated lake volume variations can be considered as a strong constraint in the water balance equation.

Additional uncertainty in back-calculated evaporation (BCE) comes from the lack of the exact coordinate of the streamflow gauge, which measures Amu Darya streamflow into the Aral Sea. The Amu Darya terminates in a massive delta with many small lakes. During the low/normal flow, water reaches directly into the East Aral Sea. However, during heavy streamflow (2005 and 2010) a significant amount of water is consumed in the delta region. Therefore, except during the flood months, the BCE can be considered as the nearest approximation of the evaporation loss from the South Aral Sea. The BCE estimates (Figure 6, magenta plot) are limited by the availability of the streamflow data (until 2010).

Precipitation is relatively a smaller direct input in the region with 80–200 mm annual range [43]. GPCP observations are closer to this range compared to GPCC (Figure 5c). Therefore, for the back-calculation of evaporation from the lake (Equation (3)), GPCP is used. Furthermore, GPCP utilizes the nearest rain-gauge stations to reduce potential biases. Figure 6 shows that most of the products severely underestimate ET compared to the BCE estimation. The PT-JPL ET, however, indicates notably different and reasonable agreement with the BCE.

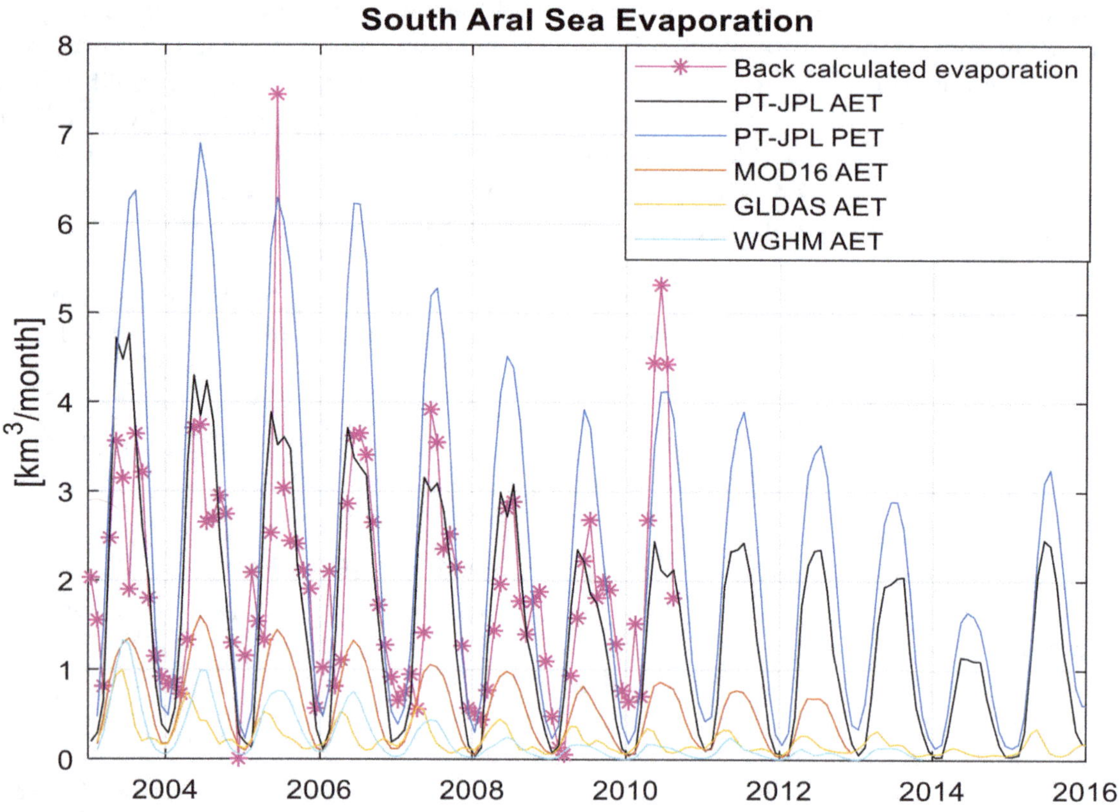

Figure 6. Evaporation from the South Aral Sea back-calculated evaporation (BCE) compared with the other ET products.

The BCE shows an even better agreement with the PT-JPL AET given a month's time lag because the BCE is estimated based on changes in the lake volume compared to its previous month. Except for the flooding months, the PT-JPL AET shows 0.83 correlation with the BCE and 0.76 cubic km RMSE (42% NRMSE). Theoretically, AET should be similar to PET over open water, but we found that they differ because a lot of the radiation/energy to drive ET gets absorbed deep into the water without actually ever evaporating any water. Therefore, the actual ET ends up being less than the potential. The MOD16 and GHMs ET showed very low seasonality in the lake volume estimation. Thus, our analysis indicates that PT-JPL ET is more consistent with the rest of the water cycle component for the evaluation of evaporation over the open water. PT-JPL has been widely and independently assessed as the top-performing global ET remote-sensing algorithm for evaporation [44–51]. However, this paper is the first to compare it with other ET products over open water. This shows that PT-JPL probably has a good representation of surface-atmosphere moisture coupling.

4.3. Amu Darya Streamflow into the Lake

As global lakes are not always well instrumented, this analysis provides insights on how well streamflow might be estimated using other water-cycle components. The study attempted to determine the Amu Darya streamflow into the South Aral Sea by two independent methods.

- A water balance-based streamflow estimate (R1, Figure 7b, green plot) is generated by combining PT-JPL ET (assuming it as actual evaporation from the lake), GPCP and South Aral Sea volumetric variations (Equation (4)). The average annual Amu Darya streamflow into the lake (except 2005 and 2010 flow) ranges between 0–1 km^3/month while the accumulated error from different datasets in Equation (4) is more than one km^3/month. Consequently, accurate estimation of the streamflow is not possible with this method. Therefore, three-monthly weighted-average (3MWA) by 0.25, 0.5, 0.25 weights, is calculated to obtain a long-term trend of the streamflow into the lake. The derived estimate (R1, Figure 7b) showed 0.71 correlation with the in situ 3MWA streamflow.

$$R1 = 3MWA\ (\Delta(VV) - P + ET) \tag{4}$$

where,

R1 = Streamflow estimated from lake water budget (green plot in Figure 7b)
3MWA = three-monthly weighted-average
ET = Evaporation from the lake (PT-JPL ET) and P = Precipitation (GPCP)

- Second streamflow (R2, Figure 7b, red plot) is calculated from the deseasonalized GRACE signal obtained from the Amu Darya basin (DGADB) (Figure 1, green polygon). An empirical relation between 3MWA of the in-situ Amu Darya streamflow and 3MWA of the DGADB is used to generate GRACE-based streamflow (R2). The Least-absolute-residuals method based two-degree polynomial curve showed a good agreement ($r^2 = 0.94$ and RMSE = 0.2 km^3) between the two. The derived curve (R2, Figure 7b) showed 0.68 correlation with the in situ 3MWA streamflow.

Furthermore, the dashed vertical lines in Figure 7a shows that the DGADB observes flood peaks into the lake about two months earlier than the in situ data (compare red and blue dashed lines). Reager et al. [52] also demonstrated that basin-scale TWS could be used to characterize regional flood potential with longer lead times in flood warnings. This two months' early warning of heavy streamflow into the Amu Darya delta could be useful for the water resource management of the region.

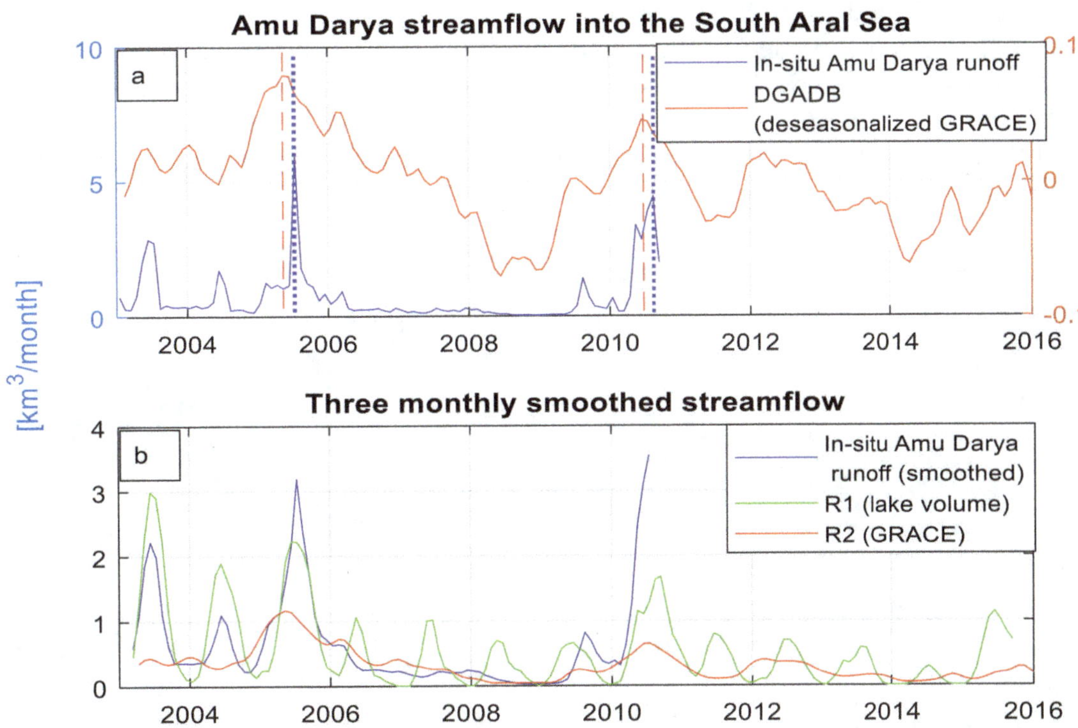

Figure 7. The Amu Darya streamflow. (**a**) Deseasonalized GRACE signal from the Amu Darya basin (DGADB) compared with the in-situ streamflow. The red vertical lines are peaks of DGADB, and the blue vertical lines are peaks of in situ streamflow; (**b**) three monthly-smoothed Amu Darya streamflow observed by in situ data compared with the two derived estimates (R1 and R2).

As discussed above, the streamflow during the flood events (2005 and 2010) was significantly absorbed in the massive Amu Darya delta. Therefore, the estimated R1 and R2 have a relatively smaller peak in 2005 and 2010 than in situ streamflow. Figure 7b suggests that R1 and R2 provide only a general idea of the long-term trend and are likely to be far from true streamflow. However, they have captured well the extreme weather events in the long-term patterns of the streamflow into the lake, i.e., 2005 and 2010 floods and 2009 and 2014 droughts. R2 produces low flow reasonably well (except in 2004), but underestimates peak flow, yet the pattern of the peak flow is captured. R1 is relatively skillful in capturing peak flows but overall provides many false signals not observed by either in situ observation or GRACE. The GRACE signal (DGADB, Figure 7a red plot) indicates that 2015 and 2016 had high seasonality, but they do not exceed 2005/2010 peak flows or 2009/2014 low flows.

4.4. The Amu Darya Basin

Figure 8 shows the spatiotemporal linear trend in the Amu Darya basin between 2003 and 2017 of (a) TWS from GRACE, (b) precipitation from GPCP, (c) MODIS NDVI and (d) ET from MOD16. The figure shows that during this period TWS went down in the Amu Darya delta and the southwestern Hindukush region. The decrease in the delta TWS can be attributed to the increasing absorption of water in the central and northeastern part of the Amu Darya basin. This area is a major irrigation belt (other than a delta region) of the basin. The figure shows that in the irrigation area water mass increased (Figure 8a), ET decreased (Figure 8d), and the precipitation trend remained relatively flat (Figure 8b), which indicates a possible increase in infiltration. Furthermore, a decrease in total annual Amu Darya streamflow into the Aral Sea from more than 10 km^3 in 2002 to less than 4 km^3 by 2009 shows that either precipitation has gone down during those years (which is not seen in GPCP dataset) or abstraction of water has increased upstream. On the other hand, the reduction of NDVI in the central part of the basin (Figure 8c) supports the assumption of increased infiltration loss. However, vegetation loss can also be attributed to increasing salt deposition in the soil profiles and desertification [8]. Forkutsa et al. [53] discussed the fact that more than 40% of the river water is lost to evaporation and infiltration from the canal. Water escapes the route to form lakes and ponds along the way. Consequently, the rise in groundwater level has brought soil salt to the surface, leading to widespread salinization [9]. It is probable that during this time frame the canal conditions are deteriorated, leading to more infiltration loss.

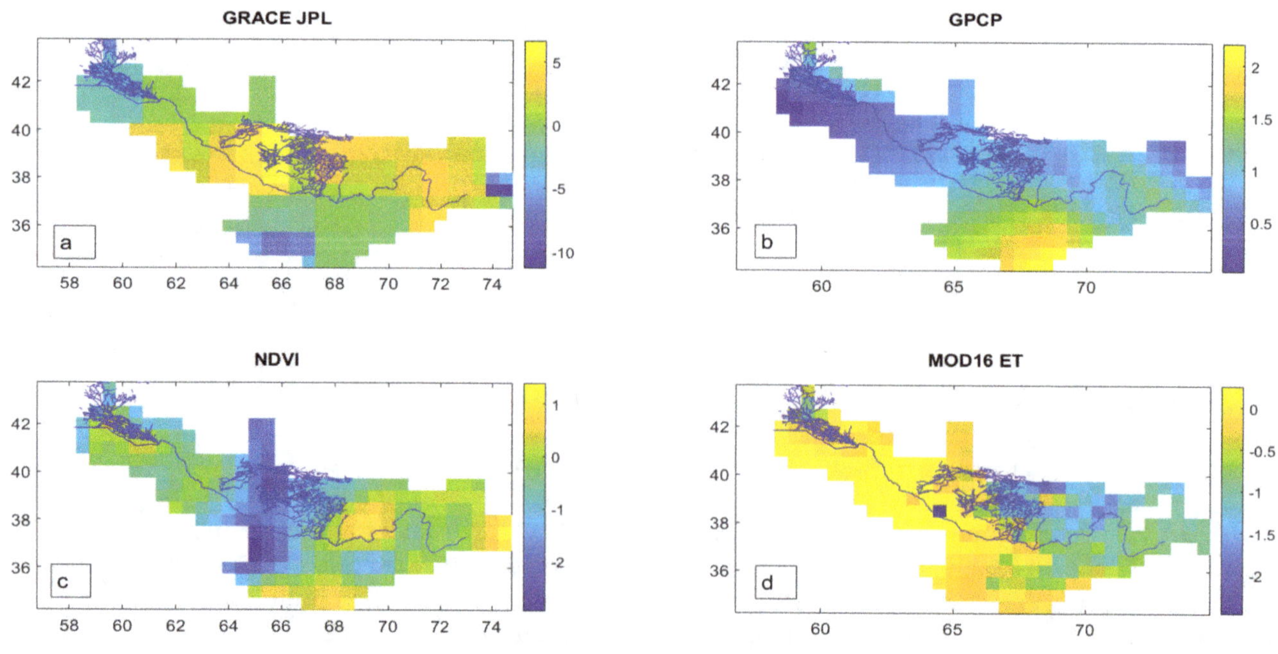

Figure 8. The Amu Darya basin: (**a**) GRACE TWS; (**b**) GPCP; (**c**) normalized difference vegetation index (NDVI); and (**d**) MODIS based ET (MOD16).

5. Discussion

Accurate monitoring of the world's lakes in a changing climate is increasingly essential, as lakes often significantly contribute to regional climatology and ecosystems. Several lakes are drying (e.g., Aral Sea, Urmia, Lake Chad) due to climate change and anthropogenic activities (e.g., through water abstractions for irrigation). These not only lead to loss of Lake Habitat and biodiversity but also can trigger several adverse impacts on the environment and life in the surrounding region, like desertification, dust storms, and the salinization of land. All these processes can eventually offset economic growth near the lakes. Similar processes can be seen in various global lakes like Lake Urmia [54,55], Lake Chad [56,57], the Dead Sea [58–60], and Lake Poopo [61,62]. Therefore, careful monitoring of the changes in lake/inland sea properties is becoming critical more than ever. Unfortunately, many of these lakes are not well instrumented. Thus exploiting the potential of integrating multi-dimensional remote-sensing data with other complementary data is essential. The following are the highlights of this research:

1. Lake level estimate: this paper suggests methods for filling gaps in the altimetry observations. These data gaps may occur due to intermission time lag or loss of altimetry ground track due to changes in the shape of the water bodies. Landsat images together with bathymetry can provide an alternative water level estimate. However, sometimes, optical images have limitations during lousy weather. In that case, GRACE signals from lakes like the Aral Sea have a potential to estimate water level. The linear regression between the TWS and water level has been explored to generate the water level from GRACE.
2. The rate of evaporation loss: most of the models/data products do not estimate evapotranspiration (ET) from inland waterbodies well, except for one. We have back-calculated the lake evaporation (BCE) by integrating altimetry-based lake volume variations, with the in situ runoff and GPCP precipitation. This study found the PT-JPL ET estimate to have the closest approximation to the BCE compared to the other existing ET products MODIS (MOD16) and hydrological models (WGHM and GLDAS). While PT-JPL has never been tested over open water bodies, our findings are consistent with multiple studies that have consistently found PT-JPL to be the top-performing ET remote-sensing algorithm over terrestrial vegetation [44,45,48,49,51,63].
3. Estimating river streamflow to the lake: the study also suggests that the GRACE signal from the Amu Darya basin can provide a long-term trend of streamflow into the lake and may predict flood events one or two months in advance. Another streamflow is estimated based on the lake water budget, which showed a good long-term progression but has some false highs. The back-calculated streamflow (R1) indicated strikingly high seasonality, which demonstrates possible seasonal groundwater infiltration into the lake, assuming error in other datasets are not seasonally biased. Nevertheless, in the absence of any in-situ streamflow, these methods can be explored.
4. Assessing the spatiotemporal variations in the water cycle of the Amu Darya basin: finally, we monitored the spatial changes of the Amu Darya basin to examine the cause of reducing streamflow. Various insights could be gained through analyzing the maps of a temporal trend in ET, TWS, NDVI, and Precipitation. The decrease in TWS in the Amu Darya delta region is mainly due to the increase in water mass in the central part of the Amu Darya basin, which is probably due to rising infiltration with the worsening of the canal system. This assumption cannot be validated due to lack of ground-based observations but is supported by the decrease in ET and NDVI in the region with the increase in TWS.
5. Future of the Aral Sea: the low Amu Darya streamflow and huge evaporation loss from the vast open body have endangered the existence of the South Aral Sea. If the present trend continues, the remnant West Aral Sea will also disappear by nearly 2032 or reach the level of its base flow. One possible solution is to drain the Amu Darya streamflow directly into the West Aral Sea to avoid evaporation loss from the vast shallow East Aral Sea. Assuming 4 km^3/year water flows

into the West Aral Sea based on the current the annual Amu Darya streamflow (without any flood), the West Aral Sea will start increasing at a rate of more than 1 km^3/year. Additionally, a dam is also required to be built between the East and West Aral Sea to stop flooding from the west when it reaches more than 28 m above MSL.

6. Concluding Remarks

In this paper, we show how the combination of various remote-sensing products can help gain more information on the fate of the Aral Sea lake. A framework that can be extended to study other endorheic lakes. The Aral Sea was well instrumented in the Soviet era, after which it gradually lost ground-based stations. However, some in situ records for streamflow to the lake (until 2010), and lake bathymetry (from the 1960s) are available. These enable us to assess current remote-sensing potential and determine challenging areas for further improvement. Our study over the Aral Sea mainly consists of using remote-sensing data to quantify changes in the lake level and volume, the rate of evaporation loss, estimating river streamflow to the lake, and assessing the spatiotemporal variations in the water cycle of both the lake and the Amu Darya basin. The study demonstrated that in addition to altimetry, optical and GRACE data can help to estimate the lake level. Also, GRACE may predict flood events one or two months in advance. Furthermore, lake evaporation was calculated using runoff, altimetry, and precipitation. In comparison to the other global ET products (for example WGHM, GLDAS, MOD16), PT-JPL ET showed the best approximation to the estimated ET for the open water. The Aral Sea basin analysis by the GRACE and other datasets suggests a rise in infiltration, potentially due to worsening of the canals.

Efforts are ongoing to monitor other world's inland lake. Given the lack of quality in situ observations, remote sensing has shown great potential. However, several areas require further improvement by the community, some of which are highlighted below:

- Higher spatial resolution GRACE signals can improve its application tremendously by reducing the impact of contributions from other hydrological compartments.
- Evaporation estimates from the waterbodies need to be better estimated. The lake's volume variations and its salinity need to be incorporated in the models.
- With the recent operation of Global Precipitation Measurements (GPM) and Soil Moisture Active Passive (SMAP) missions, precipitation, soil moisture is expected to be monitored better than before. The role of new observations in studies like that presented here needs to be further investigated.
- The upcoming Surface Water and Ocean Topography (SWOT) mission is expected to provide volumetric variations of most of the inland water bodies because of its wide swath altimetry. This can potentially advance water balance studies such as that investigated in this work.
- By increasing confidence in the quality of surface/sub-surface estimates (surface water and soil moisture), the role of groundwater dynamics can be better explored from GRACE.

The study demonstrated the potential of deriving some proxy estimates of the missing or inadequate hydrological variables by combining multi-dimensional, multi-sensor and multi-mission earth observation datasets to undertake a comprehensive analysis of the hydrological state of a basin.

Author Contributions: The study was conceptualized and investigated by A.S. She prepared the original draft and took care of modifications. A.B. is the principal investigator of the project. He consistently improved the work by important guidance, ideas, and effective discussions. J.B.F. contributed to the paper by providing the PT-JPL data and its analysis. A.B. and J.B.F. helped in the improvement of the original draft by thorough revisions. J.T.R. and A.B. provided funding for the project. All the co-authors assisted in the finalization of the manuscript.

Acknowledgments: Authors acknowledge different agencies for proving various data, including NASA Physical Oceanography Distributed Active Archive Center (PO.DAAC) for GRACE TWS, NASA Land Process Distributed Active Archive Center (LP DAAC) for NDVI, Numerical Terradynamic Simulation Group for MOD16, Cawater group for the Aral Sea in-situ data, DAHITI for altimetry data, NASA Goddard Earth Sciences Data and Information Services Center (GES DISC) for GLDAS ET, NOAA's Earth System Research Laboratory for GPCP and GPCC data. The authors are thankful to Prof. Andreas Guentner (GFZ, Potsdam) for proving WGHM data. Gregory Halverson processed the PT-JPL ET data. The research described in this paper was carried out at the Jet Propulsion Laboratory, California Institute of Technology, under a contract with the National Aeronautics and Space Administration. Financial support was also made available from NASA GRACE, and GRACE-FO (NNH15ZDA001NGRACE) and NASA Energy and Water Cycle Study, (NH13ZDA001N-NEWS) awards; J.B.F. was also supported by NASA SUSMAP. Government sponsorship is acknowledged.

References

1. Gleick, P.H.; Pacific Institute for Studies in Development, Environment; Security, Stockholm Environment Institute. *Water in Crisis: A Guide to the World's Fresh Water Resources*; Oxford University Press: New York, NY, USA, 1993; ISBN 978-0-19-507627-1.
2. Nicholson, S.E. Historical Fluctuations of Lake Victoria and Other Lakes in the Northern Rift Valley of East Africa. In *Environmental Change and Response in East African Lakes*; Lehman, J.T., Ed.; Springer Netherlands: Dordrecht, The Netherlands, 1998; Volume 79, pp. 7–35. ISBN 978-90-481-5043-4.
3. Bortnik, V.N.; Chistyaeva, S.P. Hydrometeorology and Hydrochemistry of the USSR Seas. *Aral Sea Leningr. Gidrometeoizdat* **1990**, *VII*, 196.
4. Zavialov, P.O. *Physical Oceanography of the Dying Aral Sea*; Springer Science & Business Media: Berlin, Germany, 2005; ISBN 978-3-540-22891-2.
5. Dodson, J.; Betts, A.V.G.; Amirov, S.S.; Yagodin, V.N. The nature of fluctuating lakes in the southern Amu-dar'ya delta. *Palaeogeogr. Palaeoclimatol. Palaeoecol.* **2015**, *437*, 63–73. [CrossRef]
6. Micklin, P.P. The Water Management Crisis in Soviet Central Asia. *Carl Beck Pap. Russ. East Eur. Stud.* **1991**, 131. [CrossRef]
7. Asokan, S.M.; Rogberg, P.; Bring, A.; Jarsjö, J.; Destouni, G. Climate model performance and change projection for freshwater fluxes: Comparison for irrigated areas in Central and South Asia. *J. Hydrol. Reg. Stud.* **2016**, *5*, 48–65. [CrossRef]
8. Bosch, K.; Erdinger, L.; Ingel, F.; Khussainova, S.; Utegenova, E.; Bresgen, N.; Eckl, P.M. Evaluation of the toxicological properties of ground- and surface-water samples from the Aral Sea Basin. *Sci. Total Environ.* **2007**, *374*, 43–50. [CrossRef] [PubMed]
9. Micklin, P. Introduction to the Aral Sea and Its Region. In *The Aral Sea*; Springer Earth System Sciences; Springer: Berlin/Heidelberg, Germany, 2014; pp. 15–40. ISBN 978-3-642-02355-2.
10. Roget, E.; Khimchenko, E.; Forcat, F.; Zavialov, P. The internal seiche field in the changing South Aral Sea (2006–2013). *Hydrol. Earth Syst. Sci.* **2017**, *21*, 1093–1105. [CrossRef]
11. Singh, A.; Seitz, F.; Eicker, A.; Güntner, A. Water Budget Analysis within the Surrounding of Prominent Lakes and Reservoirs from Multi-Sensor Earth Observation Data and Hydrological Models: Case Studies of the Aral Sea and Lake Mead. *Remote Sens.* **2016**, *8*, 953. [CrossRef]
12. Vörösmarty, C.J.; Fekete, B.M.; Meybeck, M.; Lammers, R.B. Geomorphometric attributes of the global system of rivers at 30-minute spatial resolution. *J. Hydrol.* **2000**, *237*, 17–39. [CrossRef]
13. Fekete, B.M.; Vörösmarty, C.J.; Lammers, R.B. Scaling gridded river networks for macroscale hydrology: Development, analysis, and control of error. *Water Resour. Res.* **2001**, *37*, 1955–1967. [CrossRef]
14. Singh, A.; Seitz, F.; Schwatke, C. Application of Multi-Sensor Satellite Data to Observe Water Storage Variations. *IEEE J. Sel. Top. Appl. Earth Obs. Remote Sens.* **2013**, *6*, 1502–1508. [CrossRef]
15. Birkett, C.; Murtugudde, R.; Allan, T. Indian Ocean Climate event brings floods to East Africa's lakes and the Sudd Marsh. *Geophys. Res. Lett.* **1999**, *26*, 1031–1034. [CrossRef]
16. Crétaux, J.-F.; Biancamaria, S.; Arsen, A.; Bergé-Nguyen, M.; Becker, M. Global surveys of reservoirs and lakes from satellites and regional application to the Syrdarya river basin. *Environ. Res. Lett.* **2015**, *10*, 15002. [CrossRef]
17. Kleinherenbrink, M.; Ditmar, P.G.; Lindenbergh, R.C. Retracking Cryosat data in the SARIn mode and robust lake level extraction. *Remote Sens. Environ.* **2014**, *152*, 38–50. [CrossRef]
18. Singh, A.; Kumar, U.; Seitz, F. Remote Sensing of Storage Fluctuations of Poorly Gauged Reservoirs and State Space Model (SSM)-Based Estimation. *Remote Sens.* **2015**, *7*, 17113–17134. [CrossRef]
19. Medina, C.E.; Gomez-Enri, J.; Alonso, J.J.; Villares, P. Water level fluctuations derived from ENVISAT Radar Altimeter (RA-2) and in-situ measurements in a subtropical waterbody: Lake Izabal (Guatemala). *Remote Sens. Environ.* **2008**, *112*, 3604–3617. [CrossRef]

20. Crétaux, J.-F.; Abarca-del-Río, R.; Bergé-Nguyen, M.; Arsen, A.; Drolon, V.; Clos, G.; Maisongrande, P. Lake Volume Monitoring from Space. *Surv. Geophys.* **2016**, *37*, 269–305. [CrossRef]
21. Schwatke, C.; Dettmering, D.; Bosch, W.; Seitz, F. DAHITI—An innovative approach for estimating water level time series over inland waters using multi-mission satellite altimetry. *Hydrol. Earth Syst. Sci.* **2015**, *19*, 4345–4364. [CrossRef]
22. Wahr, J.; Swenson, S.; Zlotnicki, V.; Velicogna, I. Time-variable gravity from GRACE: First results. *Geophys. Res. Lett.* **2004**, *31*, L11501. [CrossRef]
23. Rodell, M.; Famiglietti, J.S.; Chen, J.; Seneviratne, S.I.; Viterbo, P.; Holl, S.; Wilson, C.R. Basin scale estimates of evapotranspiration using GRACE and other observations. *Geophys. Res. Lett.* **2004**, *31*, L20504. [CrossRef]
24. Richey, A.S.; Thomas, B.F.; Lo, M.-H.; Reager, J.T.; Famiglietti, J.S.; Voss, K.; Swenson, S.; Rodell, M. Quantifying renewable groundwater stress with GRACE. *Water Resour. Res.* **2015**, *51*, 5217–5238. [CrossRef] [PubMed]
25. Watkins, M.M.; Wiese, D.N.; Yuan, D.-N.; Boening, C.; Landerer, F.W. Improved methods for observing Earth's time variable mass distribution with GRACE using spherical cap mascons. *J. Geophys. Res. Solid Earth* **2015**, *120*, 2648–2671. [CrossRef]
26. Wiese, D.N.; Landerer, F.W.; Watkins, M.M. Quantifying and reducing leakage errors in the JPL RL05M GRACE mascon solution. *Water Resour. Res.* **2016**, *52*, 7490–7502. [CrossRef]
27. Wiese, D.N.; Yuan, D.-N.; Boening, C.; Landerer, F.W.; Watkins, M.M. *JPL GRACE Mascon Ocean, Ice, and Hydrology Equivalent Water Height RL05M.1 CRI Filtered Version 2*, PO.DAAC. 2016. [CrossRef]
28. Didan, K. *MYD13C2 MODIS/Aqua Vegetation Indices Monthly L3 Global 0.05Deg CMG V006*; NASA: Washington, DC, USA, 2015.
29. Running, S.W.; Kimball, J.S. Satellite-Based Analysis of Ecological Controls for Land-Surface Evaporation Resistance. In *Encyclopedia of Hydrological Sciences*; Anderson, M.G., McDonnell, J.J., Eds.; John Wiley & Sons, Ltd.: Hoboken, NJ, USA, 2006; ISBN 978-0-470-84894-4. [CrossRef]
30. Mu, Q.; Heinsch, F.; Zhao, M.; Running, S. Development of a global evapotranspiration algorithm based on MODIS and global meteorology data. *Remote Sens. Environ.* **2007**, *111*, 519–536. [CrossRef]
31. Mu, Q.; Zhao, M.; Running, S.W. Improvements to a MODIS global terrestrial evapotranspiration algorithm. *Remote Sens. Environ.* **2011**, *115*, 1781–1800. [CrossRef]
32. Fisher, J.B.; Tu, K.P.; Baldocchi, D.D. Global estimates of the land–atmosphere water flux based on monthly AVHRR and ISLSCP-II data, validated at 16 FLUXNET sites. *Remote Sens. Environ.* **2008**, *112*, 901–919. [CrossRef]
33. Rodell, M. *GLDAS CLM Land Surface Model L4 Monthly 1.0 × 1.0 degree, Version 1*; Goddard Earth Sciences Data and Information Services Center (GES DISC): Greenbelt, MD, USA. [CrossRef]
34. Derber, J.C.; Parrish, D.F.; Lord, S.J. The New Global Operational Analysis System at the National Meteorological Center. *Weather Forecast.* **1991**, *6*, 538–547. [CrossRef]
35. Benduhn, F.; Renard, P. A dynamic model of the Aral Sea water and salt balance. *J. Mar. Syst.* **2004**, *47*, 35–50. [CrossRef]
36. Singh, A.; Seitz, F. Updated bathymetric chart of the East Aral Sea, supplement to: Singh, Alka; Kumar, Ujjwal; Seitz, Florian (2015): Remote sensing of storage fluctuations of poorly gauged reservoirs and State Space Model (SSM)-based estimation. *Remote Sens.* **2015**, *7*, 17113–17134. [CrossRef]
37. Schneider, U.; Becker, A.; Finger, P.; Meyer-Christoffer, A.; Rudolf, B.; Ziese, M. *GPCC Full Data Reanalysis Version 7.0 at 0.5°: Monthly Land-Surface Precipitation from Rain-Gauges built on GTS-based and Historic Data*; NCAR: Boulder, CO, USA, 2015. [CrossRef]
38. Adler, R.F.; Sapiano, M.; Huffman, G.J.; Bolvin, D.; Wang, J.-J.; Nelkin, E.; Xie, P.; Chiu, L.; Ferraro, R.; Schneider, U.; et al. New Global Precipitation Climatology Project monthly analysis product corrects satellite data drifts. *GEWEX News* **2016**, *26*, 7–9.
39. Adler, R.F.; Huffman, G.J.; Chang, A.; Ferraro, R.; Xie, P.-P.; Janowiak, J.; Rudolf, B.; Schneider, U.; Curtis, S.; Bolvin, D.; et al. The Version-2 Global Precipitation Climatology Project (GPCP) Monthly Precipitation Analysis (1979–Present). *J. Hydrometeorol.* **2003**, *4*, 1147–1167. [CrossRef]
40. Arsen, A.; Crétaux, J.-F.; Berge-Nguyen, M.; del Rio, R. Remote Sensing-Derived Bathymetry of Lake Poopó. *Remote Sens.* **2013**, *6*, 407–420. [CrossRef]

41. Gao, H.; Birkett, C.; Lettenmaier, D.P. Global monitoring of large reservoir storage from satellite remote sensing: Global monitoring of large reservoir storage from space. *Water Resour. Res.* **2012**, *48*, W09504. [CrossRef]
42. The Aral Sea Crisis. Available online: http://www.columbia.edu/~tmt2120/introduction.htm (accessed on 10 November 2017).
43. Issanova, G.; Abuduwaili, J.; Galayeva, O.; Semenov, O.; Bazarbayeva, T. Aeolian transportation of sand and dust in the Aral Sea region. *Int. J. Environ. Sci. Technol.* **2015**, *12*, 3213–3224. [CrossRef]
44. Chen, Y.; Xia, J.; Liang, S.; Feng, J.; Fisher, J.B.; Li, X.; Li, X.; Liu, S.; Ma, Z.; Miyata, A.; et al. Comparison of satellite-based evapotranspiration models over terrestrial ecosystems in China. *Remote Sens. Environ.* **2014**, *140*, 279–293. [CrossRef]
45. Ershadi, A.; McCabe, M.F.; Evans, J.P.; Chaney, N.W.; Wood, E.F. Multi-site evaluation of terrestrial evaporation models using FLUXNET data. *Agric. For. Meteorol.* **2014**, *187*, 46–61. [CrossRef]
46. Fisher, J.B.; Melton, F.; Middleton, E.; Hain, C.; Anderson, M.; Allen, R.; McCabe, M.F.; Hook, S.; Baldocchi, D.; Townsend, P.A.; et al. The future of evapotranspiration: Global requirements for ecosystem functioning, carbon and climate feedbacks, agricultural management, and water resources: The future of evapotranspiration. *Water Resour. Res.* **2017**, *53*, 2618–2626. [CrossRef]
47. Fisher, J.B.; Malhi, Y.; Bonal, D.; Da Rocha, H.R.; De AraãšJo, A.C.; Gamo, M.; Goulden, M.L.; Hirano, T.; Huete, A.R.; Kondo, H.; et al. The land–atmosphere water flux in the tropics. *Glob. Chang. Biol.* **2009**, *15*, 2694–2714. [CrossRef]
48. McCabe, M.F.; Ershadi, A.; Jimenez, C.; Miralles, D.G.; Michel, D.; Wood, E.F. The GEWEX LandFlux project: Evaluation of model evaporation using tower-based and globally gridded forcing data. *Geosci. Model Dev.* **2016**, *9*, 283–305. [CrossRef]
49. Miralles, D.G.; Jiménez, C.; Jung, M.; Michel, D.; Ershadi, A.; McCabe, M.F.; Hirschi, M.; Martens, B.; Dolman, A.J.; Fisher, J.B.; et al. The WACMOS-ET project – Part 2: Evaluation of global terrestrial evaporation data sets. *Hydrol. Earth Syst. Sci.* **2016**, *20*, 823–842. [CrossRef]
50. Polhamus, A.; Fisher, J.B.; Tu, K.P. What controls the error structure in evapotranspiration models? *Agric. For. Meteorol.* **2013**, *169*, 12–24. [CrossRef]
51. Vinukollu, R.K.; Wood, E.F.; Ferguson, C.R.; Fisher, J.B. Global estimates of evapotranspiration for climate studies using multi-sensor remote sensing data: Evaluation of three process-based approaches. *Remote Sens. Environ.* **2011**, *115*, 801–823. [CrossRef]
52. Reager, J.T.; Thomas, B.F.; Famiglietti, J.S. River basin flood potential inferred using GRACE gravity observations at several months lead time. *Nat. Geosci.* **2014**, *7*, 588–592. [CrossRef]
53. Forkutsa, I.; Sommer, R.; Shirokova, Y.I.; Lamers, J.P.A.; Kienzler, K.; Tischbein, B.; Martius, C.; Vlek, P.L.G. Modeling irrigated cotton with shallow groundwater in the Aral Sea Basin of Uzbekistan: I. Water dynamics. *Irrig. Sci.* **2009**, *27*, 331–346. [CrossRef]
54. Hassanzadeh, E.; Zarghami, M.; Hassanzadeh, Y. Determining the Main Factors in Declining the Urmia Lake Level by Using System Dynamics Modeling. *Water Resour. Manag.* **2012**, *26*, 129–145. [CrossRef]
55. Jeihouni, M.; Toomanian, A.; Alavipanah, S.K.; Hamzeh, S. Quantitative assessment of Urmia Lake water using spaceborne multisensor data and 3D modeling. *Environ. Monit. Assess.* **2017**, *189*, 572. [CrossRef] [PubMed]
56. Lemoalle, J.; Bader, J.-C.; Leblanc, M.; Sedick, A. Recent changes in Lake Chad: Observations, simulations and management options (1973–2011). *Glob. Planet. Chang.* **2012**, *80*, 247–254. [CrossRef]
57. Okpara, U.T.; Stringer, L.C.; Dougill, A.J. Lake drying and livelihood dynamics in Lake Chad: Unravelling the mechanisms, contexts and responses. *Ambio* **2016**, *45*, 781–795. [CrossRef] [PubMed]
58. Ali, W. Environment and Water Resources in the Jordan Valley and Its Impact on the Dead Sea Situation. In *Water Security in the Mediterranean Region*; Scozzari, A., El Mansouri, B., Eds.; Springer Netherlands: Dordrecht, The Netherlands, 2011; pp. 229–238. ISBN 978-94-007-1622-3.
59. Rawashdeh, S.A.; Ruzouq, R.; Al-Fugara, A.; Pradhan, B.; Ziad, S.H.A.-H.; Ghayda, A.R. Monitoring of Dead Sea water surface variation using multi-temporal satellite data and GIS. *Arab. J. Geosci.* **2013**, *6*, 3241–3248. [CrossRef]
60. Shafir, H.; Alpert, P. Regional and local climatic effects on the Dead-Sea evaporation. *Clim. Chang.* **2011**, *105*, 455–468. [CrossRef]

61. Satgé, F.; Espinoza, R.; Zolá, R.P.; Roig, H.; Timouk, F.; Molina, J.; Garnier, J.; Calmant, S.; Seyler, F.; Bonnet, M.-P. Role of Climate Variability and Human Activity on Poopó Lake Droughts between 1990 and 2015 Assessed Using Remote Sensing Data. *Remote Sens.* **2017**, *9*, 218. [CrossRef]
62. Zola, R.P.; Bengtsson, L. Long-term and extreme water level variations of the shallow Lake Poopó, Bolivia. *Hydrol. Sci. J.* **2006**, *51*, 98–114. [CrossRef]
63. Michel, D.; Jiménez, C.; Miralles, D.G.; Jung, M.; Hirschi, M.; Ershadi, A.; Martens, B.; McCabe, M.F.; Fisher, J.B.; Mu, Q.; et al. The WACMOS-ET project – Part 1: Tower-scale evaluation of four remote-sensing-based evapotranspiration algorithms. *Hydrol. Earth Syst. Sci.* **2016**, *20*, 803–822. [CrossRef]

8

Monitoring the Arctic Seas: How Satellite Altimetry can be used to Detect Open Water in Sea-Ice Regions

Felix L. Müller *, Denise Dettmering, Wolfgang Bosch and Florian Seitz

Deutsches Geodätisches Forschungsinstitut, Technische Universität München, Arcisstraße 21, 80333 Munich, Germany; denise.dettmering@tum.de (D.D.); wolfgang.bosch@tum.de (W.B.); florian.seitz@tum.de (F.S.)
* Correspondence: felix-lucian.mueller@tum.de

Academic Editors: Frédéric Frappart, Ole Andersen, Sergey Lebedev, Guillaume Ramillien, Deepak R. Mishra and Prasad S. Thenkabail

Abstract: Open water areas surrounded by sea ice significantly influence the ocean-ice-atmosphere interaction and contribute to Arctic climate change. Satellite altimetry can detect these ice openings and enables one to estimate sea surface heights and further altimetry data derived products. This study introduces an innovative, unsupervised classification approach for detecting open water areas in the Greenland Sea based on high-frequency data from Envisat and SARAL. Altimetry radar echoes, also called waveforms, are analyzed regarding different surface conditions. Six waveform features are defined to cluster radar echoes into different groups indicating open water and sea ice waveforms. Therefore, the partitional clustering algorithm K-medoids and the memory-based classification method K-nearest neighbor are employed, yielding an internal misclassification error of about 2%. A quantitative comparison with several SAR images reveals a consistency rate of 76.9% for SARAL and 70.7% for Envisat. These numbers strongly depend on the quality of the SAR images and the time lag between the measurements of both techniques. For a few examples, a consistency rate of more than 90% and a true water detection rate of 94% can be demonstrated. The innovative classification procedure can be used to detect water areas with different spatial extents and can be applied to all available pulse-limited altimetry datasets.

Keywords: satellite altimetry; Envisat; SARAL; unsupervised classification; K-medoids; Greenland Sea; Fram Strait

1. Introduction

The Arctic Ocean, including its peripheral seas, e.g., the Greenland Sea, is considered one of the most important components of the Earth's climate system [1]. In particular, these areas show strong responses to global warming and may affect climate conditions globally, for example, by changing the oceanic thermohaline circulation. The north polar regions are crucial contributors to the global ocean current system by carrying cold and fresh water southwards. Most of the Arctic Ocean is covered by varying extents of sea ice with open water areas and floes with different spatial extents as well as fully closed ice surfaces. The seasonal fluctuations of ice covers significantly impacts the atmosphere-ocean interaction (e.g., ice-albedo). While a closed sea ice cover prevents the ocean from heat emission, openings in the ice lead to a warming of the first atmospheric layers.

The evolution of sea ice is strongly influenced by sea surface temperature, wind, waves and ocean currents [1]. During recent decades, increasing sea surface temperatures and an enhanced warm water inflow in the Arctic Ocean resulted in decreased sea ice extent and volume [2,3]. Additionally, the Greenland ice sheet experienced strong environmental changes due to an increasing mass loss enhancing melt water influx into the Arctic Ocean [4].

The monitoring of the changing north polar ocean conditions, especially in the Greenland Sea, allows investigating interconnections between land, ocean, and atmospheric processes as well as their climate forcing. Today, remote sensing systems provide a large set of different sensors for monitoring the polar regions. Radar satellite altimetry is able to provide quantitative information about sea surface heights, significant wave heights, and dynamic ocean topography [5,6]. However, in order to derive reliable altimetry products, a careful selection of measurements from open water areas is necessary. By analyzing the radar return signal of the altimeter, the so-called waveform, information about the reflecting surface can be derived. This allows the classification of waveforms in water- and sea ice-returns and the detection of open water areas in sea ice regions. For example, calm open water areas within the footprint cause a very single-peak shape. With an along-track resolution of less than a kilometer using high-frequency data, conventional satellite altimetry missions (such as Envisat and SARAL) are able to detect small open water areas that might be missed by imaging Synthetic Aperture Radar (SAR) satellite missions (Sentinel-1A/B, Radarsat-1/2, etc.) in case a high-resolution acquisition mode is not available. Additionally, small water areas have insufficient backscatter properties to be mapped by passive microwave satellite missions (e.g., Special Sensor Microwave Imager (SSM/I) and SSM/I Sounder (SSM/IS)). However, altimeter radar echoes reflected from non-uniform scatterers, like sea ice regions, are challenging to interpret because the large surface footprint of several kilometers usually covers several ice types.

The first studies dealing with satellite altimetry in sea ice regions were published in 1980 by Dwyer and Godin [7] and in 1992 by Fetterer et al. [8]. After the launch of the ESA satellites ERS-1 and ERS-2, covering high latitudes in a repeat orbit, further studies were conducted by Laxon [5] and Laxon et al. [9]. They analyzed the potential of ERS-1 sea ice monitoring and the interannual variability of sea ice thickness by employing ERS-1 and ERS-2 altimetry data. Furthermore, Peacock [10] provides a first sea surface height determination in the Arctic ocean. In recent years, several sea ice applications have been explored e.g., the detection of openings in the ice. Connor et al. [11] applied a peakiness parameter, defined by Peacock [10] in order to detect small open water bodies in the sea ice cover using high-frequency data of Envisat. Zakharova et al. [12] continues with the development of a lead detection algorithm by using the Centre National d' Études Spatiales (CNES) and Indian Space Research Organisation (ISRO) satellite SARAL and maximum power threshold. Currently, all pulse-limited altimetry-based approaches for detecting water returns in sea ice regions use thresholds for different parameters. This has the disadvantage that the thresholds have to be set manually and individually for every altimetry mission. Furthermore, a deep knowledge about the different scatter characteristics in sea ice regions is required. Besides Zygmuntowska et al. [13] developed another approach using the waveforms shape for classifying airborne SAR altimeter echoes over the Arctic sea ice in a supervised way.

The present study proposes a new strategy to detect open water areas based on an unsupervised classification of high-frequency altimetry radar echoes. The approach is able to detect water domains with different spatial extents and can be easily applied without any deeper knowledge about surface-dependent backscatter characteristics. The method is applicable to all available pulse-limited altimeter data and is independent of mission-specific radar frequencies and characteristics. Furthermore, the results are compared to processed SAR images using the method described in [14] to obtain quantitative information about the classification performance.

The present paper is structured into three main parts. First, the study area and the applied datasets are introduced. Section 3 presents the method and processing procedure as well as the comparison process of the obtained results. Section 4 presents the classification results and provides evidence of the classification performance. At first, quantitative information considering the entire available validation dataset (Section 4.1) is derived before some visual comparisons between the SAR images and the altimetry overflights are provided. The paper finishes with a conclusion and an outlook to future research.

2. Study Area and Data Sets

This section provides an introduction to the study area and the different remote sensing datasets used for classification and validation.

2.1. Greenland Sea and Fram Strait

The study area ranges from 70.0°N to 81.5°N in latitude and from the north-east coast of Greenland to 28.0°E in longitude and covers the Greenland Sea and the Fram Strait (see Figures 1 and 2). The Greenland Sea belongs to the peripheral seas of the Arctic ocean. It connects the Fram Strait in the north, a narrow passage between north-east Greenland and Svalbard, with the Norwegian Sea as well as the Iceland Sea in the south. This area is affected by the East Greenland Current (EGC), which transports more than 90% of fragmented sea ice from the Arctic Ocean through the Fram Strait southwards [15]. Therefore, the EGC represents the main and most important freshwater outlet of the Arctic Ocean. According to Serreze and Barry [16] the Greenland Sea and the region of the Fram Strait is strongly influenced by rapid atmospheric and changing sea ice conditions as well as comparatively fast ocean currents with a mean velocity of 20–30 cm/s [17] and maxima up to 80 cm/s [18]. The sea ice state reaches from a nearly closed sea ice cover, showing straight lined and circular shaped open water bodies, leads and polynyas, up to individual ice floes ranging from a few meters to kilometers in diameter [16]. Applying open water detection to the Greenland Sea and the Fram Strait offers one the chance to sensitize the unsupervised classification method for a various number of different sea ice and ocean conditions.

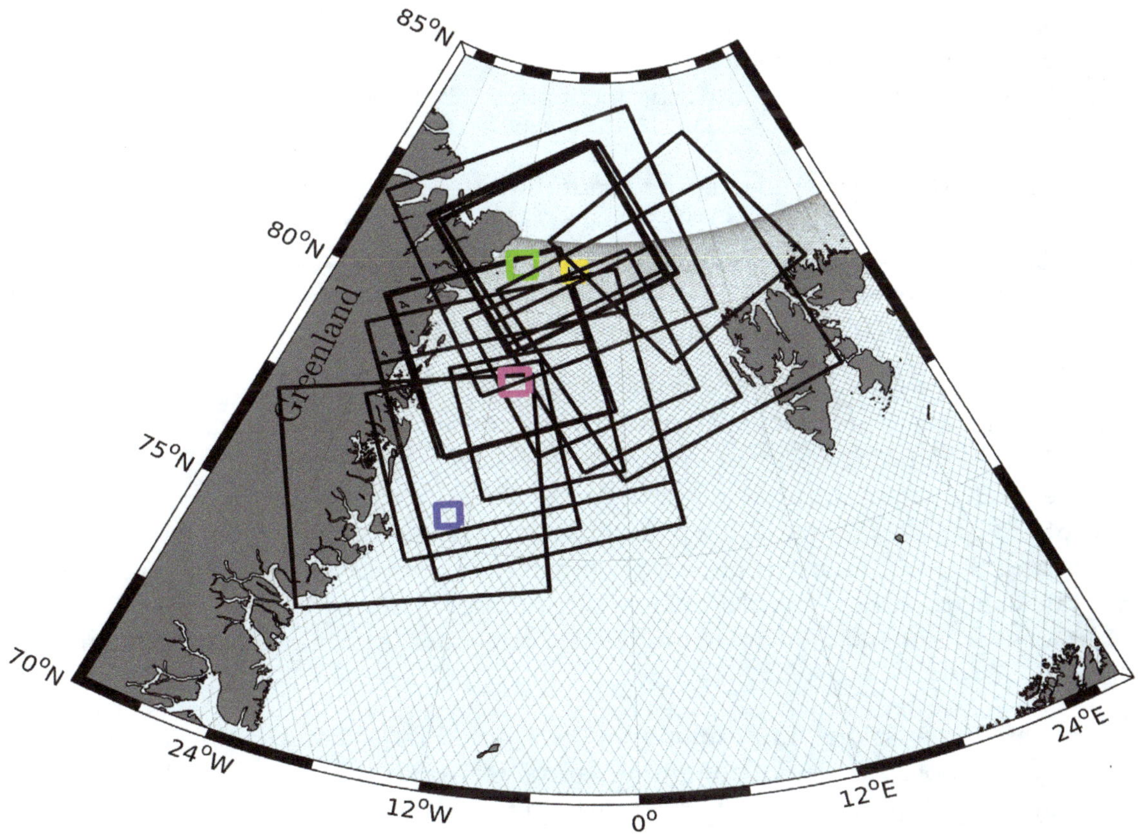

Figure 1. Black rectangles indicate locations of the SAR images from ALOS and Radarsat-2 used for comparison with Envisat classification results against the background of nominal sun-synchronous ground tracks of one Envisat cycle. The four subsets discussed in Section 4.2 are highlighted by different colors.

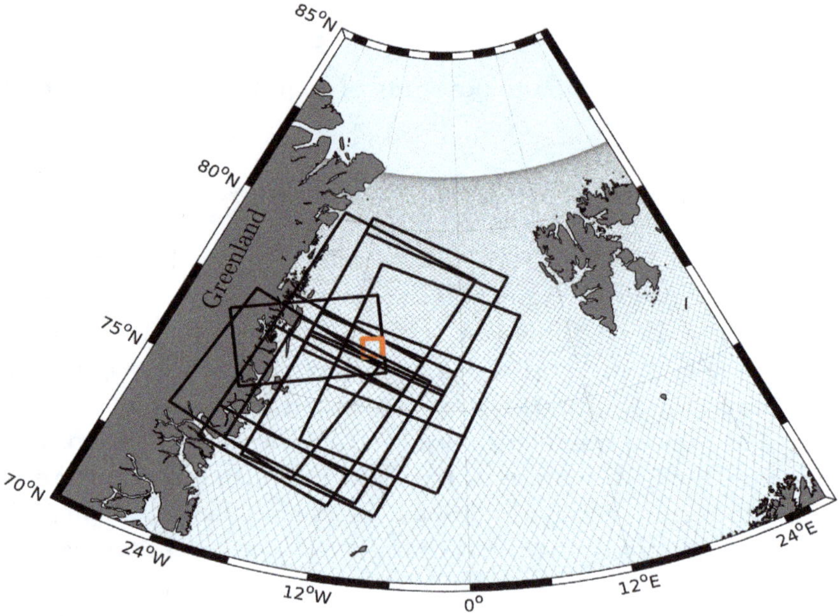

Figure 2. Black rectangles indicate locations of the SAR images from Sentinel-1A used for comparison with SARAL classification results against the background of nominal sun-synchronous ground tracks of one SARAL cycle. One subset discussed in Section 4.2 is highlighted in orange.

2.2. *Radar Altimetry Data*

In the present investigation, the high-frequency radar altimetry data of the ESA satellite Envisat and the CNES/ISRO altimetry satellite SARAL are used. Data of the missions Jason-1, Jason-2 and Jason-3 are disregarded due to their low orbit inclination (about 66°) not covering the Greenland Sea and Fram Strait.

Envisat and SARAL carry pulse-limited radar altimeters and are placed on the same 35 day repeat-orbit covering polar areas up to ±81.5° geographical latitude. Envisat was launched in March 2002 and orbits the Earth at an altitude of nearly 800 km. In October 2010, Envisat left the repeat-orbit and started to drift until in May 2012, the ESA mission was decommissioned after an unexpected signal loss. SARAL was placed in orbit in February 2013 and is still active even though in July 2016, the satellite started its drifting orbit phase without fix repeat period.

All computations and methodologies used in this study are based on official high-frequency Sensor Geophysical Data Record (SGDR) v2.1 dataset of Envisat's radar altimeter (RA-2) and the SGDR-T dataset of the AltiKa radar altimeter mounted on SARAL. In case of SARAL, data until July 2016 and in case of Envisat, data until the end of the mission are used. In this study, waveforms observed in the Greenland Sea and Fram Strait (see Section 2.1) are employed in the classification process. In order to calculate the altimeter backscatter values, different features stored in the SGDR dataset, for example, atmospheric attenuation and instrumental corrections (e.g., sigma naught calibration factor) are additionally used.

The two satellite missions differ mainly in the emitted radar bandwidth, the pulse repetition frequency and the footprint size of the illuminating area onto the surface. RA-2 emits *Ku*-band signals with an repetition frequency of 1800 pulses per second, covering an nominal elliptic area of approximately up to 10 km diameter [11] depending on the surface conditions. Before transmitting to earth, the waveforms are sampled to 18 Hz by the on-board processing. AltiKa works in the *Ka*-band, with a repetition frequency of 4 kHz, generates 40 Hz averaged waveforms and has half the antenna aperture of Envisat. This leads to a smaller footprint size up to 8 km diameter and an improved spatial resolution [19]. Beside instrumental influences, the waveform's shape is mainly affected by various surface characteristics. Detailed explanations referring to the representation of the varying waveform's shape can be found in Section 3.1.

2.3. Imaging Synthetic Aperture Radar (SAR) Data

A possible source for validating the classification results is the usage of imaging synthetic aperture radar (SAR) data. Beside the altimeter satellites, several multispectral and SAR imaging satellite missions regularly provide snapshots of periodically changing ocean conditions. In contrast to multispectral sensors, working mostly in the visible and infrared spectrum, SAR sensors are unaffected by cloudiness and lighting conditions, which makes it easier to identify appropriate scenes. However, SAR sensors are side-looking instruments, which can cause a shadowing of very flat and smooth surface structures (e.g., leads or polynyas) due to interjacent higher topography (e.g., ice floes, ridges). Additionally, the recorded backscatter values do not only depend on the surface characteristics (e.g., roughness) but also on the incidence angle of the reflected radar waves, which makes it more complex to provide information about different surface types. Furthermore, it has to be mentioned that most SAR satellites are placed on sun-synchronous orbits, which allows for a uniform capture of ice state but limits the minimum time lag between the acquisition dates of the SAR images and the altimetry measurements of Envisat and SARAL also using sun-synchronous orbits.

Aiming at a small time lag between SAR images and satellite altimetry, wide swath data are qualified best since these images cover a spatially extended area with medium pixel spacing. For this investigation, SAR images of the JAXA Advanced Land Observing Satellite (ALOS) [20], MDA's Radarsat-2 [21] and ESA's Sentinel-1A (S-1A) [22] are used. The Envisat classification results are compared with ALOS PALSAR Level 1.5 Wide Beam (WB) images offered by the Alaska Satellite Facility (ASF) DAAC and with Radarsat-2 (R-2) Scan SAR Mode data provided by ESA. The SARAL classification outcomes are compared with Level-1, S-1A extra wide swath mode data. S-1A images are made available through the ESA/Copernicus Sentinel Data Hub. Specifications, temporal availability in the target region, and information about the used imaging SAR products are listed in Table 1. To distinguish between open water pixels, appearing in near black, and sea ice pixels, appearing in bright gray, HH-polarized images are used. For more information regarding SAR polarization and the influence of different surface scattering see Dierking W. [23] and Jackson et al. [24].

Table 1. Synthetic Aperture Radar (SAR) image specifications [20–22] and altimeter satellites covering same time periods.

SAR Satellite	Band	Mode	Swath Width (km)	Pixel Size (m)	Period (mm/yyyy)	Altimeter Satellite
ALOS	L-Band	Wide Beam	250–350	100 × 100	June 2007–May 2008	Envisat
Radarsat-2	C-Band	Scan SAR Wide	500	50 × 50	June 2008–present	Envisat/SARAL
Sentinel-1A	C-Band	Extra Wide	400	40 × 40	October 2014–present	SARAL

In order to ensure similar sea ice conditions and allow for an unbiased comparison between SAR and altimetry, only images with a time lag less than about 3.5 h, with respect to the altimetry crossings, are used. The comparison is based on 16 grayscaled SAR images during the lifetime of Envisat and 19 images for SARAL. The SAR data are selected from different epochs considering a varying sea surface state with a focus on periods with various sea ice coverage. Figures 1 and 2 display the locations of all used SAR images. The scenes are mainly located in the Fram Strait and near the north-east coast of Greenland. Tables 2 and 3 list sensor and temporal information for all conducted comparison pairs. Two of the R-2 images are used for multiple satellite overflights. In the case of SARAL classification, it has to be mentioned that, due to sun-synchronous orbits and fixed revisit times of Sentinel-1A and SARAL, it is not possible to find suitable pairs for comparison that show good spatio-temporal coverage with a time gap smaller than 2 h 40 min during the study period.

Table 2. Acquisition date of the SAR images and time gap between altimetry observations and imaging data used for comparison with Envisat classification results.

SAR-Satellite	Acquisition Date	Time Gap hh-mm
ALOS	14 June 2007	02-30
ALOS	1 October 2007	02-57
ALOS	7 October 2007	01-55
ALOS	10 November 2007	03-07
ALOS	10 December 2007	02-50
ALOS	26 December 2007	02-13
ALOS	5 January 2008	02-40
ALOS	7 January 2008	01-46
ALOS	12 January 2008	02-49
ALOS	4 May 2008	01-25
R-2	4 November 2008	01-47
R-2	20 April 2009	02-09 00-29
R-2	21 April 2009	02-07
R-2	10 February 2010	03-04
R-2	14 March 2010	01-52 00-13 01-27
R-2	16 October 2010	02-04

Table 3. Sentinel-1A acquisition date of the SAR images and time gap between altimetry observations and imaging data used for comparison with SARAL classification results.

Acquisition Date	Time Gap hh-mm
23 October 2014	02-41
16 November 2014	03-34
14 November 2014	02-49
18 November 2014	02-42
3 December 2014	02-40
6 December 2014	02-58
27 December 2014	03-33
1 January 2015	02-59
15 January 2015	03-20
18 January 2015	03-27
16 March 2015	03-08
6 February 2015	03-29
6 February 2015	03-28
10 February 2015	03-22
22 February 2015	02-59
2 March 2015	02-44
9 March 2015	02-56
19 April 2015	02-53
15 May 2015	02-54

2.4. *Sea Ice Data*

Polar sea areas are affected by moving sea ice due to the influences of wind and ocean currents [16]. This results in a rapid change and high diversity of the sea surface conditions. To reach a realistic comparison of altimetry results and SAR images, the compensation for sea ice motion within the time interval between the two observation sets is required. For this purpose, daily ice vector velocity fields are exploited within the validation process. Therefore, the "Polar Pathfinder Daily 25 km EASE-Grid Sea Ice Motion Vectors, Version 3" of the National Snow and Ice Data Center (NSIDC) are employed [25]. This dataset contains zonal and meridional sea ice velocity observations of active and passive sensors as well as in situ measurements interpolated to a 25 km spacing grid referring to an azimuthal equal area map projection. This dataset covers the entire altimetry era until the end of May 2015.

The sea ice velocity data are used to shift the SAR image, respectively, the image pixel coordinates, assuming an averaged ice motion (direction and velocity) over the time interval between the altimetry measurement and the SAR image. For this purpose, only homogeneous data represented by small standard deviations in direction and velocity inside a predefined box (± 35 km) around the altimetry track are selected to compute a mean displacement vector. Sea ice velocity vectors located close to the coastlines (within 25 km) are eliminated due to erroneous ice observations [25].

The comparison is performed only in areas affected by sea ice to suppress the influence of falsely detected SAR ice pixels caused by diffuse scattering behavior due to rough swell in the open ocean. Therefore, daily "Sea Ice Concentrations from Nimbus-7 SMMR and DMSP SSM/I-SSM/IS Passive Microwave Data, Version 1" of the NSIDC [26] with a spatial resolution of 25 km $\times$ 25 km are interpolated to the altimetry high-frequency data. Observations outside the ice edge without sea ice are excluded from the comparison process.

3. Methods

This study is based on an unsupervised classification process of altimetry waveforms. Unsupervised classification algorithms group unassigned data into a predefined number of classes without any background information about the data and their sources using only "natural" and hidden intra-cluster similarities [27]. The classification is performed based on a set of features characterizing the input data. In contrast, supervised classification is based on a-priori information of a well known or labeled dataset to classify and assign the observations [28]. Examples for unsupervised classification methods are artificial neural networks (e.g., Self-Organizing Maps [29]) or partitional clustering algorithms (e.g., K-means and K-medoids [30,31]). In the present investigation, a partitional cluster algorithm, K-medoids, is used for separating a set of unlabeled waveform data into clusters indicating different waveform properties. Therefore, features have to be defined describing various waveform characteristics. Based on the clustering results, K-nearest-neighbor is applied to assign unclassified waveform data.

In this section, at first, features for describing the various waveform shapes and their characteristics are specified and explained. This is followed by the description of the methodical background of the clustering and classification process. The last part of Section 3 presents the validation approach for the classification procedure. The presented methods are applied independently to Envisat and SARAL.

3.1. Waveform Features

The shape of altimetry waveforms strongly depends on the surface characteristics within the altimeter footprint. Figure 3 shows Envisat/RA-2 and SARAL/AltiKa radar pulses reflected by ocean, leads, and sea ice. Major differences can be detected in the power magnitude and the number and shape of the signal peaks. Leads produce very narrow and peaky waveforms due to the specular scattering of calm and flat water. In contrast, radar pulses originating from ocean or sea ice surfaces are influenced by waves or interlaced and piled ice floes, respectively, leading to multi peaks and wider, noisier shapes.

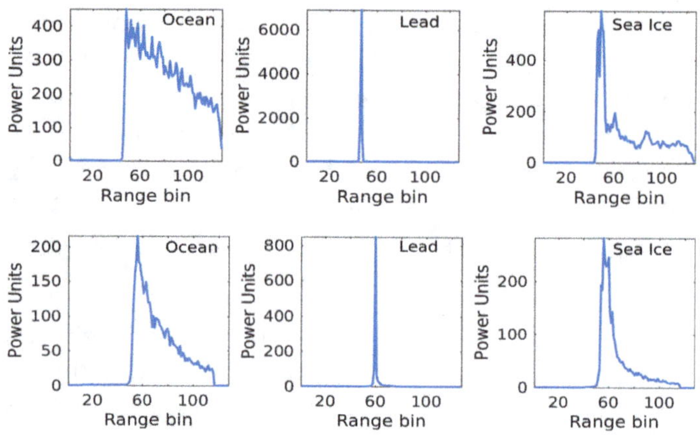

Figure 3. Waveform examples for Envisat (**top row**) and SARAL (**bottom row**) for three different surface scatterers: Ocean (**left**), Lead (**middle**), and Sea Ice (**right**).

In order to characterize a waveform and identify the main evocative surface scatterer, a number of waveform features are defined. The computed values constitute a waveform feature space that provides the input for the clustering and classification process. To increase the efficiency of the algorithm and to get a reliable open water detection, the selected features should fulfill the following conditions:

1. The features should characterize different waveform types.
2. The selected features should be stand alone and without linear dependence and major correlations among each other.
3. The feature space should be adaptable to any altimetry waveform.
4. All features should exhibit the same order of magnitude for equal weighting among each other.

In the present investigation, six features are defined to describe the waveforms mainly focusing on the reflected radar pulse shape (width) and the recorded power intensity. These features are applicable for each pulse-limited altimeter waveform, i.e., for Envisat as well as for SARAL.

- Waveform maximum (Wm)
  The waveform height is described by the maximal power of the returning radar pulse σ_{max}. It provides information about the backscatter of calm or rough surface conditions. To compute σ_{max} for Envisat as well as SARAL, the maximum waveform power and mission specific rectifications are applied by using instrumental and atmospheric corrections from the provided datasets (see Section 2.2).

- Trailing edge decline (Ted)
  The trailing edge decline is computed by fitting an exponential function, considering an exponential decay of AltiKa waveforms, from the waveform power maximum to the last bin. The estimated decay rate is used to characterize the decline of the trailing edge after the maximum.

- Waveform noise (Wn)
  This feature quantifies the trailing edge scattering. It is computed as median absolute deviations of the trailing edge fitting (see Ted) residuals. This parameter is very small for single peak waveforms (leads) and moderate for oceans.

- Waveform width (Ww)
  The number of bins where the power is equal to zero provides information about the waveform's width.

- Leading edge slope (Les)
  The leading edge slope is obtained by subtracting the first bin position containing more than 30%

of the power maximum from the bin position of the maximum power. The difference provides relative information about the width and steepness of the leading edge independent of the absolute position of the leading edge, i.e., the range.

- Trailing edge slope (Tes)
  In contrast to the leading edge slope, the trailing edge slope is obtained by subtracting the last bin position containing more than 30% of the maximum from the bin position of the maximum power. This difference provides similar information to Ted in the case of single-peak waveforms but supports the identification of strong specular peaks in front of an ocean-like trailing edge.

The selected features in the present investigation show a varying order of magnitudes, which results in an irregular weighting in the clustering algorithm. In order to comply with condition 4 (see above) a standardization has to be processed. Before conducting the unsupervised classification procedure, the features are reduced by subtracting their average and divided by their standard deviation (standard-score).

The features are calculated for RA-2 and AltiKa waveforms in the same way. In the case of SARAL, the maximum power is limited to 1250 counts. Power counts above this limit are not recorded due to too high backscatter values that cannot be resolved by the tracking window [12]. The waveforms are cut without a clear maximum peak in the radar echo, which makes it impossible to compute all features (e.g., leading edge slope) and to constitute the complete feature space. These waveforms, which are not flagged in the SGDR dataset, are skipped from the further classification process. Furthermore, all waveforms are neglected, for which no reliable computation of the defined features is possible (e.g., if trailing edge fitting is impossible with 95% confidence).

3.2. Clustering

Within the clustering process a representative subset of all waveforms from a single mission will be used to define waveform groups, so-called clusters, that will later be used to also classify all remaining observations. In a first step, this reference model has to be created. For this purpose, a set of several waveforms, containing a majority of all possible scatter types, has to be selected. To this end, waveform data covering an area in the central Greenland Sea within bounds of 15°W/10°E longitude and 68°N/80°N latitude are used. To cover as many sea ice types as possible, the epoch is selected at the beginning of the melting period in early summer from April to May [32]. For Envisat, Cycle 57 (2007, containing about 307,000 waveforms), and for SARAL, Cycle 12 (2014, ca. 670,000 waveforms) are selected.

To group the reference data, a K-medoids cluster algorithm is implemented that clusters unsupervised data into K clusters. K-medoids performs a distance minimization between the features and the most centrally located feature (medoids) based on the feature space itself. Thereby, K-medoids is more robust to outliers and noise in contrast to K-means, which tries iteratively to estimate an optimal partition of unlabeled data by minimizing the distances between the coordinates of a mean cluster center (centeroids) and the features. However, in contrast to K-medoids, K-means integrates every value of the feature space into the arithmetic average [27].

At first, K-medoids randomly chooses K medoids of the feature space and computes the distances to every feature. In the next steps, the algorithm rearranges every single feature until there is no motion within the K clusters and the minimal distances to the medoids are found. However, the clustering result depends on the initial randomly chosen medoids. This is why the algorithm is repeated several times and the best solution is selected by analyzing the final sum of all distances within the clusters. This leads to high computational efforts by employing large input datasets, but it is considerable that the clustering has to be performed just once per altimetry mission. To reduce the computation times, the algorithm examines only a random sample of cluster members during each medoids updating step. The size of the sample is set by default to 0.1% of the total number of data points. The iteration terminates if the medoids are stabilized.

Partition clustering algorithms require an initialization of the number of clusters K. In the present investigation, K is chosen empirically after several test runs by evaluating the best segmentation results [31]. Indicators for defining an appropriate K are, for example, the analysis of the sum of all distances within the clusters and, additionally, a visual analysis of all clusters. In order to obtain a clear partitioning of waveforms, it is useful to set K larger than the desired number of the three surface types, the present investigation is looking for, namely, calm open water, ocean or sea ice conditions. Figure 4 shows the clustering for 30 classes based on derived waveform features of about 307,000 Envisat waveforms (the clustering results for SARAL waveforms can be found in the Supplementary section, Figure S1). The displacements between the points and the medoids are computed using Euclidean distances.

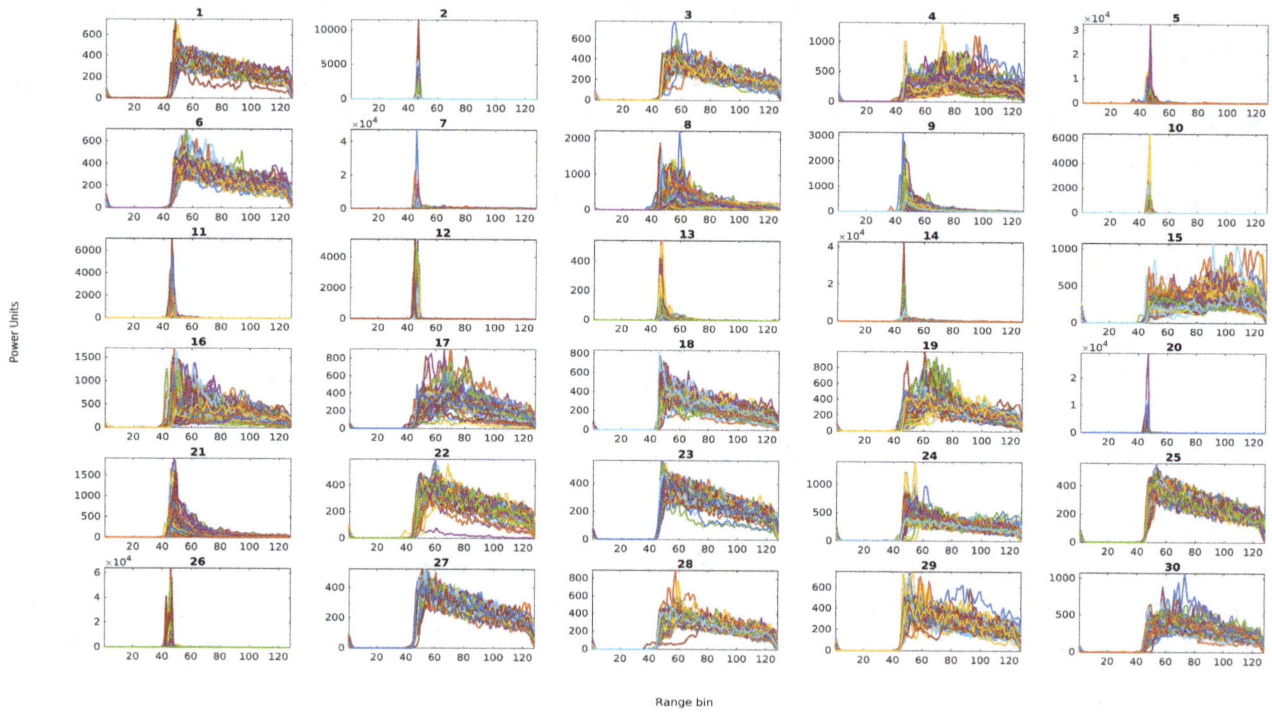

Figure 4. Envisat waveform clusters (K = 30) after K-medoids clustering showing segmented waveforms (every twenty fifth per cluster).

After running K-medoids, each cluster has to be assigned to one surface condition. The 30 clusters need to be manually condensed to three main classes indicating ocean, sea ice, and lead/polynya returns. This is done based on the feature statistic per cluster (see Figure 5) and knowledge on the physical backscattering behavior of different surfaces. It is well known that radar returns from dominant scatterers (i.e., a lead with a calm, mirror-like surface) cause single-peak waveforms with high power and narrow shape. Radar echoes nearly entirely reflected by sea-ice show a more diffuse scattering, weak power and no clear peaks. Using these relationships and transferring them to the cluster statistics enable a nearly unambiguous assignment. However, questionable clusters with ambiguous feature properties remain and are labeled as "undefined".

Figure 5 indicates the cluster assignment by different colors. Lead and polynya returns (clusters 2, 10–12, 20, and 26) are characterized by a very narrow and peaky shape and high maximum power values. In contrast, ocean returns (clusters 1, 3, 6, 18, 22, 23, 25, 27–29) are wider and show a greater trailing edge decay. Waveforms belonging to the ice class (clusters 5, 7–9, 13, 14, 16, 21, 24, and 30) are between these two groups. They are defined by a smaller trailing edge decay and slope as well as bigger power values than ocean returns. However, there are clusters (4, 15, 17, and 19), that cannot clearly be assigned to one surface type. As an example, cluster 19 shows an ocean like behavior, but is characterized by an indistinct leading edge as well as a steeper ice-like trailing edge. Undefined waveform classes show, apart from more noise in the cluster itself, no clear signature or trend to the underlying feature space or to the three main surface classes.

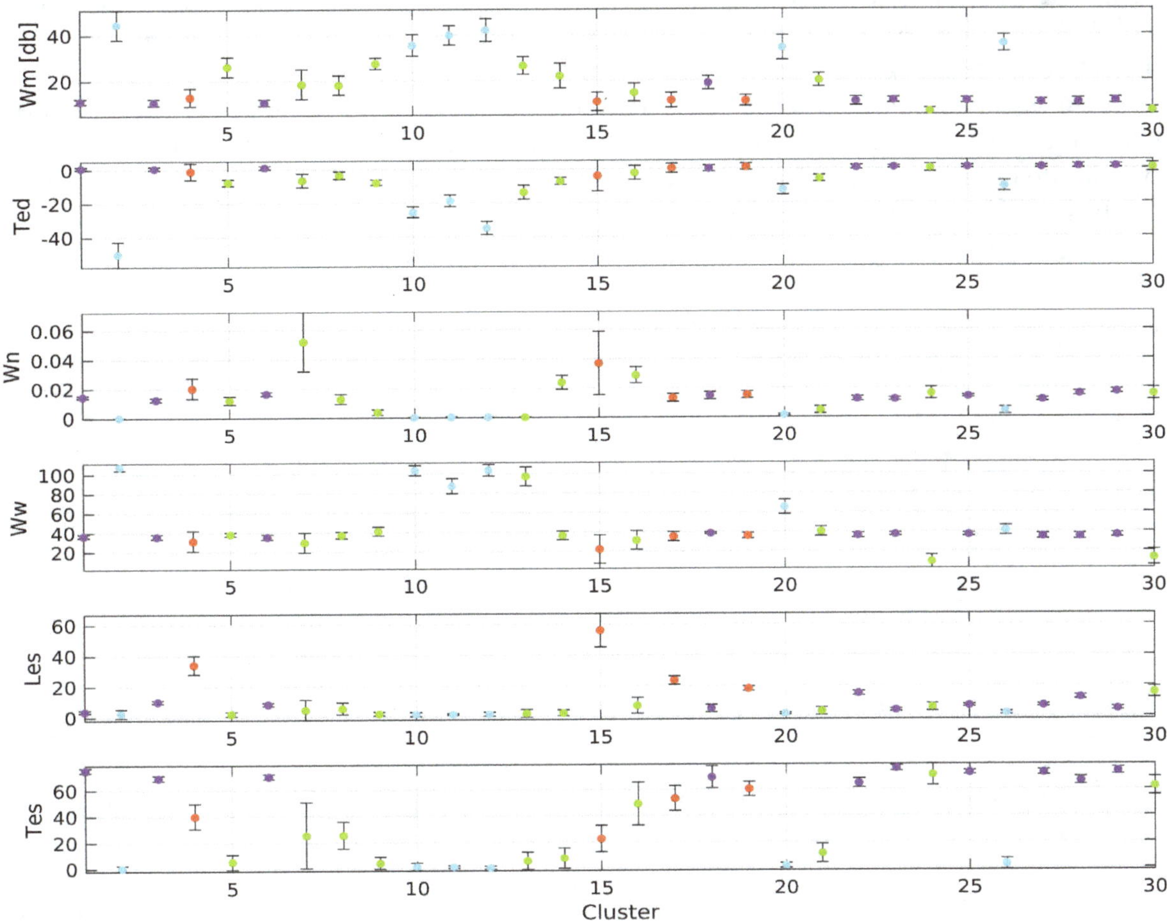

Figure 5. Means and standard deviations of waveform features (see Section 3.1) per cluster. Four classes are illustrated using different colors: "lead/polynya" (cyan), "ocean" (purple), "ice" (green), and "undefined" (red).

3.3. Classification

The waveform model created by the clustering (Section 3.2) can now be used to classify all waveforms. For this purpose the K-Nearest Neighbor (K-NN) classifier is employed. In general, K-NN belongs to the memory-based classifiers and does not require a stochastic model [28]. Basically, K-NN searches for the closest distance between a query point and a given input model. Similar to the K-medoids algorithm, the K-NN uses the euclidean distance. However, K has a completely different meaning than in the K-medoids algorithm. The K is now defined as the number of neighbors used for the classification. The cluster assignment of a specific waveform is done based on the majority of clusters of these K nearest neighbors. K must be set before the classification process starts.

In the present study, K is estimated by performing a 10-fold cross-validation. Therefore the reference model used for the clustering and already assigned to the clusters is divided into 10 randomly sorted, but equally sized subsets and validated against each other. This means, that every subset is used as a test sample and the remaining subsets as training sets. In order to find an appropriate K for K-NN, the cross-validation is performed based on different numbers of neighbors. Figure 6 shows the mean misclassification error as a percentage of the 10-fold cross-validation in the case of SARAL and Envisat. Similar errors can be expected for the classification of the remaining unlabeled waveforms. The minimum error defines the optimal number of neighbors. SARAL displays less variability and a smaller misclassification rate than Envisat. The K-NN method seems to be more stable with clustered SARAL than with Envisat waveforms, which can be explained due to less variability in the AltiKa waveforms and a more robust waveform clustering. For SARAL a minimum error rate is obtained with K = 20 (1.93 ± 0.05%). In the case of Envisat, nearest-neighbor number K = 44 (2.3 ± 0.08%) is used, providing a good balance between low error and variance.

The misclassification rate in connection with the defined number of neighbors gives information about the K-NN prediction error based on the reference model and class labels. This parameter can be used to estimate the internal precision of the classification approach. In this study, a minimal error of about 2% has to be expected from the methodology itself.

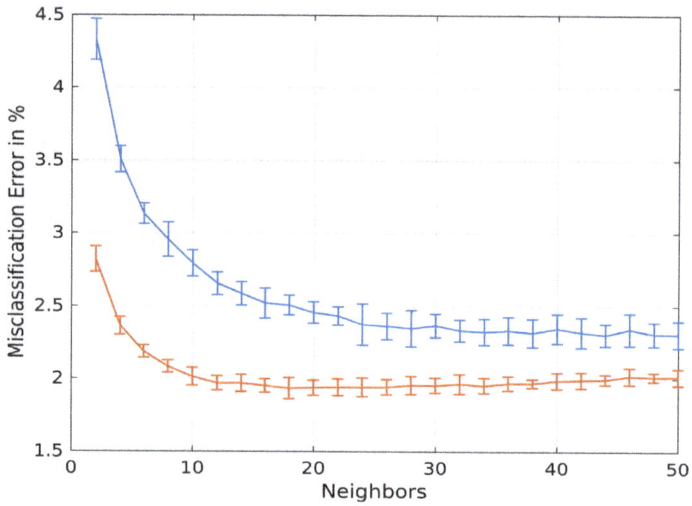

Figure 6. Misclassification error and its standard deviation for SARAL (red) and Envisat (blue) with varying number of K neighbors as computed by 10-fold cross-validation.

After defining an appropriate K, the remaining waveforms are applied to K-NN. In the end of the classification process every waveform is labeled by a certain cluster and, consequently, assigned to a specific surface type.

3.4. Validation Approach

In order to conduct an external validation for the waveform classification, a comparison with independent SAR images is performed. For this purpose, the defined waveform classes of ocean, lead/polynya, sea ice are assigned to water (ones) and non-water (zeros) observations. Undefined waveforms classes are also labeled with zeros. In order to provide quantitative information about the classification performance, it is necessary to compare the results to an external dataset. For this purpose, imaging SAR data are used, as they regularly provide snapshots of different sea surface states in the study area.

Before performing an automatic comparison between SAR and the classification results, the SAR images are pre-processed by using the ESA toolbox SNAP, version 4.0.0 for Sentinel-1A as well as Radarsat-2 and the MapReady toolbox, version 3.1.22 for ALOS image data offered by ASF. Basically, the following standard routines are applied to the imaging SAR data:

1. Thermal noise removal (only S-1A)
2. Radiometric calibration
3. Speckle filter (only Radarsat-2)
4. Delay Doppler terrain correction
5. Reprojection to Lambert Azimuthal Equal Area map projection
6. Converting backscatter values to db
7. Datatype conversion in *uint8*

On the pre-processed SAR images, linear and circular shaped black and near black areas indicate openings inside the sea ice cover generated by a smooth surface and specular reflection of the radar waves. To automatically extract these areas, the SAR images have to be converted into binary pixel values by applying several image processing tools. The applied approach is described in detail by Passaro et al. [14]. Briefly summarized, the images undergo a noise and minimum filtering in order to emphasize dark pixel regions, followed by an adaptive thresholding that considers local illumination changes. Finally, a mathematic morphological closing operation is applied to the black and white coded images to link fragmented open water regions. To control the effect of the morphological closing operation a structure element (kernel) or convolution matrix is needed. Regarding linear and circular shapes of open water areas, an octagon with various size, considering the nominal pixel spacing of the SAR images, is employed. In the case of ALOS, the octagon size is six pixels around the center pixel, and in the case of Sentinel-1A and Radarsat-2, a kernel size of 12 pixels is used. Moreover, the image coordinates are shifted to compensate for sea ice-motion, for the acquisition time difference between altimetry and SAR (see Section 2.4). In a last step, the locations of the altimetry returns are interpolated to the SAR pixel locations by using nearest neighbor method.

4. Results and Discussion

In this study, 15,025 Envisat and 19,919 SARAL observations are investigated for which SAR image classification results are available for validation. 31.2% of the Envisat waveforms and 15.0% of the SARAL returns are assigned to water classes. Furthermore, 4.7% of Envisat and 14.2% of SARAL waveforms are set to undefined and defined as non-water returns. For a quantitative rating, 19 comparison pairs for SARAL and Envisat, respectively (see Section 2.3), are used. The results of this comparison are presented in the following section. Afterwards, examples are displayed to illustrate and discuss the functionality of the validation approach.

4.1. Automatic Comparison to SAR Images

As mentioned above, the automatic comparison process only relies on observations in areas with a semi-closed sea ice layer. This allows one to reduce false SAR classifications outside the ice edge due to an unreliable SAR image processing. Table 4 provides the numbers of measurements assigned to water and ice by the two observation techniques and, therefore, allows for an assessment of the altimetry classification performance. The absolute number of water and ice detections are listed column-wise for the altimetry classification results and row-wise for the SAR open water detection. The table shows that 1124 of the 15025 Envisat observations are identified as water by both, altimetry and SAR, whereas 837 locations are assigned to non-water by altimetry and to water by SAR. Assuming the SAR to be the ground truth validating the altimetry water detection, four dependencies are derived to rate the classification results. The total consistency rate, $P(CR)$ is computed by summing up the bold values and dividing them by the total number of comparison points. In addition, three conditional frequencies are derived: The true water detection rate ($P(Alt|SAR)$) is computed by dividing the "correct" altimetry water detections by the total number of SAR water observations, whereas the false water detection rate ($P(Alt|\overline{SAR})$) is defined as the relation between the water altimetry detections not confirmed by SAR and the total number of SAR ice detections. Moreover, the percentage of correctly classified water returns $P(SAR|Alt)$ represents the "correct" water altimetry detections in relation to the total number of open water detections by altimetry.

In the case of Envisat, a consistency rate of 70.7% is reached. In detail, nearly 60% of SAR water detections are truly classified by Envisat ($P(Alt|SAR) = 0.57$) in contrast to below 30% of SAR ice observations that are falsely assigned to water areas by Envisat ($P(Alt|\overline{SAR}) = 0.27$). However, only about a quarter of all Envisat open water detections are also classified by SAR ($P(SAR|Alt) = 0.24$).

The comparison between SARAL and Sentinel-1A water detection yields a higher consistency rate of 76.9% but a smaller true water detection rate of less than 30% ($P(Alt|SAR) = 0.28$). At the same time, the false water detection rate $P(Alt|\overline{SAR})$ is very small and yields only 12.3%. Moreover, the correctly classified water return rate $P(SAR|Alt) = 33\%$ is better as for Envisat.

It has to be noticed that for the interpretation of these numbers it is important to consider that inconsistencies are not only due to altimetry classification but that the SAR open water detection as well as the sea ice-motion correction also contribute to the error budget. For example, most of the SARAL comparisons take place during the sea ice maximum between January and mid March, when the pack ice is very close and exhibits only small openings in the ice, which makes it challenging to be detected by the SAR image processing.

Analyzing the absolute water detection numbers of Envisat versus SAR images, it is remarkable that the number of open water points differs by 2732 between the SAR detection and the Envisat classification. The Envisat classification identifies significantly more open water areas than the SAR processing (factor of nearly 2.4). In the case of SARAL, a transposed situation can be found. This can be explained by different SAR sensor characteristics and an insufficient pixel resolution as well as an imprecise SAR image processing, including an unreliable sea ice-motion correction. Additionally, the altimeters are affected by off-nadir returns, which can cause an enhanced number of open water detections. In the case of Envisat, a larger footprint size than SARAL intensifies off-nadir effects.

Overall, it is important to understand, that the classification performance numbers of SARAL and Envisat are not directly comparable with each other. The underlying different instrumental, sensor, and spatio-temporal conditions differ too strongly to provide qualitative information that would allow for a comparative assessment of the two altimetry satellites. More details related to the impacts of SAR and altimetry processing on the quantitative comparison process can be found in Section 4.2.

Analyzing, for example, $P(CR)$ of Envisat and SARAL, the quantitative comparison confirms the reliability of the altimetry-based classification method and a good performance of their results. However, it has to be kept in mind that a data comparison of two totally different Earth observation techniques for open water detection in a very dynamic study area is not possible without a variety of uncertainties and inaccuracies. In order to provide a better impression of the difficulties of a quantitative comparison approach, the next section shows a couple of examples in a visual comparison.

Table 4. 2D contingency tables based on Envisat—ALOS/R-2 (top) and SARAL—S-1A (bottom) comparisons. The table shows the number of points classified as water/ice from altimetry (Alt) with the corresponding classification from SAR.

	Alt (Water)	**Alt (Ice)**	**Σ**
SAR (water)	**1124**	837	1961
SAR (ice)	3569	**9495**	13064
Σ	4693	10332	15025
	Alt (Water)	**Alt (Ice)**	**Σ**
SAR (water)	**987**	2600	3587
SAR (ice)	2007	**14325**	16332
Σ	2994	16925	19919

4.2. Visual Comparison

Using different SAR image subsets, this section will provide some visual comparisons between open water classification by altimetry and SAR images. The images were selected in order to indicate possible difficulties due to uncertainties in the SAR image or altimetry processing as well as sea ice motion correction. Figures 7 and 8 show five visual examples before (left) and after (right) image processing. The altimetry measurement locations are superimposed on the SAR image. Cyan colored altimetry observations identify open water classifications. These regions are plotted in white in the binary coded SAR images (right column). Figure 7a–f display Envisat-ALOS and Figure 7g,h Envisat—Radarsat-2 comparisons. Figure 8 shows a visual example of one SARAL—Sentinel-1A comparison. Metadata information, i.e., acquisition date and applied sea ice motion correction on each comparison is provided next to the images (visualized classification results without class assignment can be found in supplementary Figures S2–S6). Moreover, Table 5 displays quantitative comparison results.

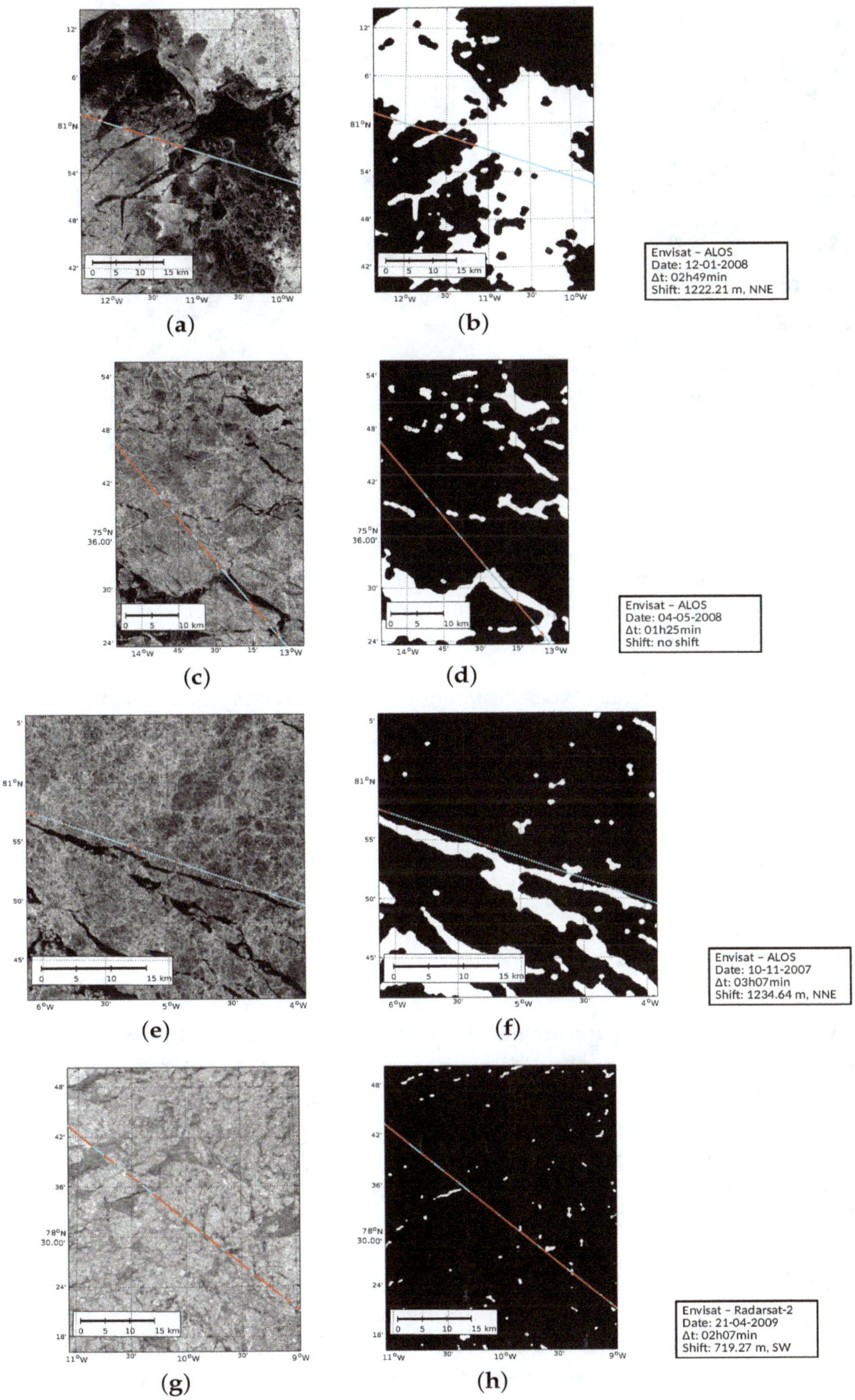

Figure 7. Examples of open water detection from Envisat against ALOS (**a**–**f**) and Radarsat-2 (**g**,**h**) before (**left**) and after SAR image processing (**right**) with open water indicated in white. Boxes provide additional image and processing information. Red: ice detection, cyan: open water detection. The geographical locations of the image subsets are displayed in Figure 1; from top to bottom in green, blue, yellow and magenta.

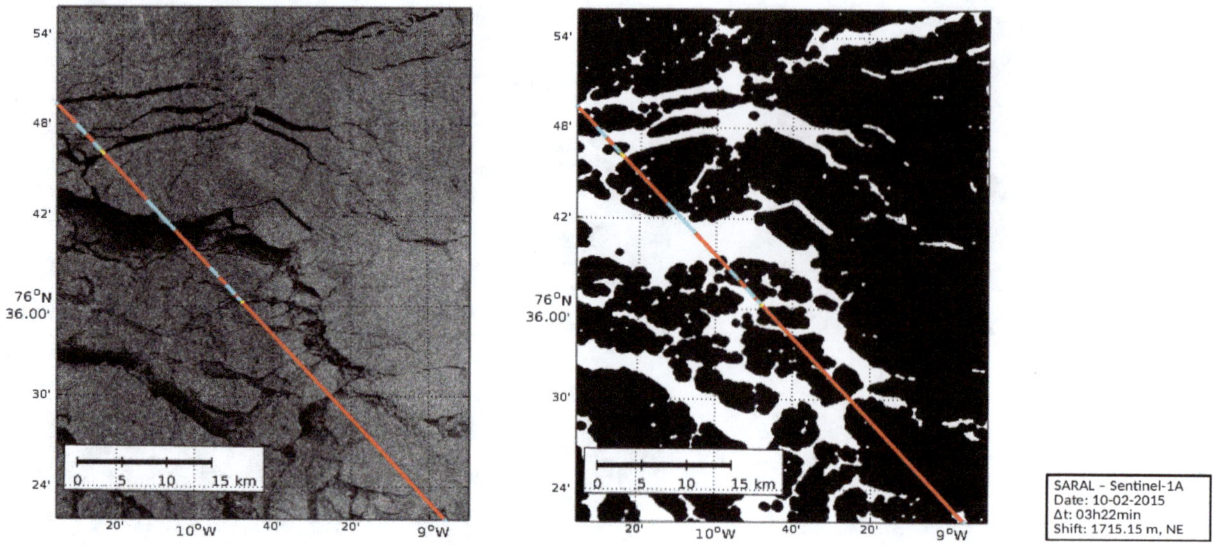

(a) (b)

Figure 8. Example (orange highlighted in Figure 2) of open water detection from SARAL against Sentinel-1A before (**a**) and after SAR image processing (**b**) with open water indicated in white. Box provides additional image and processing information. Red: ice detection, cyan: open water detection, yellow: saturated AltiKa observations.

Table 5. Table providing percentage statistical information about conditional $P(X|Y)$ and consistency $P(CR)$ rates of visual examples discussed in Section 4.2. "SAR" and "Alt" indicate imaging SAR and altimetry. Ice detections are indexed by overline character marked shortcuts.

Subset	$P(CR)$	$P(Alt|\overline{SAR})$	$P(SAR|Alt)$	$P(Alt|SAR)$
Figure 7b	90.63%	15.91%	91.86%	94.05%
Figure 7d	85.10%	16.04%	64.58%	88.57%
Figure 7f	12.61%	88.99%	2.02%	100.00%
Figure 7h	76.22%	22.36%	0.00%	0.00%
Figure 8b	72.07%	12.41%	40.68%	25.53%

The first example (Figure 7a) demonstrates very good accordance between altimetry and SAR classification.The L-band image displays different sea ice and open water conditions. From West to East, various sized open water areas ranging from 200 m up to 3.5 km are visible. A large region appearing in dark reaches from the image center at 11°W to the eastern edge. It indicates a mixture of differently sized ice floes interrupted by open water sections. Analyzing the colored altimetry observations, the open water detection is in good accordance with the grayscaled as well as with the binary coded SAR image (Figure 7b). The quantitative comparison yields a consistency rate $P(CR)$ of 90.63%. Moreover, the altimetry classification approach provides a true classification rate close to 100% ($P(Alt|SAR) = 0.94$). This comprises small leads as well as larger areas of open water.

An almost perfect accordance between altimetry classification and SAR images can also be observed in Figure 7c, detecting an expanded lead in the southern image part and some small leads in the central part of the image. However, comparing Figure 7c and d to the altimetry classification results at 75°36.00′N, it is clearly visible that SAR image processing is not always able to segment very narrow lead fragments. This might happen because of a poor spatial pixel resolution of the SAR sensor (100 m), an insufficient identification of the ice-water transition, or a too restrictive threshold level in the SAR image processing. This deficiency in the automatic SAR image processing results in reduced performance in the quantitative comparison with a total consistency rate of about 85% and a true water detection rate of 88.6%.

However, there are also problems related to the altimetry observation technique. The example displayed in Figure 7e,f is characterized by a very long (ca. 47 km) and in most parts narrow lead located parallel to the satellite track. The SAR image is shifted about 1.2 km in the northeastern direction, assuming a steady sea ice motion. Even if the altimeter track is still located northwards at the off-nadir position of the lead, almost all measurements are classified as open water. As a consequence, in the quantitative comparison, just 2% of all Envisat open water detections are confirmed by SAR classification although they can be visually connected with the dominant lead in the image center. The overall consistency rate yields only 12.6%—probably due to the fact that the altimetry classification approach is not able to separate off-nadir water returns from nadir water returns or the mean sea ice motion correction is not enough to consider the total sea ice drift.

Additional discrepancies between altimetry and SAR classification can occur in areas with new, very thin ice coverage. Figure 7g shows those areas, appearing light gray in C-band, only a little darker than the surrounding older ice. These areas are correctly set to ice by the SAR image processing (see Figure 7h) because of the small brightness differences between the thin and surrounding ice types. In contrast, the altimetry returns within these areas are falsely classified and interpreted as calm open water since they show a very narrow and single-peaked lead/polynya-like shape. One explanation for this mis-interpretation is the dominant scattering of all flat and specular surfaces. Connor et al. [11] found that strong reflective surfaces, for example, leads/polynyas, can also affect the waveform shape if covered by very thin ice. A distinction from open water is not possible based on the altimetry waveform's shape. Since the ice is very thin, the retracked ranges should represent the water level well enough, even if the classification is wrong.

Related to the comparison process itself, uncertainties in sea ice motion correction can reduce the quantitative consistency rate. Figure 8a,b are corrected by ice-motion considering a time difference of more than 3 h. Analyzing Figure 8b, it can be shown that only 25.53% of the SAR detected ice openings are well identified by the altimetry data. A visual image inspection suggests that the applied ice motion correction is too small to completely compensate for the effect of the time lag.

Further challenging issues using SARAL SGDR-T data are so-called saturated waveforms. Zakharova et al. [12] pointed out that leads or strongly reflecting surfaces can exceed the maximum permissible power count value of 1250. The waveforms are cut and feature no clear peak due to a saturated power tracking window. Figure 8 highlights saturated SARAL observations in yellow. They are mainly located near small and calm open water areas, producing very high backscatter returns. In the classification process, they are omitted because of an unknown maximum peak position. In general, saturated SARAL waveforms are mainly traceable within the sea ice edge, but can provide evidence about the location of further open water areas. However, just 0.14% (i.e., 288 waveforms) of the comparison data are affected by a saturated power tracking window.

The present section shows a number of challenging and unavoidable impacts on the validation of the waveform classification process. Considerable parts of the inconsistencies do not originate from the altimetry classification but from the SAR classification or the ice-motion correction. In order to adequately rate the quantitative comparison results, it is necessary to keep these effects in mind.

5. Conclusions and Outlook

The present paper introduces an unsupervised classification approach based on pulse-limited multi-mission altimetry data to detect open water areas in a largely sea ice covered region. The study demonstrates the successful application of the clustering of pulse-limited altimeter waveforms for the automatic identification of open ocean, sea ice, lead and polynya observations. The approach is based on known partition cluster strategies (i.e., K-medoids) and memory-based classification methods (i.e., K-nearest-neighbor). A 10-fold cross-validation for the assessment of the precision of the classification method is performed. It indicates an internal misclassification error of about 2% for Envisat and SARAL. The algorithm is applicable to every pulse-limited altimetry satellite mission without requiring any deeper knowledge about mission specific details. Moreover, it can be assumed that the developed approach also works for SAR altimetry waveforms if the waveform feature space is adapted adequately. Additionally, the presented method can be adapted to a number of open water detection or waveform classification tasks, e.g., for the identification of lake returns [33] or in inundation areas.

In order to evaluate the classification results, a comparison with SAR images is performed. In contrast to previous studies, the present validation relies not only on visual and manually selected examples, but also on a larger set of images and an automated comparison procedure. The comparison procedure allows for a quantitative assessment of the classification performance by assigning the altimetry observations to open water and sea ice returns and checking them against processed SAR images that indicate sea ice and open water areas. We reach consistency rates of 70.7% for Envisat and 76.9% for SARAL. However, it has to be underlined that the quantitative comparison results of Envisat and SARAL are not directly comparable because of significant differences in the underlying sensor and instrument characteristics of the available SAR missions.

When interpreting the comparison results, different sources of inconsistencies have to be considered, e.g., effects from the altimetry data and their classification procedure and uncertainties in SAR image processing as well as in the ice-motion correction. The Fram Strait and the Greenland Sea are one of the most dynamic areas on Earth. Fast changing sea ice conditions due to short, periodic melting and refreezing as well as rapid climate change make it hard to provide a high reliability in the comparison as well as in the altimetry classification results. Local phenomena, such as melt ponds (i.e., open water pools on the sea ice surface) and their impacts on the open-water detection, have to be investigated. Over specific sea ice types, altimetry waveforms show ambiguities, which prevents a clear attribution to sea ice or open water returns. In particular, specular thin and flat ice produces very specular returns resembling open water returns. In contrast, big ice floes or landfast ice can imitate ocean-like returns due to similarities in ocean surface roughness and reflectivity.

Further improvements of the classification method are possible. In particular, saturated SARAL waveforms have to be included in the classification process. In addition, the application of more recent sea ice motion data in combination with Sentinel-1B data could lead to a better spatio-temporal ratio within the validation process.

A reliable classification is an indispensable requirement for a meaningful estimation and an efficient computation of sea surface heights in the Arctic by retracking only open water waveforms. In addition to Envisat and SARAL, more pulse-limited (e.g., ERS-1/2) as well as delay-doppler altimetry data (e.g., CryoSat-2, Sentinel-3A) may be employed in the classification process and, thus, contribute to the generation of a long-term sea level record for the Arctic ocean.

Acknowledgments: The authors thank the following institutions and agencies for providing their data and software products under the terms of the GNU General Public License. ESA for operating and managing Envisat, Sentinel-1A and the SNAP Toolbox. CNES for providing SARAL data. JAXA/METI and ASF DAAC for making ALOS PALSAR L1.5 data and MapReady Toolbox available as well ESA for maintaining MDA Radarsat-2 data in framework of Third Party Mission program. This work was supported by the German Research Foundation (DFG) through grants, BO1228/13-1 and DE2174/3-1. The publication is funded by the Technical University of Munich (TUM) within the framework of the Open Access Publishing Program. We thank five anonymous reviewers for their valuable comments that helped to improve the manuscript.

Author Contributions: Felix L. Müller developed the classification and validation methods, conducted the data analysis and wrote the majority of the paper. Denise Dettmering supervised the present study, contributed to the manuscript writing and helped with the discussions of the applied methods and results. Wolfgang Bosch initiated the study. Florian Seitz supervised the research. Both were involved in the writing process and discussed the methods presented in the manuscript as well.

References

1. Comiso, J. *Polar Oceans from Space*; Atmospheric and Oceanographic Sciences Library, Springer: New York, NY, USA, 2010.

2. Stroeve, J.C.; Markus, T.; Boisvert, L.; Miller, J.; Barrett, A. Changes in Arctic melt season and implications for sea ice loss. *Geophys. Res. Lett.* **2014**, *41*, 1216–1225.
3. Polyakov, I.V.; Beszczynska, A.; Carmack, E.C.; Dmitrenko, I.A.; Fahrbach, E.; Frolov, I.E.; Gerdes, R.; Hansen, E.; Holfort, J.; Ivanov, V.V.; et al. One more step toward a warmer Arctic. *Geophys. Res. Lett.* **2005**, *32*, L17605.
4. Sasgen, I.; van den Broeke, M.; Bamber, J.L.; Rignot, E.; Sørensen, L.S.; Wouters, B.; Martinec, Z.; Velicogna, I.; Simonsen, S.B. Timing and origin of recent regional ice-mass loss in Greenland. *Earth Planet. Sci. Lett.* **2012**, *333–334*, 293–303.
5. Laxon, S.W. Sea-Ice Altimeter Processing Scheme at the EODC. *Int. J. Remote Sens.* **1994**, *15*, 915–924.
6. Rio, M.H.; Hernandez, F. A mean dynamic topography computed over the world ocean from altimetry, in situ measurements, and a geoid model. *J. Geophys. Res. Oceans* **2004**, *109*, doi:10.1029/2003JC002226.
7. Dwyer, R.; Godin, R. *Determining Sea-Ice Boundaries and Ice Roughness Using GEOS-3 Altimeter Data*; Technical Report; NASA Wallops Flight Center: Wallops Island, VA, USA, 1980.
8. Fetterer, F.M.; Drinkwater, M.R.; Jezek, K.C.; Laxon, S.W.C.; Onstott, R.G.; Ulander, L.M.H. Sea Ice Altimetry. In *Microwave Remote Sensing of Sea Ice*; American Geophysical Union: Washington, DC, USA, 2013; pp. 111–135.
9. Laxon, S.; Peacock, N.; Smith, D. High interannual variability of sea ice thickness in the Arctic region. *Nature* **2003**, *425*, 947–950.
10. Peacock, N.R. Sea surface height determination in the Arctic Ocean from ERS altimetry. *J. Geophys. Res.* **2004**, *109*, C07001.
11. Connor, L.N.; Laxon, S.W.; Ridout, A.L.; Krabill, W.B.; McAdoo, D.C. Comparison of Envisat radar and airborne laser altimeter measurements over Arctic sea ice. *Remote Sens. Environ.* **2009**, *113*, 563–570.
12. Zakharova, E.A.; Fleury, S.; Guerreiro, K.; Willmes, S.; Rémy, F.; Kouraev, A.V.; Heinemann, G. Sea Ice Leads Detection Using SARAL/AltiKa Altimeter. *Mar. Geod.* **2015**, *38*, 522–533.
13. Zygmuntowska, M.; Khvorostovsky, K.; Helm, V.; Sandven, S. Waveform classification of airborne synthetic aperture radar altimeter over Arctic sea ice. *Cryosphere* **2013**, *7*, 1315–1324.
14. Passaro, M.; Müller, F.L.; Dettmering, D. Lead Detection using Cryosat-2 Delay-Doppler Processing and Sentinel-1 SAR images. *Adv. Space Res.* **2017**, under review.
15. Rudels, B.; Friedrich, H.J.; Quadfasel, D. The Arctic Circumpolar Boundary Current. *Deep-Sea Res. Part II* **1999**, *46*, 1023–1062.
16. Serreze, M.; Barry, R. *The Arctic Climate System*; Cambridge Atmospheric and Space Science Series; Cambridge University Press: Cambridge, UK, 2014.
17. Bersch, M. On the circulation of the northeastern North Atlantic. *Deep-Sea Res. Part II* **1995**, *42*, 1583–1607.
18. Woodgate, R.A.; Fahrbach, E.; Rohardt, G. Structure and transports of the East Greenland Current at 75°N from moored current meters. *J. Geophys. Res. Oceans* **1999**, *104*, 18059–18072.
19. Verron, J.; Sengenes, P.; Lambin, J.; Noubel, J.; Steunou, N.; Guillot, A.; Picot, N.; Coutin-Faye, S.; Sharma, R.; Gairola, R.M.; et al. The SARAL/AltiKa Altimetry Satellite Mission. *Mar. Geod.* **2015**, *38*, 2–21.
20. Japan Aerospace Exploration Agency (JAXA). *ALOS Data Users Handbook, Revision C*; Japan Aerospace Exploration Agency: Chofu, Tokyo, Japan, 2008.
21. MDA. *RADARSAT-2 Product Description*; Report RN-SP-52-1238, Issue 1/13; MacDonald, Dettwiler and Associates Ltd.: Vancouver, BC, Canada, 2016.
22. Sentinel-1 Team. *Sentinel-1 User Handbook*; GMES-S1OP-EOPG-TN-13-0001, Issue Draft; European Space Agency: Paris, France, 2013.
23. Dierking W. Sea Ice Monitoring by Synthetic Aperture Radar. *Oceanography* **2013**, *26*, doi:10.5670/oceanog.2013.33.
24. Jackson, C.R.; Apel, J.R. *Synthetic Aperture Radar: Marine User's Manual*; US Department of Commerce, National Oceanic and Atmospheric Administration, National Environmental Satellite, Data, and Information Serve, Office of Research and Applications: Washington, DC, USA, 2004; pp. 81–115.
25. Tschudi, M.; Fowler, C.; Maslanik, J.; Stewart, J.S. *Polar Pathfinder Daily 25 km EASE-Grid Sea Ice Motion Vectors, Version 3*; Subset: Greenland Sea, Date Accessed: 16.11.2016; National Snow and Ice Data Center: Boulder, CO, USA, 2016.

26. Cavalieri, D.; Parkinson, C.; Gloersen, P.; Zwally, J.H. *Sea Ice Concentrations from Nimbus-7 SMMR and DMSP SSM/I-SSMIS Passive Microwave Data, Version 1*; Subset: Greenland Sea, Date Accessed: 16.11.2016; NASA National Snow and Ice Data Center Distributed Active Archive Center: Boulder, CO, USA, 1996.
27. Xu, R.; Wunsch, D.C. *Clustering*; IEEE Press Series on Computational Intelligence; IEEE Press: Hoboken, NJ, USA; Wiley: Piscataway, NJ, USA, 2009.
28. Hastie, T.; Tibshirani, R.; Friedman, J. *The Elements of Statistical Learning*; Springer: New York, NY, USA, 2009; Volume 2, pp. 337–387.
29. Kohonen, T. *Self-Organizing Maps*; Springer Series in Information Sciences, Springer: Berlin/Heidelberg, Germany, 2012.
30. Celebi, M. *Partitional Clustering Algorithms*; EBL-Schweitzer, Springer International Publishing: Cham, Switzerland, 2014.
31. Kaufman, L.; Rousseeuw, P.J. *Finding Groups in Data: An Introduction to Cluster Analysis*; Wiley Series in Probability and Mathematical Statistics; Wiley: New York, NY, USA, 1990.
32. Kvingedal, B. Sea-Ice Extent and Variability in the Nordic Seas, 1967–2002. In *The Nordic Seas: An Integrated Perspective*; American Geophysical Union: Washington, DC, USA, 2013; pp. 39–49.
33. Göttl, F.; Dettmering, D.; Müller, F.L.; Schwatke, C. Lake Level Estimation Based on CryoSat-2 SAR Altimetry and Multi-Looked Waveform Classification. *Remote Sens.* **2016**, *8*, 885.

9

The Drifting Phase of SARAL: Securing Stable Ocean Mesoscale Sampling with an Unmaintained Decaying Altitude

Gérald Dibarboure [1,*], Alain Lamy [1], Marie-Isabelle Pujol [2] and Ghita Jettou [2]

1 Centre National d'Etudes Spatiales, 31400 Toulouse, France; alain.lamy@cnes.fr
2 Collecte Localisation Satellites, 31520 Ramonville Saint-Agne, France; mpujol@cls.fr (M.-I.P.); gjettou@cls.fr (G.J.)
* Correspondence: gerald.dibarboure@cnes.fr

Abstract: The French/Indian altimeter project Satellite with ARGOS and AltiKa (SARAL) completed its nominal 3-year mission on the historical European Remote-sensing Satellite (ERS) orbit in Spring 2016. In order to extend the lifetime of the satellite as much as possible, the agencies in charge of SARAL decided to initiate a so-called drifting phase where the satellite altitude is no longer maintained. In this paper we describe how the ocean mesoscale sampling capability of SARAL has been preserved during the drifting phase by initiating it at a specific altitude: the optimal starting point was approximately 1 km above the historical ERS/ENVIronment SATellite (ENVISAT) orbit. This strategy secured the ocean mesoscale sampling capability of SARAL drifting phase (DP) for 6 years or more. We also generalize these findings: any altimeter could follow SARAL's drifting phase strategy if their maneuvering capability is limited. Lastly, we explain how an altimetry mission or an entire altimeter constellation could be operated without any form of altitude control: some specific altitude bands (e.g., near 1230 km) guarantee a high-quality mesoscale sampling for years or decades even if the altitude is not maintained.

Keywords: altimetry; orbit decay; drifting orbit; geodetic orbit

---

1. Introduction and Context

The Satellite with ARGOS and AltiKa (SARAL) mission was developed by CNES (Centre National d'Etudes Spatiales) and ISRO (Indian Space Research Organization). Launched in 2013, the SARAL satellite was primarily a gapfiller for the Advanced Research and Global Observation Satellite (ARGOS) localization system and a Ka-band altimeter technology demonstrator (AltiKa). Verron et al. [1] and Vincent et al. [2] recall that SARAL/AltiKa was initially designed as a response to the requirements expressed by operational oceanography users to bridge the gap between the ENVIronment SATellite (ENVISAT) and the Sentinel-3 altimeter missions. An important objective of AltiKa was to complement the altimeter constellation and to help resolve large ocean mesoscale variability, i.e., scales ranging from 150 to 500 km and 15 to 50 days [3]. SARAL/AltiKa was instrumental in providing the ocean mesoscale sampling needed by operational systems such as AVISO [4] and the Copernicus Marine Environment Monitoring Service (CMEMS) [5]. Indeed, according to Chelton et al. [6] and Pascual et al. [7], two operational altimeters are required to monitor ocean mesoscale variability in delayed time, and up to four are needed in near real time.

Furthermore, Verron et al. [1,8] and Bonnefond et al. [9] recently reported that the Ka-band demonstrator exhibited unprecedented altimeter precision for topography over the ocean, rivers and lakes, and even over sea and land ice, as well as for wind and wave. This precision in turn led to innovative user-products and new scientific findings (e.g., [10–15]). Thanks to this demonstrated scientific value, CNES and ISRO decided to extend the SARAL mission beyond the 3-year nominal duration that was successfully completed in Spring 2016.

As reported by Bonnefond and Willis [16], SARAL exhibited an increasing need to reduce the stress on its momentum wheels in order to extend the lifetime of the mission. This could be achieved by limiting the orbit maintenance maneuvers. The agencies eventually decided that the best strategy was to stop all orbit control maneuvers and to let the altitude decay naturally. This strategy is also known as a drifting phase (DP). It was initiated in July 2016.

The main consequence of a DP was that the SARAL subsatellite track would rapidly drift away from the historical ground track formerly used by the European Remote-sensing Satellite (ERS) and ENVISAT missions. At this point it is important to highlight that the term "drift" refers to the evolution of the nadir track (i.e., position of the altimeter measurement profile): SARAL DP is therefore very different from the ENVISAT mission extension where the orbit was drifting but the nadir ground track was relatively stable and composed of 431 orbits per 30-day cycle.

Sandwell et al. [17] have shown that using a drifting ground track could be extremely useful for the marine geodesy community: the unprecedented precision of AltiKa could help resolve uncharted sea mounts and sea floor topography. Improving marine geodesy products requires a so-called "geodetic altimeter sampling", i.e., a sampling pattern where the subsatellite tracks over a long period (e.g., one year) create a very dense grid with a resolution of 8 km or less. In practice, this can be achieved by keeping precisely the satellite on a so-called "geodetic orbit", i.e., an orbit with a very long exact repeat cycle. This strategy was used for the Jason-1 Geodetic-Mission (GM) phase or the Jason-2 Long Repeat Orbit (LRO) phase. For SARAL DP, there is no longer an exact repeat because the orbit altitude decays continuously, yet the geodetic sampling of SARAL DP is naturally dense, albeit somewhat random (discussed in Sections 3.4 and 4.1).

However, putting SARAL on such a drifting ground track raised the question of the new mesoscale sampling capability (still a high priority for the mission). Dibarboure et al. [18] have shown that the geodetic orbit used by Cryosat-2 is very poor when it comes to mesoscale monitoring: the orbit has a monthly linear scanning pattern that leaves most of the ocean unobserved over a typical ocean mesoscale decorrelation period of 15 days. On the other hand, Dibarboure et al. [19] demonstrated that it could be possible to find geodetic orbits that were compatible with mesoscale monitoring: this was achieved by including intermediate sub-cycles (defined in Section 2.1) that maximize the ocean mesoscale sampling over a period of 15 to 20 days. But these studies were both carried out in the context of a well-maintained altitude, and the mesoscale sampling properties of decaying orbits has not been studied so far.

In this context, the objective of this paper is to demonstrate that it is possible to secure high-quality ocean mesoscale sampling and good marine geodesy sampling for 6 years or more even with an unmaintained decaying altitude. Sections 2 and 3 describe how the continuous evolution of the orbit sub-cycles could be predicted to find mesoscale-friendly altitude bands where a decaying altitude does not alter the altimeter sampling capability. In Section 4, the anticipated results are confirmed with in-flight data from the first 18 months of SARAL DP. Lastly Section 5 generalizes the SARAL DP findings to other altimeter missions.

2. Methodology Used to Prepare the Drifting Phase of SARAL

This section gives an overview of the methodology used in 2016 to define the drifting phase of SARAL: Section 2.1 describes how to infer the quality of mesoscale sampling for a given altitude, and Section 2.2 shows how we defined conservative scenarios to predict the altitude decay of SARAL DP. These two essential inputs are then used in Section 3 to predict the evolution of the ocean mesoscale sampling throughout the drifting phase, and to select the best starting point.

2.1. *Quality of Mesoscale Sampling and Orbit Sub-Cycles*

To measure the quality of the mesoscale sampling for a given altitude, we analyze the distribution of the satellite tracks in the longitude/time plane, following the methodology of Dibarboure et al. [19] for Jason-1: in essence, the quality of mesoscale sampling is inferred by gauging when altimeter tracks get too close in space or in time. Indeed, a single altimeter cannot resolve all the mesoscale space and time scales [6], so it is essential to avoid duplications between adjoining tracks: if the satellite tracks are too close, it means that a different region is not properly observed. Here, "too close" is measured in the longitude/time plane because all satellite tracks have the same geometry. Furthermore, the metric is based on mesoscale covariance models used by Ducet et al. [20] to build a multi-altimeter gridded map of the ocean mesoscale. The covariance models are a statistical approximation of the size of mesoscale features in space and time [21,22], i.e., a good description of when altimeter tracks are too close and correlated.

Note that because SARAL is the only satellite of the 2016+ altimeter constellation that is near the historical ERS/ENVISAT orbit, we only optimize its own sampling: we do not try to maximize the combined sampling of the entire altimeter constellation: the ground track geometry and repeat cycles of other altimeters are very different so the constellation-wide gain would be very small.

Figure 1 shows the distribution of the satellite tracks for the ERS/ENVISAT altitude originally used by SARAL (approximately 780 km). Each black dot is one satellite track. The vertical alignment of the black dots corresponds to the 35-day exact repeat cycle of this orbit. The grey circles are 150 km by 15 days. This is an approximation of the decorrelation scale of mesoscale eddies at mid-latitudes.

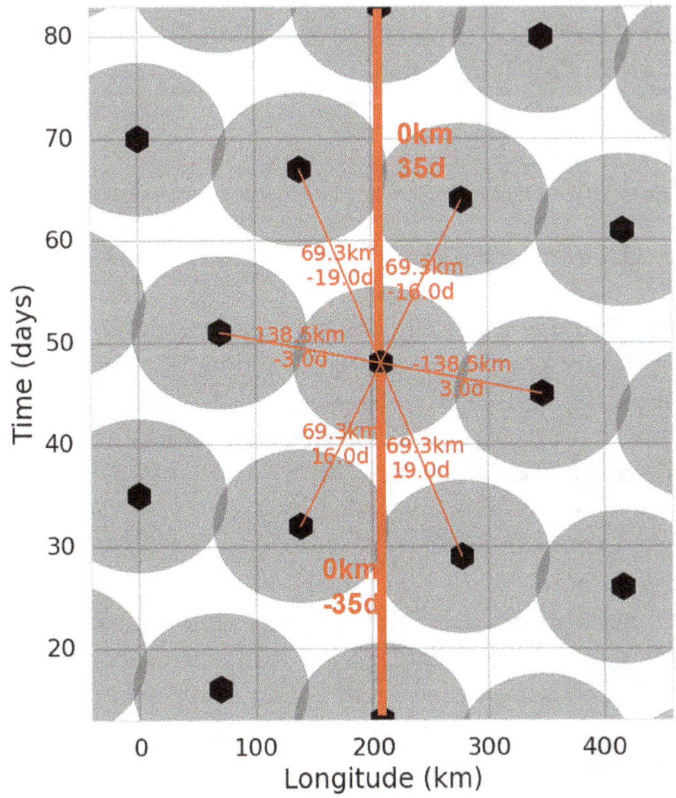

Figure 1. Longitude of Advanced Research and Global Observation Satellite (ARGOS) and AltiKa (SARAL/AltiKa) ascending tracks at 30°N as a function of time (black dots). The grey circles are 150 km by 15 days (typical decorrelation scale of ocean mesoscale). The red lines highlight the main orbit sub-cycles (major: 3 days, 16 days, minor: 19 days) as well as the 35-day repeat cycle of the European Remote-sensing Satellite (ERS)/ENVIronment SATellite (ENVISAT)/SARAL orbit.

In other words, if two grey circles overlap, then the corresponding satellite tracks are too close in space or in time: their measurements are highly correlated and in turn other regions of the space/time plane are completely unobserved. As expected from Le Traon and Dibarboure [23], the ERS/ENVISAT orbit is very good for mesoscale observation: the overlap between grey circles is almost non-existent as they are well-scattered in the longitude/time plane.

In addition to the true repeat cycle (highlighted with two thick vertical red lines), the diagonal alignment of grey circles in Figure 1 (thin red lines) shows the presence of orbital sub-cycles (SC). Orbital sub-cycles are defined by Rees [24] as a period of near-repeat for Earth remote-sensing satellites. The influence of SC in the context of altimetry missions is explained by Dibarboure et al. [19].

For the ERS/ENVISAT/SARAL altitude there are two dominant sub-cycles: 3 days and 16 days. In practice, dominant sub-cycles could also be defined by their global coverage: a sub-cycle corresponds to the time needed to collect a coverage that is globally homogeneous (e.g., 1000 km resolution after 3 days, or 180 km after 16 days). The 1002 tracks of the ERS/ENVISAT/SARAL repeat cycle are composed of two interleaved 16-day periods and one 3-day period. A 19-day sub-cycle is also visible in Figure 1: it is composed of a 16-day and a 3-day sub-cycle.

2.2. Predicting the Altitude Loss

Section 3.1 will show that the quality of the mesoscale sampling changes with the altitude and the associated sub-cycles. To determine how the sampling will evolve throughout the drifting phase, it is necessary to know how the altitude is likely to change in 3 years or more.

Like all low-earth orbit satellites, SARAL experiences an atmospheric drag force in a direction opposite to the direction of its motion. This force slows down the satellite, which in turn decreases its altitude. The drag force is given by Equation (1):

$$F_d = \frac{1}{2} \cdot \rho \cdot v \cdot C_b \tag{1}$$

where ρ is the atmospheric density, v is the satellite speed, and C_b the ballistic coefficient. Note that if the orbit is circular, the speed of the satellite is constant and it is a function of the orbit altitude. For the sake of simplicity, we here ignore that the satellite drag actually depends on the speed relative to the atmosphere which in turn creates a modulation of the drag force with latitude: the effect on the ground track geometry is extremely small and an analysis on the longitude of ascending node only requires us to know the mean satellite drag over a long period of time. The ballistic coefficient C_b is given by Equation (2) where C_x is the drag coefficient of the satellite, S is the area of the cross section surface in the direction of flight, and m is the mass of the satellite. In practice the coefficients used for SARAL in this study were estimated through an analysis of all the maneuvers performed by ISRO during the 3-year nominal mission.

$$C_b = \frac{C_x \cdot S}{m} \tag{2}$$

To get a projection of the altitude decay, it is necessary to have a prediction of the atmospheric density. The density ρ as a function of the altitude h around a reference altitude h_0 can be approximated by a simple exponential with variable scale height H (Equation (3)). Here, the scale height depends on the solar radio flux which is a proxy of the upper atmosphere density [25]. In practice, it is more accurate to use more sophisticated atmosphere models (e.g., MSIS2000). These models also use solar activity coefficient to parameterize the atmospheric density at all altitudes.

$$\rho = \rho_0 \cdot e^{\frac{h-h_0}{H}} \tag{3}$$

In this study, we used the solar flux predictions from the National Aeronautics and Space Administration (NASA) Marshall Space Flight Center [26]. The solar flux prediction used to prepare the SARAL DP in 2016 is shown in Figure 2a. Using the atmosphere model and the measured ballistic coefficient of SARAL, we obtain the altitude prediction in Figure 2b.

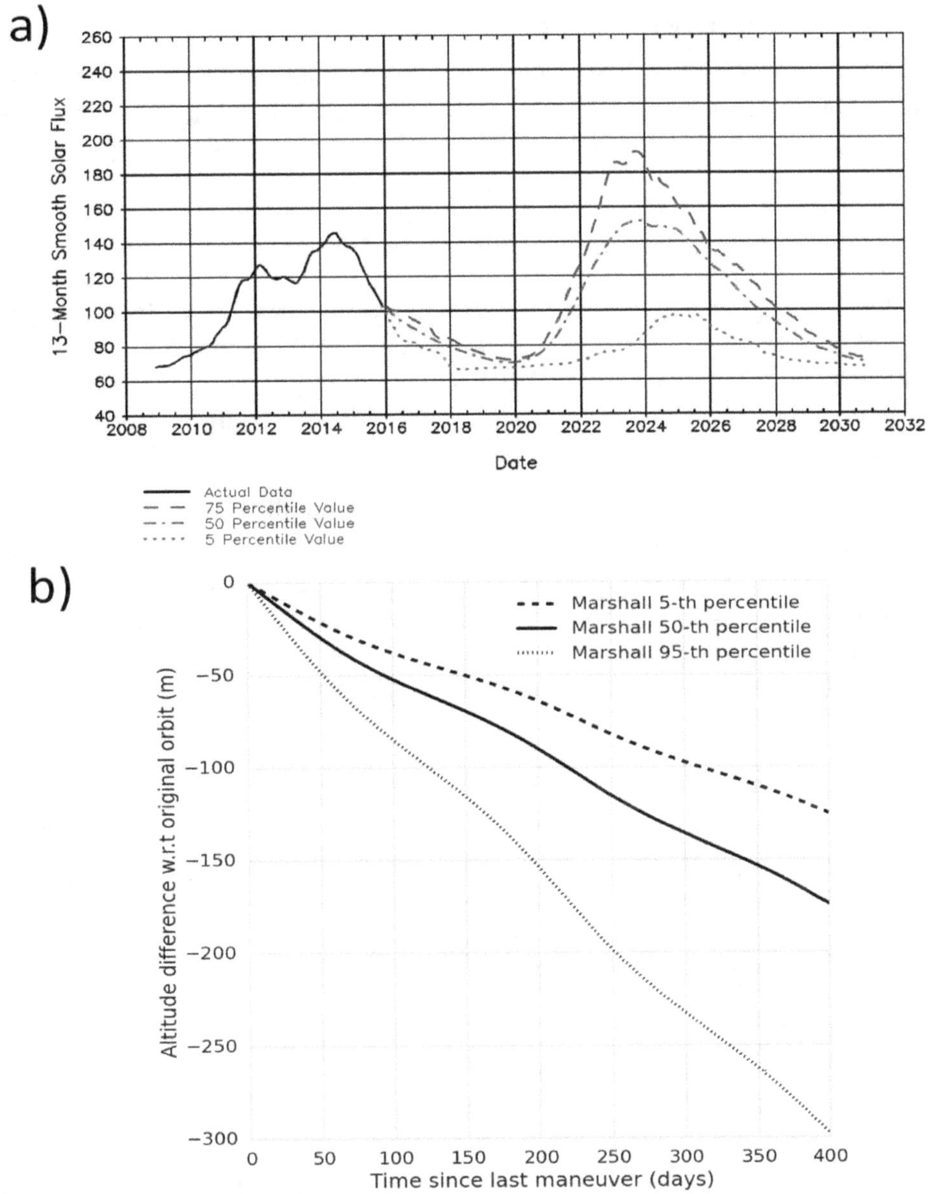

Figure 2. Panel (**a**) shows the solar flux prediction (unit: sfu, i.e., 10^{-22} $W{\cdot}m^{-2}{\cdot}Hz^{-1}$) used as an input (from the National Aeronautics and Space Administration (NASA) Marshall Space Flight Center). Panel (**b**) shows the altitude decay with respect to the original ERS/ENVISAT orbit for three solar flux scenarios: 5-th percentile (weak solar activity), 50-th percentile (median solar activity) and 95-th percentile (very intense solar activity).

The altitude decay is almost linear over one year, albeit modulated by the solar flux. To that extent, panel (b) shows three predictions: one for the 5-th percentile (weak solar activity, weak altitude decay), one for the 50-th percentile (median solar activity) and one for the 95-th percentile (strong solar activity).

The average altitude decay ranges from 150 m to 300 m per year. To prepare the SARAL drifting phase, we used a conservative value of 300 m/year (95-th percentile solar flux scenario): with this strategy it was very likely that the altitude would decrease more slowly. In other words, the drifting phase SARAL DP could be as long as three times our conservative estimate from Section 3.3 (the observed solar fluxes where indeed weaker and close to the 5-th percentile value, as discussed in Section 4.4).

3. Predictions Obtained to Define the Drifting Phase of SARAL

Based on the inputs from Section 2, the following sections explore the behavior of an unmaintained orbit strategy for SARAL, and the consequence for its mesoscale sampling capability. These predictions helped define the final strategy of the SARAL drifting phase. The observations after 18 months of SARAL DP are discussed in Section 4.

3.1. Diversity of the Mesoscale Sampling Near the ERS Altitude

The SARAL DP strategy was designed for a baseline of 3 years (i.e., as long as nominal lifespan of SARAL) with a backup scenario as long as 9 years (e.g., scenario where SARAL remains operational as long as the TOPEX/Poseidon mission). Because of the unmaintained decaying altitude and our conservative value of 300 m/year, the entire drifting phase could span over an altitude range of 1 km (baseline) to 3 km (backup).

But the mesoscale sampling properties over such a range of altitude can be very different: Figure 3 shows the position of SARAL tracks in the longitude/time plane for 5 different altitudes near the ERS orbit. Panel a in Figure 3 is 2 km above: it exhibits an exact 13-day repeat cycle and the 3-day sub-cycle that exists for all orbits in this altitude range. In panel (a), mesoscale sampling is not as good as for the ERS altitude because the 13-day repeat cycle is slightly too short: the grey circles are overlapping in the vertical direction. This weakness is the same as for the TOPEX orbit [23].

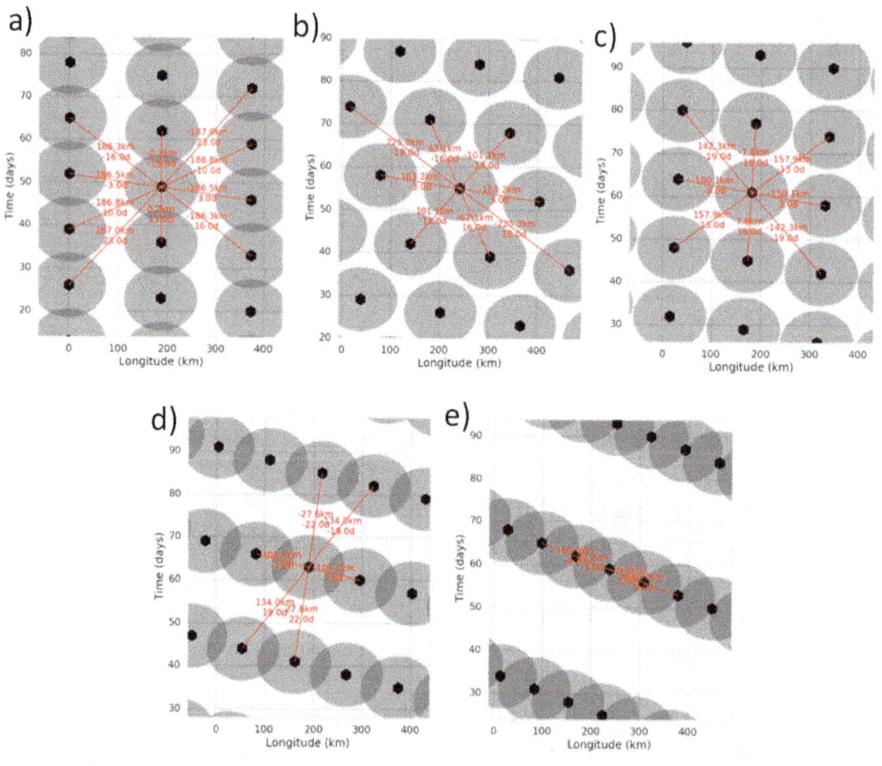

Figure 3. Same as Figure 1 for 5 altitudes near the historical ERS orbit. Panel (**a**) is approximately +2 km with respect to the ERS orbit. Panel (**b**) is approximately +1 km. Panel (**c**) is approximately +0.5 km. Panel (**d**) is approximately −1.5 km. Panel (**e**) is approximately −2.5 km. The alignment of gray circles (i.e., mesoscale-relevant orbit sub-cycles) is very different even though these 5 altitudes are relatively close to the ERS orbit.

Panel b in Figure 3 is 1 km above the ERS orbit. Here, the distribution of altimeter tracks in the longitude/time plane is different: the grey circles are rotated clockwise with respect to panel (a). The 13-day exact repeat of panel (a) has morphed into a 13-day sub-cycle and the 16-day sub-cycle has appeared (the 16-day SC is a composition of the 13-day SC and one 3-day SC). As a result, the mesoscale sampling is excellent at this altitude: the altimeter tracks are as far away from one another as possible, and there is no overlap between them.

Figure 3c is 500 m above the ERS altitude. The grey circles are rotated clockwise with respect to panel b in Figure 3, but the sampling capability is still excellent. The 16-day sub-cycle is very strong (almost vertical, i.e., a true repeat cycle), and the 19-day sub-cycle starts to appear due to the clockwise rotation (composition of a 16-day SC and a 3-day SC).

Panel d and panel e in Figure 3 are 1.5 km and 2.5 km below the SARAL altitude, respectively. At these altitudes, the sub-cycles are longer than 20-days, and the grey circles are largely overlapping: they are separated by less than 70 km and 3 days. The overlaps highlight the regions of the space and time plane that are correlated (as defined in Section 2.1) with two or more altimeter measurements in black. In panel (d) and panel (e), there is a large amount of overlap between adjoining circles; the visible consequence is that there are large bands of the space/time plane which are not observed whatsoever (not covered by grey circles). In essence, the nadir tracks are too close in space or in time and the resulting duplication between altimeter measurements creates large blind regions elsewhere.

The poor mesoscale sampling in panel (d) and (e) is due to the clockwise rotation of the grey circles in the longitude/time plane: they now align diagonally (very strong 3-day SC). This poor sampling pattern is typically observed on CryoSat-2 [18].

To summarize, the 3 examples above the ERS/ENVISAT orbit are good for mesoscale sampling, and it is preferable to stay between the current altitude and +1 km because the sampling at +2 km is not as good. The other 2 examples below the ERS/ENVISAT orbit are progressively worse and should be avoided because their SC are too long for mesoscale. But these are only a handful of examples, and the satellite altitude will decay continuously and change the sub-cycles.

3.2. The Continuous Evolution of Sub-Cycles (SC)

To understand how the orbit sub-cycles would change when there is no station-keeping maneuvers, we performed a long simulation starting 2-km above the ERS altitude and using the 300 m/year altitude decay obtained in Section 2.2. We chose to keep a constant altitude decay to show how a 15-year SARAL DP (starting in summer 2016) would look like at all altitudes.

Figure 4 shows the longitude of all subsatellite tracks as the altitude decays. We highlighted the altitudes studied in the 5 panels of Figure 3. The timeline goes from top to bottom (altitude decaying with time, 1 km of altitude is approximately equivalent to 3 years) so here the rotation of the grey circles discussed in Section 3.1 is counterclockwise: with this broader scale the rotation looks like horizontal parabolas (highlighted as colored arrows).

This continuous parabola-shaped effect is created by the modulo between the duration of a satellite revolution and the Earth rotation as seen by the satellite: a tiny change in the altitude also reduces the revolution period and the new modulo rotates the altimeter tracks in the longitude/time plane.

This result is important for SARAL DP: all sub-cycles appear, strengthen, become a true repeat cycle, weaken, and then disappear. More importantly this phenomenon occurs in a continuous way and the large parabolas are approximately 1-km large in altitude: in other words, if we find one orbit with good sampling properties, then there is also a 3-year time-frame where the sampling will be good because the sub-cycles change only progressively.

Incidentally, Figure 4 shows that larger sub-cycles are composed of smaller sub-cycles. Most SC can be decomposed as 1-day plus a given multiple of the 3-day sub-cycle (13, 16, 19, 22), but there are also larger sub-cycles composed of shorter sub-cycles (e.g., the 35-day repeat of ERS is 16 + 16 + 3 or 16 + 19). This rather complex composition effect is interesting because if two altitudes are attractive and separated by 1–2 km, then all intermediate altitudes will be good for mesoscale because the grey circle rotation seen in Figure 3 will ensure that the composition of intermediate sub-cycles is also good.

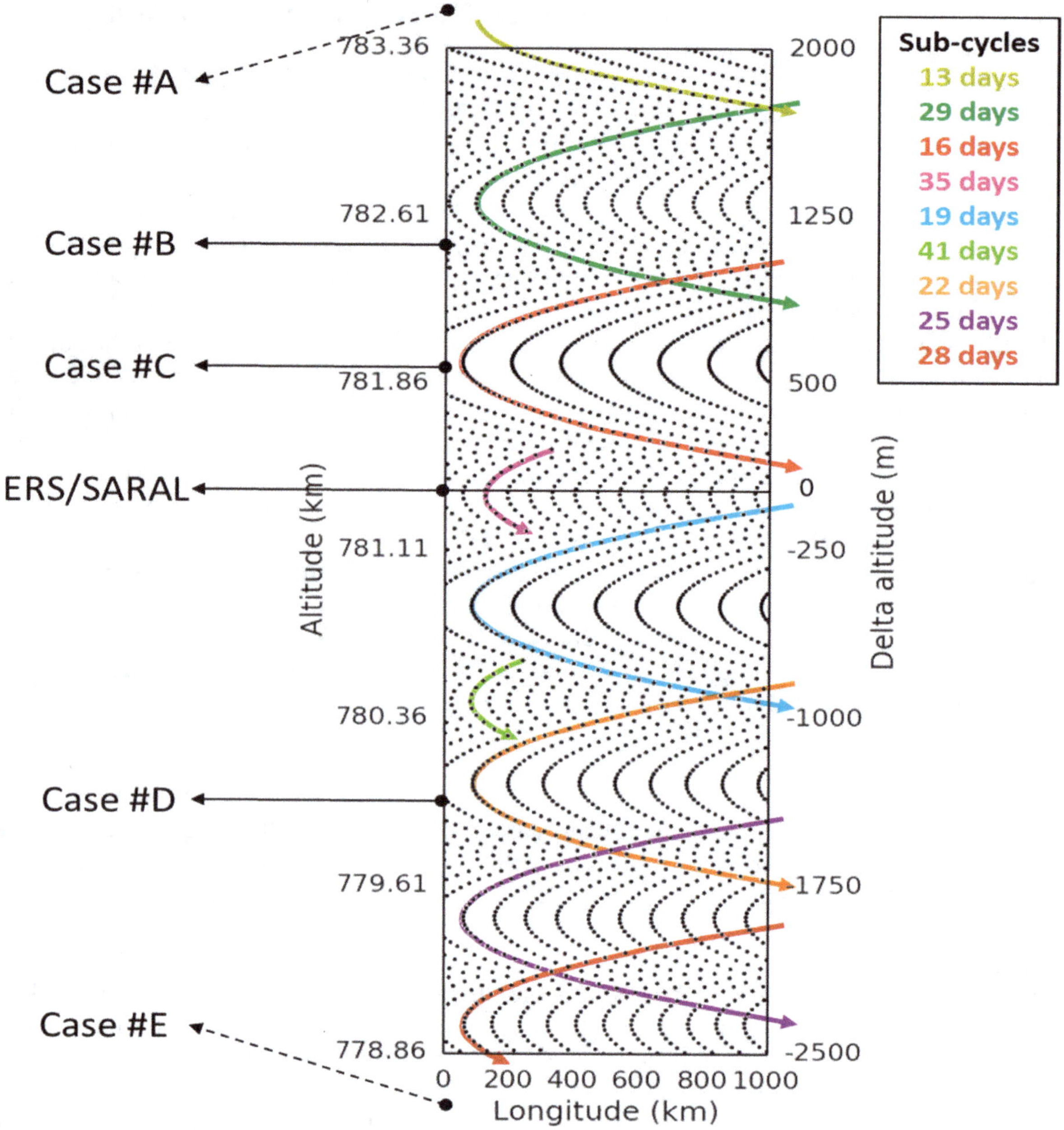

Figure 4. Simulated longitude of the AltiKa ascending tracks over a 15-year period. The simulation starts 2 km above the ERS orbit and it decreases by approximately 300 m per year. The continuously decreasing altitude transforms the orbit repeat cycle and sub-cycles (highlighted as colored parabolas): the 3-day sub-cycle of this altitude range generates composite sub-cycles increasing by 3 days (e.g., 13, 16, 19, 22, 25, 28), and the intermediate sub-cycles compose higher level sub-cycles (e.g., 29-day from 13 and 16, or 35 from 16 and 19). The circle alignment of the 5 panels in Figure 3 is explained by the colored parabolas.

As a result, the mesoscale-friendly altitudes aggregate in small bands that should be considered for SARAL DP. This aggregation was already observed but not explained by Dibarboure et al. [19] when they were looking for a good geodetic orbit for Jason-1.

3.3. Finding a Good Starting Point for the SARAL Drifting Phase (DP)

Assuming a baseline duration of 3 years, SARAL DP requires a 1-km altitude range where the sub-cycles are good for mesoscale. Figure 3 shows that sub-cycles ranging from 13 to 19 days are good although 15 to 17 days is better. Lastly the altitude range should be as close to the ERS/ENVISAT altitude as possible in order to avoid a strong maneuver when the drifting phase starts.

With all these constraints, Figure 4 shows that the best starting point for SARAL DP is approximately 1 km above the ERS/ENVISAT orbit: Figure 3b,c showed that the starting and intermediate altitudes yield an excellent sampling for mesoscale. After 3 years, SARAL/AltiKa is back at the ERS/ENVISAT altitude where the sampling capability is still good. From 3 to 6 years of drifting phase, the sampling is decent but sub-optimal due to the long 19-day SC. After 6 years, the sampling degrades because the main SC is too long. Figure 5 gives an overview of the dominant sub-cycles for all altitude as well as a qualitative index for mesoscale sampling.

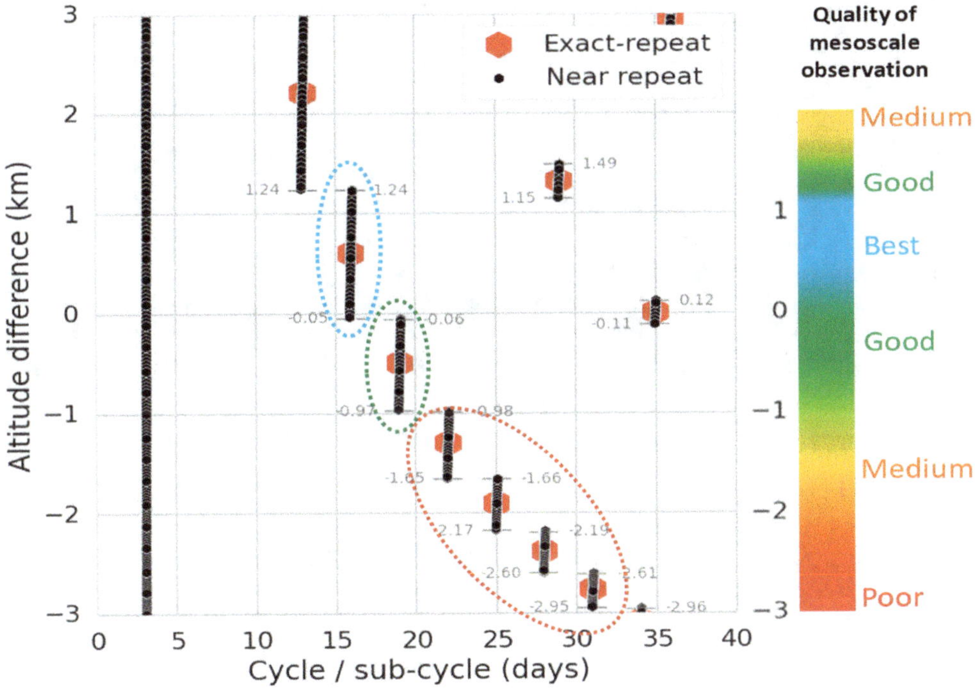

Figure 5. Overview of the dominant sub-cycles for a 6-km altitude range near the historical ERS orbit. The quality of mesoscale observation is controlled by some sub-cycles (15 to 19 days is better as shown in Figure 3) so some altitude ranges are better than others (colored ellipses and right-hand-side color bar).

3.4. Relaxed Altitude Control or No Control Whatsoever?

Assuming that SARAL/AltiKa is operated at least 3 years on the DP orbit, one might raise the following question: is it better to keep a relaxed altitude control or to have no altitude control whatsoever? E.g., one maneuver every 6–12 months to stay within 150–300 m of the altitude with an optimal sampling pattern for ocean mesoscale. As discussed above, such a yearly maneuver would be good but not essential for mesoscale for at least 3 years. But in this section we show that it would be very detrimental to the geodetic sampling.

Indeed, Zhang and Sandwell [27] recall that geodesists use altimeter data collected over a very long period (typically one year or more). They use these 1D altimeter datasets (subsatellite track) to compute a series of 2D gridded parameters such as geoid models, mean sea surface models, bathymetries, and gravity anomaly fields. For CryoSat-2 and Jason-1 GM (geodetic mission), the yearly cross-track resolution of altimeter tracks is of the order of 8 km at the equator. Thus, SARAL DP becomes exceedingly useful if it can collect a regular geodetic grid of 8 km after one year and a 4 km grid after two years. Zhang and Sandwell [27] emphasize the benefit of a drifting phase longer than 1.3 years, i.e., a cross-track resolution better than 6 km, to reveal a large number of uncharted seamounts, and important tectonic features.

Figure 6a shows a simulation of the zonal distance (in km) of all altimeter tracks (equator crossing, expressed as a distance with respect to the Greenwich meridian) after one year of SARAL DP: each track is a small black tick and the histogram shows how many tracks (i.e., how many black ticks) are located in each 8-km bin. Due to the uncontrolled altitude decay, the distribution is good albeit random with 0 to 3 satellite tracks per 8-km bin.

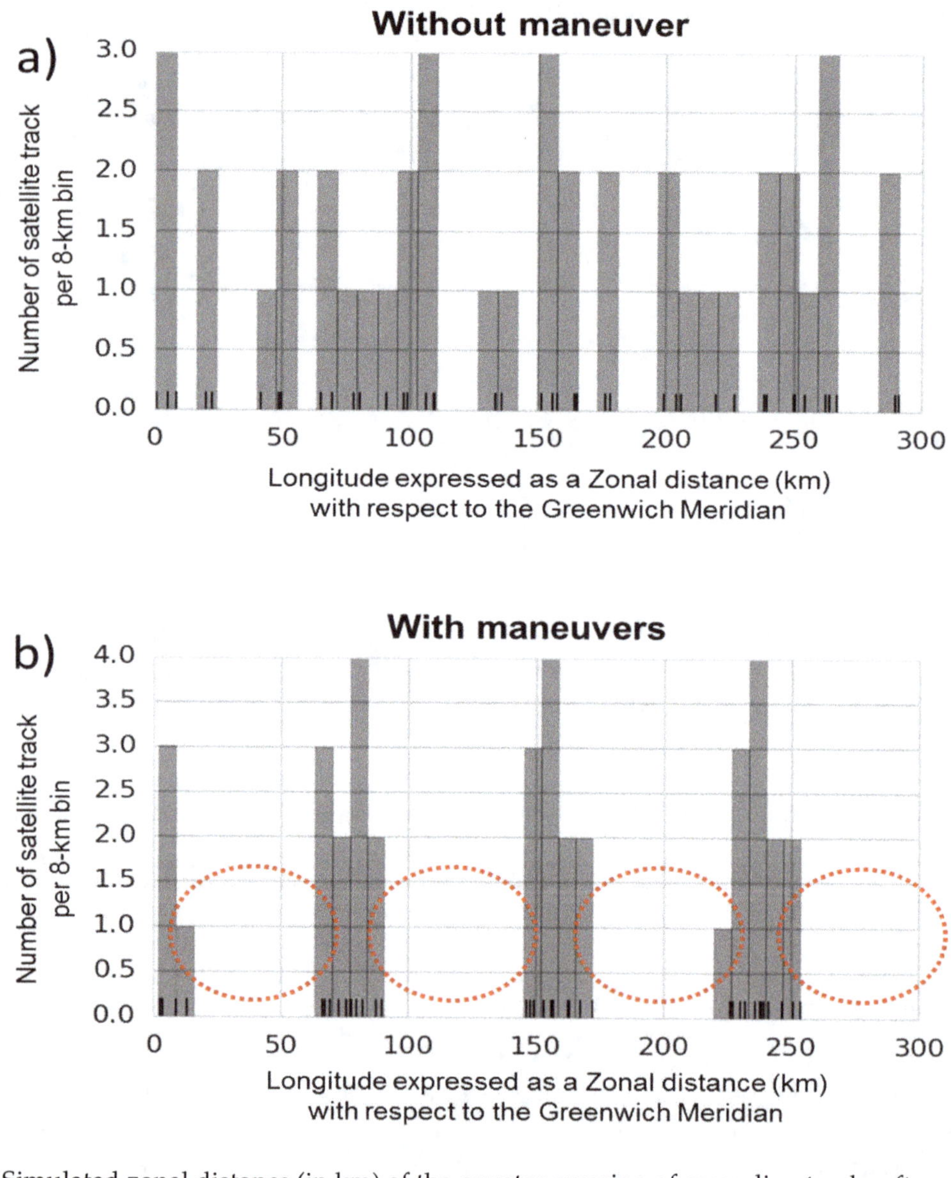

Figure 6. Simulated zonal distance (in km) of the equator crossing of ascending tracks after one year of AltiKa drifting phase (DP) (black ticks) over a 2.7° longitude region. The grey histogram shows the number of altimeter tracks in 8-km bins. Panel (**a**) is a simulation where the satellite altitude is not maintained: the distribution of altimeter tracks is relatively random with 0 to 3 tracks in each 8-km bin). Panel (**b**) is a simulation where the orbit altitude decay is mitigated with one maneuver in the middle of the drifting phase: this maneuver single-handedly aggregates all altimeter tracks in 30-km bands (typically 2 to 4 tracks per 8-km bin), leaving all the other regions completely unobserved (red circles).

In other words, this prediction shows that one year of SARAL DP does not yield a perfect 8-km geodetic grid since some grey bins are empty and other bins have up to 3 satellite track samples. But on average, all regions are reasonably well covered, albeit with a somewhat random and irregular pattern. This kind of random distribution was predicted for all regions and all solar flux scenarios.

In contrast, if a relaxed altitude control is performed, Figure 6b shows a very different picture. In this scenario a single maneuver is performed after 5 months of simulation in order to pull back SARAL to the altitude where the drifting phase was initiated. As a result, all the satellite tracks are put together in small bands of 30-km, leaving 50% of the globe unobserved even after one year. Assuming this pattern is repeated every 6 months, the geodetic dataset collected by AltiKa over the global ocean is less attractive than in Panel (a) of Figure 6, where the random distribution ensures that observation gaps are neither large nor systematic.

To summarize, as soon as the drifting phase starts, it is better for geodesy users to stop all maneuvers altogether and to let the satellite drift. Assuming that emergency debris avoidance must be performed, it is better to decrease the altitude than to increase it.

4. Observations after 18 Months of SARAL DP

The altitude of SARAL was successfully increased by 1 km in Summer 2016. The satellite has been operated on a slowly decaying altitude for approximately 18 months at the time of writing. This section gives an overview of the results obtained with in-flight data. Section 4.1 gives an overview of the geodetic sampling collected so far, and Sections 4.2 and 4.3 present some results on ocean mesoscale.

4.1. Geodetic Sampling

As expected from Section 3.4 and Figure 6a, the most important change of the drifting phase is the unprecedented cross-track resolution: Figure 7 shows the position of descending SARAL subsatellite tracks during the 35-day repeat phase (panel a) and during the first 15 months of SARAL DP (panel b). The longitude resolution of the historical ERS orbit was of the order 50 km at mid-latitudes. The cyclic sampling left very large uncovered regions, even after Spring 2015 when SARAL departed up to 10 km away from its nominal track for a brief period.

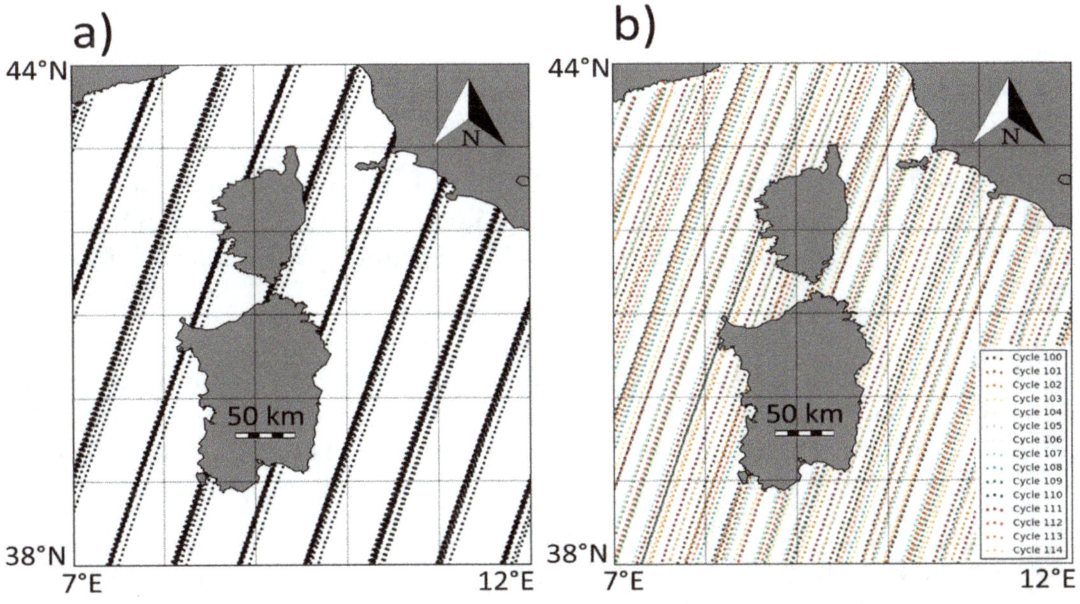

Figure 7. Position of the AltiKa measurements (1 Hz rate of the Geophysical Data Record product) from descending passes over a period of 15 months in the Mediterranean Sea. Panel (**a**) is for the 35-day repeat phase in 2015 and panel (**b**) is for the first 15 months of the drifting phase.

In contrast, the SARAL DP coverage is very dense (panel b, Figure 7): after 15 months, the grid resolution is approximately 4 km on average in this region. As discussed in Section 3.4, the geodetic grid is also somewhat random and very irregular: it can locally range from 1 km to 15 km.

Sandwell et al. [28] report that this grid is now used to enhance the smallest scales of geoid models and bathymetry fields in the open ocean. As more AltiKa measurements are collected, the local resolution will keep increasing, albeit still in an irregular way.

4.2. Mesoscale Sampling

Because the altitude of SARAL was increased at the beginning of the drifting phase, its mesoscale sampling has been modified as predicted in Section 3.3. Figure 8 shows the distribution of actual SARAL/AltiKa ascending nodes in the longitude/time plane (same metric as the simulations of Section 2 and Figure 1, but based on in-flight data). Panel (a) is for the 35-day repeat phase and panel (b) is for December 2017 after 18 months of SARAL DP. Each black dot is one satellite track, and the grey circles are their area of influence (high correlation) for large ocean mesoscale.

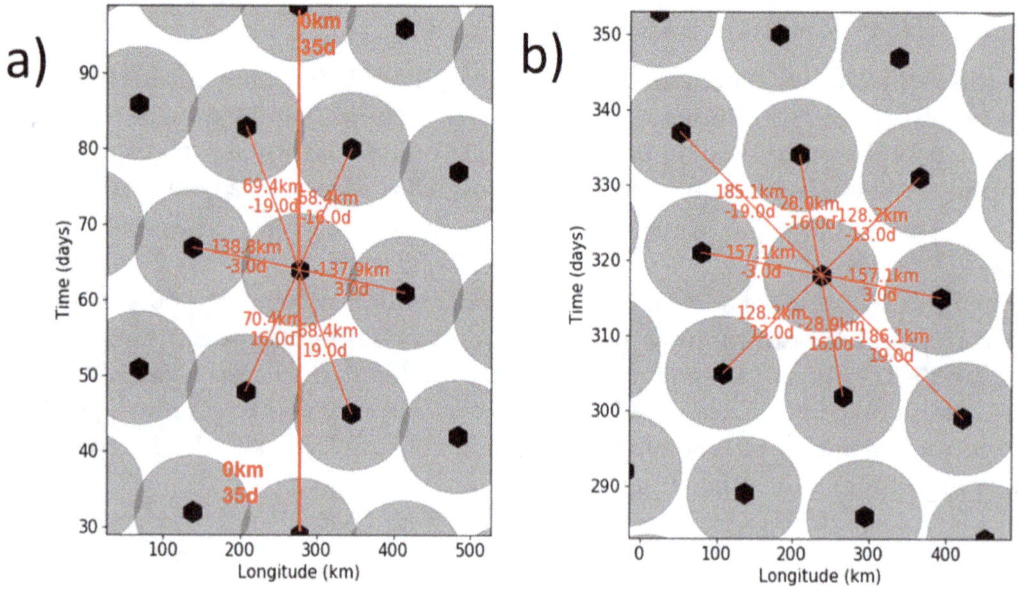

Figure 8. Same as Figure 1 for actual AltiKa data. Panel (**a**) is for the 35-day repeat phase in 2015 and panel (**b**) is December 2017 during the drifting phase.

Figure 8b shows that the distribution of SARAL DP tracks is consistent with the simulations from Figure 3, and more specifically with case #B. After 18-months the new phase of SARAL/AltiKa still provides a good sampling capability where each track/measurement minimizes the correlation with adjoining tracks.

More interestingly, the weight of SARAL/AltiKa in the Copernicus Marine Environment Monitoring Service [5] is routinely monitored in the frame of the operational production of multi-altimeter maps [22]: Figure 9 shows the evolution of the mean contribution of each altimeter used by the Optimal Interpolator. The interpolator is derived from Ducet et al. [20] and Le Traon et al. [21]. The global contribution of each satellite is measured using Degrees of Freedom of Signal following the methodology of Dibarboure et al. [29].

Figure 9a shows the contribution of SARAL/AltiKa in an idealized configuration where only the sampling change is taken into account. From September 2015 to June 2016, SARAL is on the historical 35-day orbit and its contribution is of the same order as Jason-2, i.e., substantially more than CryoSat-2. The weaker score of CryoSat-2 is consistent with findings from Dibarboure et al. [18]. Still, SARAL is slightly below Jason-2 by a couple of points on average. In contrast, after July 2016, SARAL DP is strictly equal to Jason-2: changing the altitude by +1 km slightly improved the sampling capability of AltiKa as predicted in Section 3.3.

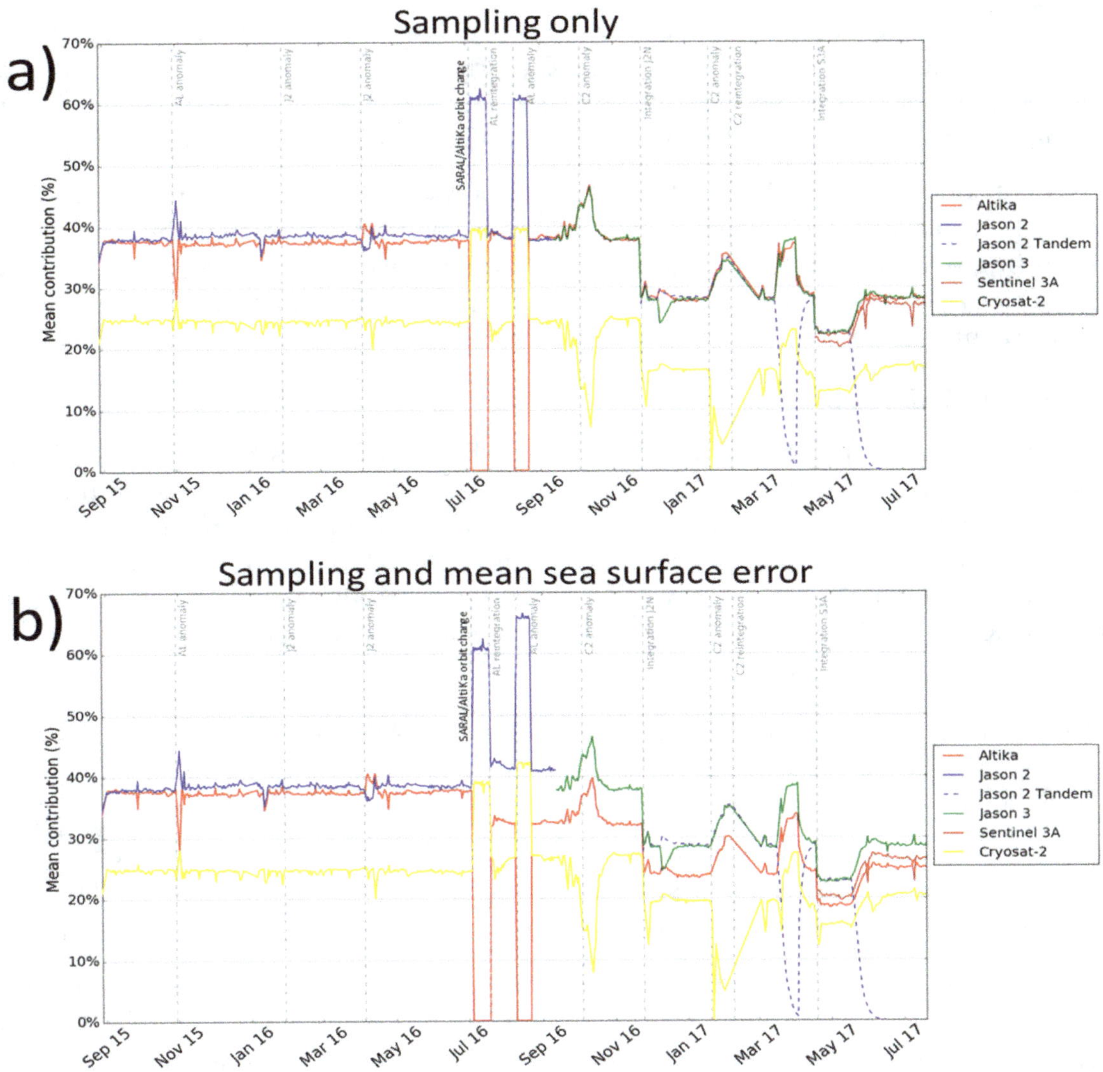

Figure 9. Relative contribution (%) of each altimeter to the multi-altimeter maps of the sea-level thematic assembly center of Copernicus Marine Environment Monitoring Service (CMEMS) from September 2015 to July 2017. Panel (**a**) measures the impact of SARAL DP sampling change only (the measurement error is assumed to be constant between the 35-day and the drifting phase). Panel (**b**) accounts for both the sampling change the increased error (sub-optimal mean sea surface model along the new uncharted orbit).

4.3. Impact of the Mean Sea Surface Error

An important component of the accuracy of altimetry products is the reference surface used to transform the Level-2 altimeter sea surface height (SSH) into Level-3 sea-level anomalies (SLA). Dibarboure et al. [18,19] have shown the difference between repeat orbits such as the 35-day orbit originally used for SARAL, and geodetic or non-repeat orbits such as SARAL DP.

The former can use a precise mean sea surface height (also known as mean profile) based on more than 2 decades of SSH from ERS-1 to SARAL: it is a very precise local estimate of the mean sea surface (MSS). In contrast, non-repeat missions rely on a global gridded MSS model which is slightly less accurate.

Pujol et al. [30] have measured the error of recent gridded MSS models using independent measurements from SARAL DP, HY-2A and Sentinel-3A. The error along uncharted tracks was shown to explain up to 30% of the total SLA variance for wavelengths ranging from 30 km to 100 km.

To that extent, Figure 9a is only an idealized view of the change between the 35-day phase and the drifting phase, because it does not account for the larger SLA error when a gridded MSS reference must be used in place of the precise 20-year mean profile. In contrast, Figure 9b takes into account both aspects of the orbit change: the slightly increased sampling capability and the additional MSS error.

The overall contribution of AltiKa decreases by approximately 6 points and in turn the relative contributions of other missions increase. Still, SARAL remains a strong contributor of the CMEMS system, including when both Jason-3 and Jason-2 plus Sentinel-3A are active. This is notably visible during the CryoSat-2 or Jason-2 data outages where the contribution of AltiKa exhibits some peaks as high a +5 to 7 points.

4.4. Evolution of the SARAL Altitude

After 18 months of drifting phase, the altitude of SARAL has decreased by less than 150 m. This is approximately the 5th percentile curve of the simulations from Figure 2, i.e., three times less than our conservative simulations of Figure 4 which was based on the 95th percentile scenario of the 2016 predictions. Thanks to a particularly weak solar activity, SARAL remains on the upper range of the desirable altitudes of Figure 5 (blue region, 16-day sub-cycle). In other words, the mesoscale sampling capability of SARAL/AltiKa will remain good for many years, and substantially longer than predicted by the conservative simulations.

5. Discussion: Using an Unmaintained Altitude with Other Altimeters

5.1. Generalizing the SARAL DP Strategy

SARAL/AltiKa is the first altimeter to be operated on an unmaintained decaying altitude. But other altimeters could use the same strategy in the future. There are essentially three reasons to use this kind of orbit in the context of radar altimetry:

- Platform ageing: if controlling the satellite's altitude becomes difficult or risky, it might be better to simply stop maneuvers altogether after the satellite is put on a well-chosen initial altitude.
- Simpler operations: keeping the satellite on its nominal orbit requires periodic operations (e.g., every 2 to 4 weeks for ERS and Jason altitudes). Letting the altitude decay naturally in a mesoscale-friendly altitude range might reduce the manpower needed for routine operations. While it is not significant for a single altimeter above 900 km, this strategy could be attractive for lower orbits (e.g., 500 km) or for altimeter constellations (e.g., 5+ satellites) because more manpower is required, and because each maneuver creates a small data gap or temporary degradation (e.g., discontinuity in precise orbit determination). Limiting the maneuvers to debris avoidance could reduce the operation cost of such a low-altitude constellation.
- Saving fuel: most of the satellite fuel is used during launch (to reach the nominal orbit) and for de-orbitation. Using a decaying altitude and avoiding station keeping maneuvers would save only a small amount of fuel. But for some satellites that experience major launch events, it is sometimes necessary to consume more fuel than anticipated (e.g., CryoSat-2 was injected relatively far away from its nominal altitude). In this situation, using a decaying altitude might be a backup scenario able to extend the satellite lifespan by saving fuel for the final de-orbitation.

To summarize, the SARAL DP decaying altitude strategy could be used for other altimeters in the future: as a new mission phase when the satellite suffers from ageing anomalies, as a backup orbit after launch anomalies, or for a dedicated mission/constellation put specifically at an altitude where the maintenance effort is limited.

In the following section, we expand the methodology and the results of Section 3 to determine automatically all the altitude ranges that are compatible with an SARAL DP strategy.

5.2. Methodology Addition

In Figure 3, we observed very different ocean mesoscale sampling properties. We used the overlap between grey circles as a way to detect poor orbits. This can be automated with the very simple correlation model (Equation (4)) used by Dibarboure et al. [19] for Jason-1.

$$C(dx, dt) = e^{-k\cdot\left(\frac{dx^2}{r_x^2}+\frac{\partial t^2}{r_t^2}\right)} \quad (4)$$

Here r_x and r_t are the spatial and temporal decorrelation radiuses of Section 2.1, and k is arbitrarily set to yield a correlation of 0.5 on the edge of the r_x/r_t circle. By computing $C(dx, dt)$ where dx and dt is the space and time distance between two altimeter tracks, it is possible to approximate the correlation between the datasets. Sensitivity tests show that the type of function and parameter k do not significantly impact the following results.

For a given altitude, we compute the position of all altimeter tracks in the space/time place over 50 days (i.e., complete mesoscale decorrelation), we pick one arbitrary track and we search for all possible neighbors in a space/time bubble. Then we keep the 5 non-aligned tracks with the highest correlation as they define the sub-cycles for this altitude. Pruning aligned tracks from this list is important: to illustrate, with a 3-day SC, we want to detect only the closest track aligned in this direction because 6, 9 and 12 days are only repeats of this SC (they are not separate sub-cycles because they all follow the same alignment).

This computation gives not only the list of dominant sub-cycles (e.g., 3 and 16 for the ERS/ENVISAT sub-cycles) but also all minor sub-cycles (e.g., 19-day SC visible in Figure 1) and the true repeat cycle (e.g., 35 days for ERS/ENVISAT). For each sub-cycle, we also have the correlation which indicates if this sub-cycle yields independent or correlated satellite tracks. If the correlation is too high for any SC, then the altitude is not attractive to observe mesoscale. If the correlation is low for all SC, then it is a good orbit candidate: it is worthy of more in-depth analyses not discussed in this paper.

5.3. Finding Good Altitude Ranges

We performed this computation for all altitudes from 300 km to 1500 km with an increment of 30 m. We used a threshold of C = 0.5 to reject mesoscale-unfriendly altitudes. Then we aggregated all valid orbits in altitude bands to determine how large these bands were.

Figure 10 shows the result near the ERS/ENVISAT altitude. Panel (a) of Figure 10, shows the correlation C for each altitude (each sub-cycle is displayed with a different color), and panel (b) of Figure 10, shows the valid (green) and invalid (red) regions based on our threshold (if any sub-cycle has a correlation C > 0.5 the altitude is in red). Panel (b) also highlights the 5 cases shown in Figure 3. This representation illustrates how the qualitative overlap discussed in previous sections is now quantified with this metric. The main green region in Panel (b) corresponds to the altitude selected by CNES and ISRO for SARAL DP.

5.4. Mesoscale-Friendly Altitude Bands for Unmaintained Altimeters

Figure 11 shows all the valid altitude bands found for the ERS/ENVISAT inclination. We performed sensitivity tests with the inclination of various altimetry missions (GEOSAT, Jason, Sentinel-3) that yielded similar results: the good/bad altitude ranges are translated but their properties are essentially the same.

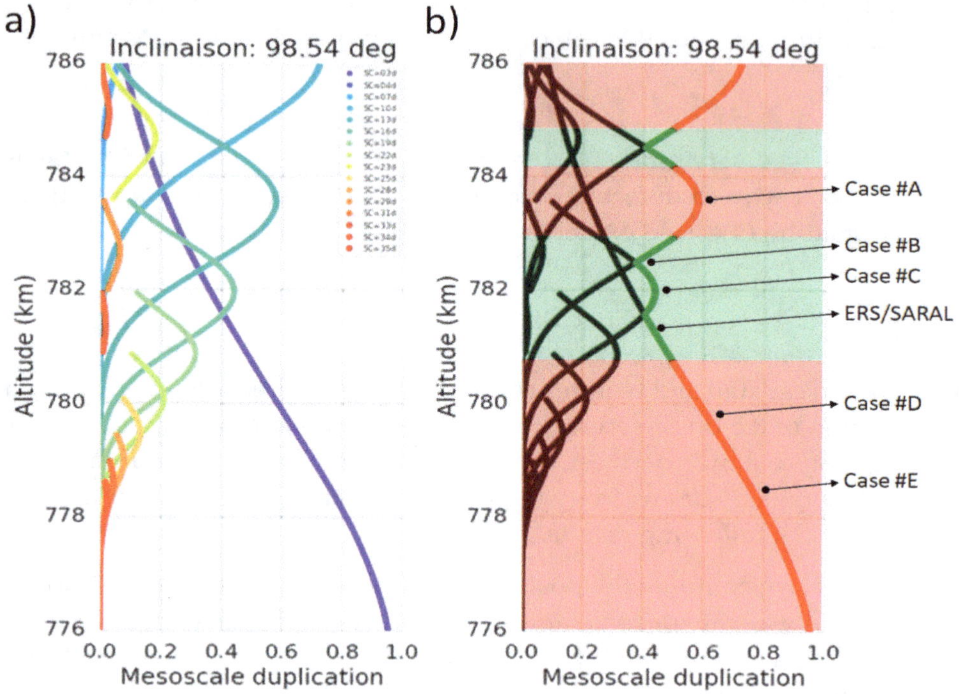

Figure 10. Quality of the mesoscale observation as a function of the altitude. For each altitude, we analyze the alignment of altimeter tracks in the space/time plane. Then we compute their correlation for ocean mesoscales. Each alignment is a sub-cycle and we can determine which sub-cycle contributes positively or negatively to the quality of mesoscale observation at this altitude (panel (**a**)). In panel (**b**), we separate favorable (green) from undesirable (red) altitudes by looking at the maximal correlation of each orbit sub-cycle. This metric corresponds to the overlap between the grey circles in Figure 1. The 5 orbit cases shown in Figure 3 and the ERS/ENVISAT altitude are highlighted in panel (**b**).

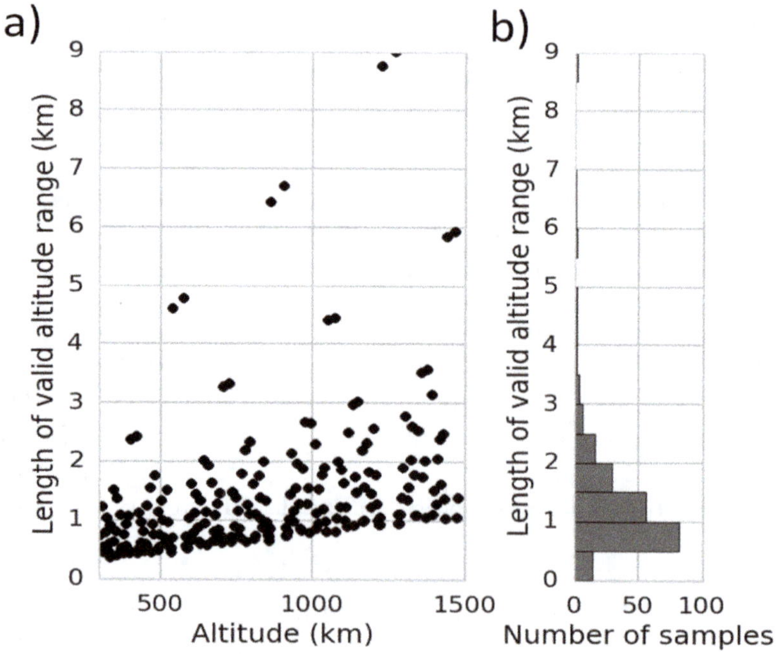

Figure 11. Length of the valid altitude bands (green regions from Figure 10b) for all altitudes and the inclination of the ERS orbit. Each dot in panel (**a**) is one green altitude range where the mesoscale correlation is always lower than 0.5. Panel (**b**) shows the associated distribution.

On average, they are rather thin (from 500 m to 1500 m) but a small fraction can be 2 to 3 km large. This finding is consistent with observations from Dibarboure et al. [19]: their orbit candidates for the so-called Extension-Of-Life phase of Jason-1 were found to be aggregated in thin altitude bands. In the case of SARAL DP there is a 2-km sweet spot just near the historical ERS/ENVISAT orbit: these larger bands are uncommon.

Furthermore Figure 11 also shows the existence of altitude bands as large as 9 km. These exceptional occurrences always come in pairs. This is explained by Figure 12: the longest valid altitude ranges are generally located symmetrically near the altitude of a very short repeat cycle. The black rectangles of panel (a) of Figure 12, on each side of the purple bell-shaped curve highlight are where valid altitude ranges are located. Panel (b) of Figure 12, is a zoom of the plain black rectangle of panel (a): it shows that the composition of a very short sub-cycle and multiple mesoscale-friendly sub-cycles minimizes ocean mesoscale duplication (i.e., the maximal correlation between adjoining satellite tracks). The validity of the altitude range is shown in green in panel (c) of Figure 12, the largest valid altitude band range from 1227 to 1236 km. The dashed black rectangle of panel (a) also contains a second valid range (essentially an upside-down mirror of panel (b)). The same dual/mirror pattern can be observed for most of the long valid ranges in Figure 11a.

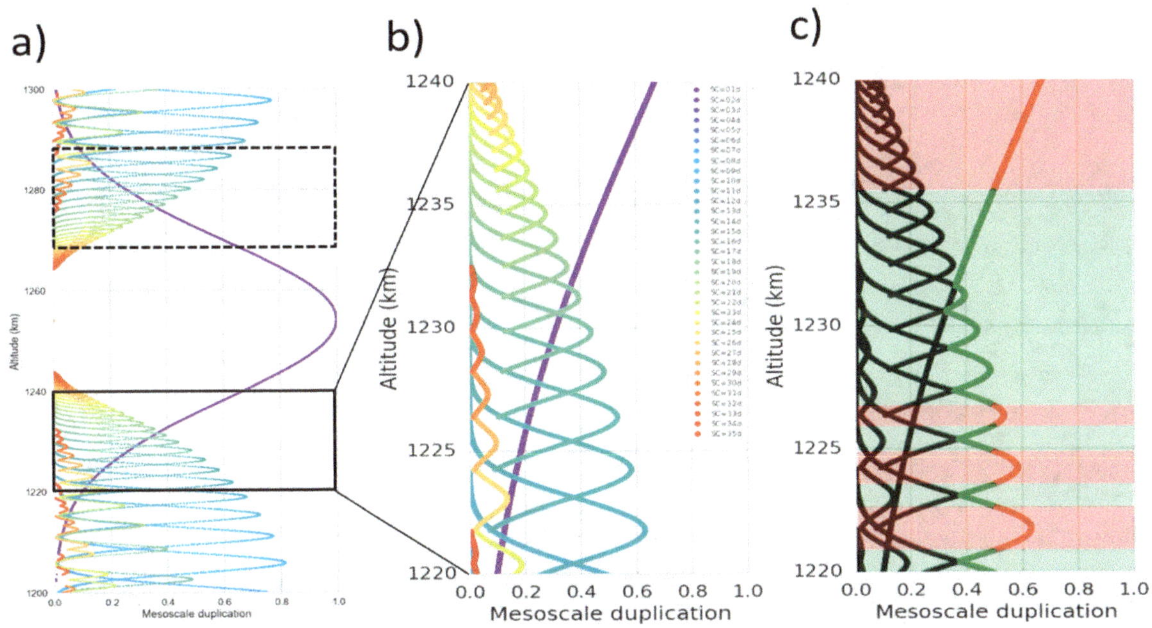

Figure 12. Same as Figure 10a from 1200 to 1300 km (panel (**a**)) and near 1230 km (panel (**b**)). In panel (**c**), we separate favorable (green) from undesirable (red) altitudes by looking at the maximal correlation of each orbit sub-cycle for the altitude range of panel (**b**). The altitude range in panel (**b**) and panel (**c**) corresponds to the plain rectangle of panel (**a**) and where the longest valid altitude range was found.

The 9-km bands are located above 1200 km where the atmospheric drag is very weak. To that extent, a dedicated altimeter mission parked at this altitude would have a great mesoscale sampling for decades. In practice, 1200 km could be too high for at least three reasons: aggressive radiation environment for the satellite, difficult to de-orbit at the end of the mission, unfavorable signal-to-noise ratio for the altimeter (e.g., possibly incompatible with a very small satellite).

In contrast, one could envision a low-cost, low-maintenance altimeter near 600 km for the following reasons:

- The lower altitude would better protect the satellite from radiations.
- The satellite de-orbitation would be easier (e.g., good for a small satellite).

- Altitudes lower than 700 km are attractive because the space debris density is lower: in turn one could anticipate fewer collision avoidance maneuvers (simpler operations)
- Lower altitudes yield a slightly better signal-to-noise ratio for the altimeter (less electrical power needed, i.e., also good for a small satellite).

Nevertheless, at 600 km the atmospheric drag increases exponentially (Figure 13). Even if there is a 4.5 km valid altimeter band that could host such as dedicated mission, the altitude of a SARAL-looking satellite would decay approximately 10 times faster than the 300 m/year of SARAL. To be viable, the platform should be specifically designed to minimize the atmospheric drag with an aerodynamic profile (e.g., smaller and compact solar arrays like Cryosat-2). The lifespan of the mission could also be extended by a handful of maneuvers (e.g., every couple of years) that would reset the altitude decay and start over a new maneuver-free cycle.

Such a low-cost and low-maintenance altimeter mission would be relevant in the context of low-cost constellations (e.g., 4 to 10 satellites): if all satellites are located at the same altitude in one sun-synchronous orbit plane, the altitude of all altimeters decays in a coherent way. With this scenario, the sampling capability of the entire constellation remains good throughout the entire altitude decay. In essence, this concept is a fire-and-forget constellation of small altimetry satellites whose orbit maintenance is limited to debris avoidance.

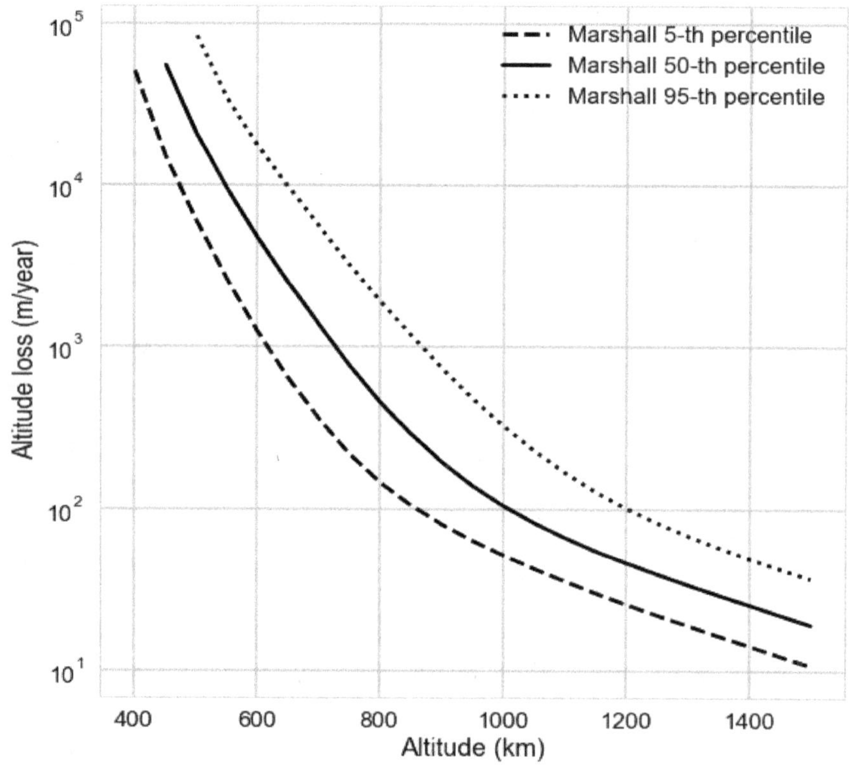

Figure 13. Approximate altitude loss (m/year) for an arbitrary satellite as a function of the initial altitude (km). Three scenarios are provided: three solar flux scenarios: 5-th percentile (weak solar activity, dashed), 50-th percentile (median solar activity, plain) and 95-th percentile (intense solar activity, dotted).

6. Conclusions

In this paper we showed that it is possible for altimeter missions to keep a good mesoscale sampling for years even if their altitude is not actively maintained and even if it decays due to the atmospheric drag. This strategy has been successfully applied on SARAL/AltiKa during the so-called drifting phase that was initiated in early July 2016 by ISRO and CNES.

Eighteen months later, SARAL DP has demonstrated the benefits of this strategy: reducing the maneuvers has made it possible to extend the satellite lifetime (less stress on momentum wheels), and the new DP orbit provides an enhanced sampling capability of ocean mesoscale, as well as precious measurements to improve marine geodesy references fields (e.g., bathymetry, geoid or mean sea surface).

For SARAL, the optimal starting point of the drifting phase was approximately 1 km above the historical ERS/ENVISAT orbit. Thanks to a continuously evolving combination of the 13, 16 and 19-day sub-cycles, the mesoscale sampling capability of SARAL DP has been secured for 6 years or more.

More generally, this strategy can be used for any altimeter (e.g., Jason-3 or Sentinel-3), for instance if the satellite maneuvering capability is limited by onboard ageing anomalies. As long as the satellite is put at the top of a favorable altitude range, the decaying altitude will not degrade the mesoscale sampling capability of the altimeter. It is possible to find mesoscale-friendly altitudes at all altitudes from 500 km to 1500 km.

We also identified a small list of larger mesoscale-friendly altitude bands where one could park a full-fledged drifting altimetry mission, or an entire altimeter constellation. In these sweet spots, the quality of mesoscale sampling is guaranteed for a very long time even if the altitude is not maintained. Near 1230 km, there are two very large altitude bands (9 km each) where an unmaintained altimeter would yield a good and stable mesoscale sampling for decades. At 600 km two large altitude ranges (4.5 km each) makes it possible to envision a low-cost/low-maintenance small altimeter satellite (good SNR, easy de-orbitation, orbit control limited to debris avoidance).

Author Contributions: G.D. carried out the mesoscale and sub-cycle analysis as well as mesoscale sampling simulations. A.L. performed the analysis of SARAL maneuvers and he computed the prediction of altitude decay. M.-I.P. performed the mesoscale sampling analysis on in-flight data using CMEMS/DUACS metrics. G.J. provided the geodetic coverage analysis of in-flight-data.

Acknowledgments: The SARAL/AltiKa mission is an achievement of the fruitful cooperation between ISRO and CNES. We acknowledge the support of the SARAL/AltiKa science and project teams during the preparation of the SARAL drifting phase.

References

1. Verron, J.; Sengenes, P.; Lambin, J.; Noubel, J.; Steunou, N.; Guillot, A.; Picot, N.; Coutin-Faye, S.; Sharma, R.; Murthy, D.R.; et al. The SARAL/AltiKa altimetry satellite mission. *Mar. Geod.* **2015**, *38* (Suppl. 1), 2–21. [CrossRef]
2. Vincent, P.; Steunou, N.; Caubet, E.; Phalippou, L.; Rey, L.; Thouvenot, E.; Verron, J. AltiKa: A Ka-band Altimetry Payload and System for Operational Altimetry during the GMES Period. *Sensors* **2006**, *6*, 208–234. [CrossRef]
3. Morrow, R.; Seo, H.; Farrar, J.T.; Fu, L.L.; Le, P.-Y. Traon Ocean Eddies and Mesoscale Variability. In *Satellite Altimetry Over Oceans and Land Surfaces*; CRC Press: Boca Raton, FL, USA, 2017; pp. 315–342.
4. AVISO. SSALTO/DUACS User Handbook: (M)SLA and (M)ADT Near-Real Time and Delayed Time Products. CNES Document SALP-MU-P-EA-21065-CLS. 2016. Available online: https://www.aviso.altimetry.fr/en/data/product-information/aviso-user-handbooks.html (accessed on 2 July 2018).
5. CMEMS. Copernicus Marine Environment Monitoring Service–Sea Level Thematic Assembly Center Product User Manual. Available online: http://marine.copernicus.eu/documents/PUM/CMEMS-SL-PUM-008-017-036.pdf (accessed on 2 July 2018).
6. Chelton, D.B.; Schlax, M.G. The accuracies of smoothed sea surface height fields constructed from tandem altimeter datasets. *J. Atmos. Ocean. Technol.* **2003**, *20*, 1276–1302. [CrossRef]
7. Pascual, A.; Boone, C.; Larnicol, G.; Le Traon, P.Y. On the quality of real-time altimeter gridded fields: Comparison with in situ data. *J. Atmos. Ocean. Technol.* **2009**, *26*, 556–569. [CrossRef]

8. Verron, J.; Bonnefond, P.; Aouf, L.; Birol, F.; Bhowmick, S.A.; Calmant, S.; Conchy, T.; Crétaux, J.-F.; Dibarboure, G.; Faugère, Y. The benefits of the Ka-band as evidenced from the SARAL/AltiKa altimetric mission: Scientific applications. *Remote Sens.* **2018**, *10*, 163. [CrossRef]
9. Bonnefond, P.; Verron, J.; Aublanc, J.; Babu, K.N.; Bergé-Nguyen, M.; Cancet, M.; Chaudhary, A.; Crétaux, J.-F.; Frappart, F.; Laurain, O. The Benefits of the Ka-Band as Evidenced from the SARAL/AltiKa Altimetric Mission: Quality Assessment and Unique Characteristics of AltiKa Data. *Remote Sens.* **2018**, *10*, 83. [CrossRef]
10. Rémy, F.; Flament, T.; Michel, A. Envisat and SARAL/AltiKa observations of the Antarctic ice sheet: A comparison between the Ku-band and Ka-band. *Mar. Geod.* **2015**, *38*, 510–521. [CrossRef]
11. Smith, W.H.F. Resolution of Seamount Geoid Anomalies Achieved by the SARAL/AltiKa and Envisat RA2 Satellite Radar Altimeters. *Mar. Geod.* **2015**, *38*, 644–671. [CrossRef]
12. Valladeau, G.; Thibaut, P.; Picard, B.; Poisson, J.C.; Tran, N.; Picot, N.; Guillot, A. Using SARAL/AltiKa to improve Ka-band altimeter measurements for coastal zones, hydrology and ice: The PEACHI prototype. *Mar. Geod.* **2015**, *38* (Suppl. 1), 124–142. [CrossRef]
13. Arsen, A.; Crétaux, J.F.; Abarca del Rio, R. Use of SARAL/AltiKa over mountainous lakes, intercomparison with Envisat mission. *Mar. Geod.* **2015**, *38* (Suppl. 1), 534–548. [CrossRef]
14. Bhowmick, S.A.; Modi, R.; Sandhya, K.G.; Seemanth, M.; Balakrishnan Nair, T.M.; Kumar, R.; Sharma, R. Analysis of SARAL/AltiKa wind and wave over Indian Ocean and its real-time application in wave forecasting system at ISRO. *Mar. Geod.* **2015**, *38* (Suppl. 1), 396–408. [CrossRef]
15. Bhowmick, S.A.; Basu, S.; Sharma, R.; Kumar, R. Impact of Assimilating SARAL/AltiKa SWH in SWAN Model During Indian Ocean Tropical Cyclone Phailin. *IEEE Trans. Geosci. Remote Sens.* **2016**, *54*, 1812–1817. [CrossRef]
16. Bonnefond, P.; Willis, J. Minutes of the 2015 Ocean Surface Topography Science Team Meeting. Available online: http://www.aviso.altimetry.fr/fileadmin/documents/OSTST/OSTST_2015_Meeting_Report.pdf (accessed on 2 July 2018).
17. Sandwell, D.; Dietmar Müller, R.; Smith, W.; Garcia, E.; Francis, R. New global marine gravity model from CryoSat-2 and Jason-1 reveals buried tectonic structure. *Science* **2014**, *346*, 65–67. [CrossRef] [PubMed]
18. Dibarboure, G.; Renaudie, C.; Pujol, M.I.; Labroue, S.; Picot, N. A demonstration of the potential of Cryosat-2 to contribute to mesoscale observation. *Adv. Space Res.* **2012**, *50*, 1046–1061. [CrossRef]
19. Dibarboure, G.; Schaeffer, P.; Escudier, P.; Pujol, M.-I.; Legeais, J.F.; Faugère, Y.; Morrow, R.; Willis, J.K.; Lambin, J.; Berthias, J.P.; et al. Finding desirable orbit options for the "Extension of Life" phase of Jason-1. *Mar. Geod.* **2012**, *35*, 363–399. [CrossRef]
20. Ducet, N.; Le Traon, P.Y.; Reverdin, G. Global high resolution mapping of ocean circulation from the combination of TOPEX/POSEIDON and ERS-1/2. *J. Geophys. Res. (Oceans)* **2000**, *105*, 19477–19498. [CrossRef]
21. Le Traon, P.Y.; Faugère, Y.; Hernandez, F.; Dorandeu, J.; Mertz, F.; Ablain, M. Can we merge GEOSAT Follow-On with TOPEX/POSEIDON and ERS-2 for an improved description of the ocean circulation? *J. Atmos. Ocean. Technol.* **2003**, *20*, 889–895. [CrossRef]
22. Pujol, M.-I.; Faugère, Y.; Taburet, G.; Dupuy, S.; Pelloquin, C.; Ablain, M.; Picot, N. DUACS DT2014: The new multi-mission altimeter dataset reprocessed over 20 years. *Ocean Sci. Discuss.* **2016**, *12*, 1067–1090. [CrossRef]
23. Le Traon, P.Y.; Dibarboure, G. Mesoscale mapping capabilities from multiple altimeter missions. *J. Atmos. Ocean. Technol.* **1999**, *16*, 1208–1223. [CrossRef]
24. Rees, W.G. Orbital subcycles for Earth remote sensing satellites. *Int. J. Remote Sens.* **1992**, *13*, 825–833. [CrossRef]
25. Priester, W. Solar activity effect and diurnal variation in the upper atmosphere. *J. Geophys. Res.* **1961**, *66*, 4143–4148. [CrossRef]
26. Hathaway, D.H.; Wilson, R.M.; Reichmann, E.J. A synthesis of solar cycle prediction techniques. *J. Geophys. Res.* **1999**, *104*, 22375–22388. [CrossRef]
27. Zhang, S.; Sandwell, D.T. Retracking of SARAL/AltiKa radar altimetry waveforms for optimal gravity field recovery. *Mar. Geod.* **2017**, *40*, 40–56. [CrossRef]
28. Sandwell, D.; Müller, D.; Matthews, K.; Smith, W. Imaging Small-scale Seafloor and Sub-seafloor Tectonic Fabric Using Satellite Altimetry. In Proceedings of the 2017 AGU Fall Meeting, New Orleans, LA, USA, 11 December 2017.

29. Dibarboure, G.; Pujol, M.I.; Briol, F.; Traon, P.L.; Larnicol, G.; Picot, N.; Mertz, F.; Ablain, M. Jason-2 in DUACS: Updated system description, first tandem results and impact on processing and products. *Mar. Geod.* **2011**, *34*, 214–241. [CrossRef]
30. Pujol, M.-I.; Schaeffer, P.; Faugère, Y.; Raynal, M.; Dibarboure, G.; Picot, N. Gauging the improvement of recent mean sea surface models: A new approach for identifying and quantifying their errors. *J. Geophys. Res. Oceans* **2018**. [CrossRef]

10

River Levels Derived with CryoSat-2 SAR Data Classification

Eva Boergens [1,*], Karina Nielsen [2], Ole Baltazar Andersen [2], Denise Dettmering [1] and Florian Seitz [1]

1 Deutsches Geodätisches Forschungsinstitut der Technischen Universität München (DGFI-TUM), Arcisstraße 21, 80333 Munich, Germany; denise.dettmering@tum.de (D.D.); florian.seitz@tum.de (F.S.)
2 Division of Geodesy, DTU Space, National Space Institute, DK-2800 Kongens Lyngby, Denmark; karni@space.dtu.dk (K.N.); oa@space.dtu.dk (O.B.A.)
* Correspondence: eva.boergens@tum.de

Abstract: In this study we use CryoSat-2 SAR (delay-Doppler synthetic-aperture radar) data in the Mekong River Basin to estimate water levels. Compared to classical pulse limited radar altimetry, medium- and small-sized inland waters can be observed with CryoSat-2 SAR data with a higher accuracy due to the smaller along track footprint. However, even with this SAR data the estimation of water levels over a medium-sized river (width less than 500 m) is still challenging with only very few consecutive observations over the water. The target identification with land–water masks tends to fail as the river becomes smaller. Therefore, we developed a classification approach to divide the observations into water and land returns based solely on the data. The classification is done with an unsupervised classification algorithm, and it is based on features derived from the SAR and range-integrated power (RIP) waveforms. After the classification, classes representing water and land are identified. Better results are obtained when the Mekong River Basin is divided into different geographical regions: upstream, middle stream, and downstream. The measurements classified as water are used in a next step to estimate water levels for each crossing over a river in the Mekong River network. The resulting water levels are validated and compared to gauge data, Envisat data, and CryoSat-2 water levels derived with a land–water mask. The CryoSat-2 water levels derived with the classification lead to more valid observations with fewer outliers in the upstream region than with a land–water mask (1700 with 2% outliers vs. 1500 with 7% outliers). The median of the annual differences that is used in the validation is in all test regions smaller for the CryoSat-2 classification results than for Envisat or CryoSat-2 land–water mask results (for the entire study area: 0.76 m vs. 0.96 m vs. 0.83 m, respectively). Overall, in the upstream region with small- and medium-sized rivers the classification approach is more effective for deriving reliable water level observations than in the middle stream region with wider rivers.

Keywords: satellite altimetry; inland water; CryosSat-2 SAR; Mekong Basin; water level time series; classification; stack data

1. Introduction

River waters are crucial as suppliers of water for irrigation and fresh water for drinking. However, river floods can destroy crops, settlements, and infrastructure. For this reason, it is essential to monitor the water level of river systems. However, an increasing number of in situ gauges have been derelicted since the 1980s [1], or the data is not publicly available. Therefore, it is more and more important to close this data gap with satellite altimetry.

In recent years many studies have been published that apply satellite pulse-limited altimetry over rivers of various sizes (e.g., [2–7]). CryoSat-2, launched in 2010, is the first satellite carrying a delay-Doppler altimeter [8] that operates in three measuring modes: the classical pulse-limited low resolution (LR) mode, the delay-Doppler synthetic-aperture radar (SAR) mode, and the SAR interferometric (SARin) mode.

Compared to conventional radar altimeters, delay-Doppler SAR measurements have a higher along-track resolution and a smaller footprint. This improves the observation of water levels of inland water bodies like lakes (e.g., [9–11]) or rivers (e.g., [12,13]). The advantage of SAR altimetry observations is that they are especially effective for measuring smaller inland waters like rivers. However, CryoSat-2 has a long repeat time of 369 days compared to 35 days for Envisat and SARAL, and 10 days for Topex/Poseidon, Jason-1, Jason-2, and Jason-3. This restricts the estimation of water level time series over lakes and even more over rivers. The advantage of the long repeat time is the very dense spatial distribution of observations. This spatial distribution enables better river monitoring in the continuous progression. Unlike lakes, rivers can change their water levels rapidly over their course which makes a denser spatial distribution of observations desirable.

To derive water levels from lakes or rivers it is necessary to identify the water returns of the altimeter. A land–water mask can be used for this purpose, like the mask provided by the World Wildlife Fund (https://www.worldwildlife.org/pages/global-lakes-and-wetlands-database). Such a mask is constant over time; therefore, it neither accounts for the seasonal variations of the water extent nor inter-annually shifting river and lake banks. Extracting dynamic land–water masks from optical remote sensing images is difficult in the tropics, which includes the study area of the Mekong Basin, since cloud-free optical data is only available during the dry season with low water level. Moreover, SAR images with sufficient spatial resolution are only available from the launch of Sentinel-1 in 2014. Although a high accuracy land–water mask is provided by the Mekong River Commission (http://portal.mrcmekong.org/map_service) for the study area with an accuracy of 30 m, this accuracy might not be sufficient for medium- and small-sized rivers. Additionally, the mask has no seasonal variations included.

Within this study, an innovative classification approach for SAR altimetry data is developed and validated, which is independent of the accuracy and availability of land–water masks. With this approach, reliable river levels, even for small and medium-sized rivers, are derived. Compared to classical approaches based on land–water masks, the number of valid measurements is increased and the water level precision is improved.

For the classical pulse-limited altimeter data, classification has been done successfully for the last decade worldwide (e.g., [14,15]). Even very small water areas in wetlands have been classified successfully with Envisat data by [16].

In these classifications, the shape of the waveform is used to discriminate between different reflecting surfaces. Also, CryoSat-2 SAR data has been classified based on the SAR waveform for lakes [11], lakes and rivers [17], or ice [18]. This study takes a step further and uses not only the waveform but also the range-integrated power (RIP) for a classification of the altimeter measurements in water and non-water returns over the Mekong River Basin in Southeast Asia. The RIP is only available for delay-Doppler SAR altimetry and gives additional insight to the reflective surface that the waveform alone could not provide [19].

The unsupervised *k-means* algorithm is employed for the classification [20] as not enough reliable training data is available for a supervised classification. The *k-means* algorithm is a widely used unsupervised clustering algorithm and has been used for altimetry classification before (e.g., [11]).

This paper is structured as follows: First, an introduction is given about the study area of the Mekong River Basin in Section 2, afterwards more information of the CryoSat-2 SAR data is given in Section 3. The classification and the used features are described in Section 4.1 followed by an explanation of the water level estimation in Section 4.2. The results and validations are presented in Section 5. These results are discussed in Section 6 and an outlook is given. The paper ends with the conclusions in Section 7. An overview over all relevant processing steps of this study is given in Figure 1.

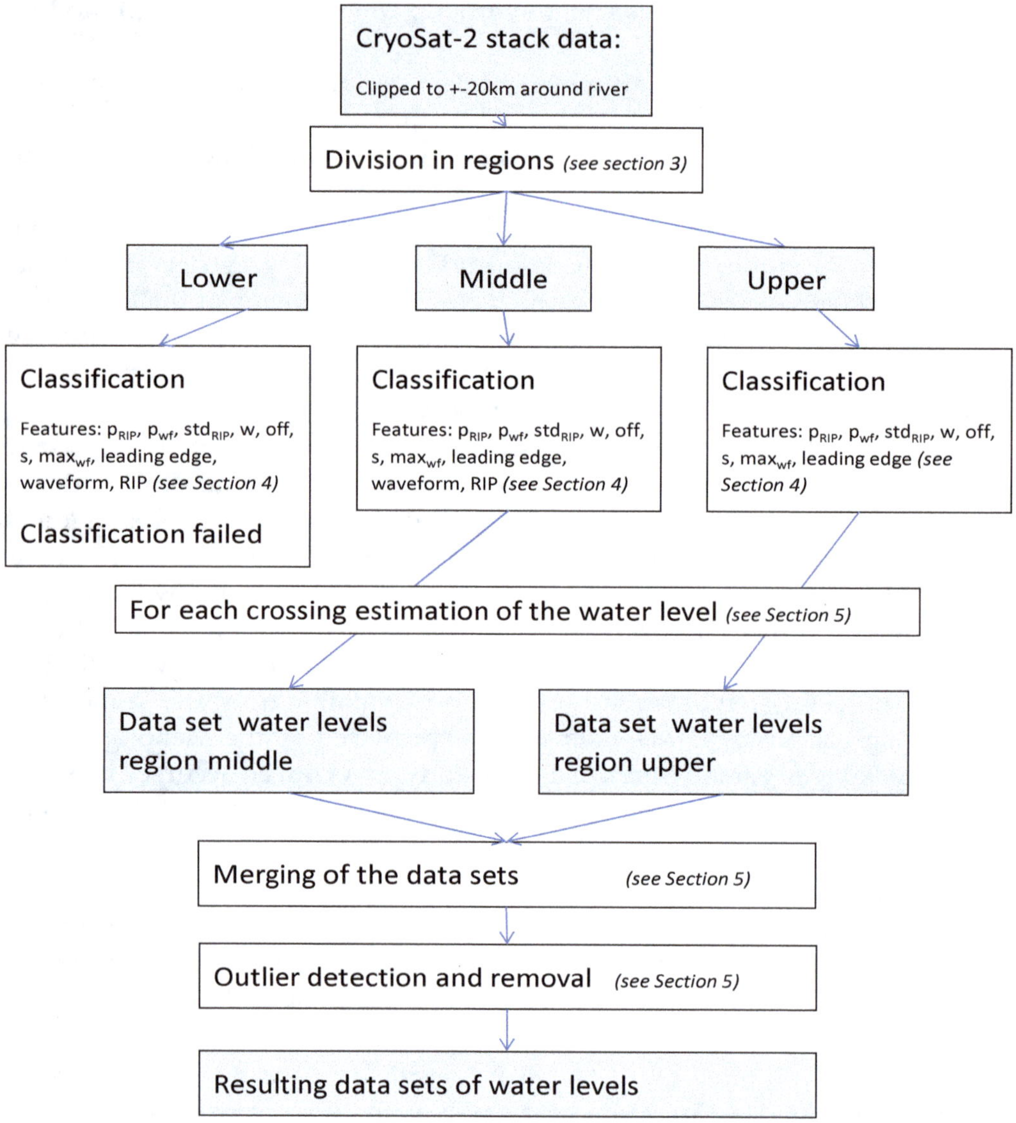

Figure 1. Processing steps used in this study for extracting water levels from the CryoSat-2 synthetic-aperture radar (SAR) data. RIP: range-integrated power.

2. Study Area

The Mekong River Basin in Southeast Asia (China, Myanmar, Thailand, Laos, Cambodia, and Vietnam) is investigated in this study, with a focus on the part of the basin south of the Chinese border. Further upstream the river gorge is too narrow for reliable altimeter observations. Downstream, the study area is limited by by the confluence with the Tonle Sap River from where the river is under a tidal influence. The tributaries, namely the large left bank side tributaries in Laos, are investigated as well. The hydrology of the Mekong Basin is primarily influenced by the precipitation on the Tibetan Plateau and the south-eastern monsoon [21].

In the Mekong River Basin the river width varies between 20 m to more than 2 km. Most of the tributaries and the upstream reach of the main river stream are less than 100- m-wide small rivers. The medium rivers, which are less than 500- m but more than 100- m-wide, are the main tributaries for the upstream main river. In the downstream reach of the river, before it splits into the delta, the river has a width of over 2 km (see also Figure 2 for a map of the basin).

The Mekong River and its tributaries flow through different topographic regions (Figure 2). The main river upstream from Vientiane and the left bank tributaries in Laos are surrounded by mountainous areas with steep banks where the rivers have a greater slope and have a width smaller than 500 m or even less than 100 m. Downstream of Vientiane and up to the Mekong Falls the river widens and flows with less slope over the Khorat plateau. Below the Mekong Falls the river is surrounded by seasonal wetlands and widens to more than 1 km. For further processing we defined three overlapping data masks according to these regions (Figure 2). The regions are determined by the roughness of a topography model and the absolute height. Afterwards a margin around each subregion allows for an overlap.

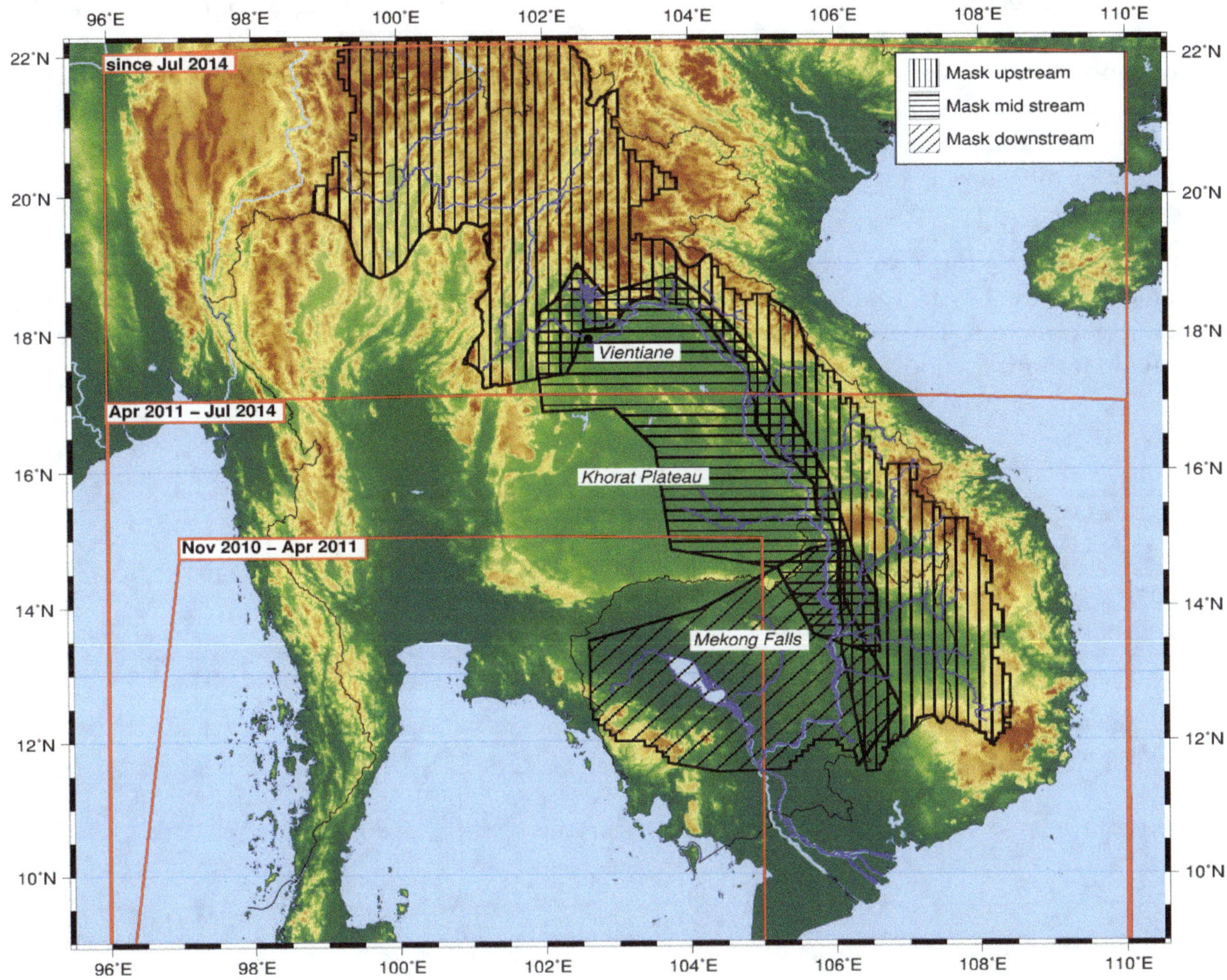

Figure 2. Map of the study area with the regional masks (black areas with different hachures) and the SAR mode mask with their validity period (red boxes).

3. Data

3.1. CryoSat-2 SAR Data

In this study we use delay-Doppler SAR altimeter data measured by CryoSat-2 between 2010 and 2016. CryoSat-2 measures in three different modes (LR, SAR, and SARIn mode), which are set in a geographical mask (https://earth.esa.int/web/guest/-/geographical-mode-mask-7107): The LRM is active mostly over the oceans and the interior of the ice sheets of Antarctica and Greenland, whereas the SAR mode measures over sea ice and other selected regions, and SARin focuses mostly on glaciated regions [22]. This mask has changed over the life time of the satellite. The entire study area of the Mekong River Basin has only been measured in SAR mode since July 2014 (see Figure 2 for the extent of the SAR mode mask). In SAR mode the along-track foot print size is reduced to 300 m while it remains at 14 km in the across-track direction [23].

The delay-Doppler SAR altimeter measures a point on the surface several times from different looking angles [24]. Unlike pulse limited altimetry, which is only able to measure the returning signal from the whole foot print, SAR altimetry discretizes the returning signal corresponding to 300- m-wide stripes of the footprint in along track direction. The signals from each of these stripes are recorded in the so-called single-look waveform. All single-look waveforms for the same surface area are collected in the stack matrix.

In Figure 3, two exemplary stack matrices are presented. The first (a) is measured over the Tonle Sap lake and the second (b) over a medium-sized river in the upstream region of the Mekong River. Each row is a single-look waveform. The integration of this matrix over all single-looks results in the multi-look SAR waveform (in Figure 3 with integration over each row of the stack) hereafter referred to as the waveform. The integration over the range bins results in the range-integrated power (RIP). In Figure 3 this corresponds to the integration over the columns. Detailed information on the delay-Doppler measurements is described in [8].

Here, we use the CryoSat-2 baseline C SAR Level 1b data provided by ESA GPOD SARvatore (https://gpod.eo.esa.int/) for the period 2010–2016. SARvatore is the official ESA platform for the dissemination of the CryoSat-2 level 1b SAR data. The data is processed on user request with different processing parameters. The most important data processing parameters of the data in this study are: the applied Hamming window, fft zero-padding, 20-Hz data, and stack data in the output file.

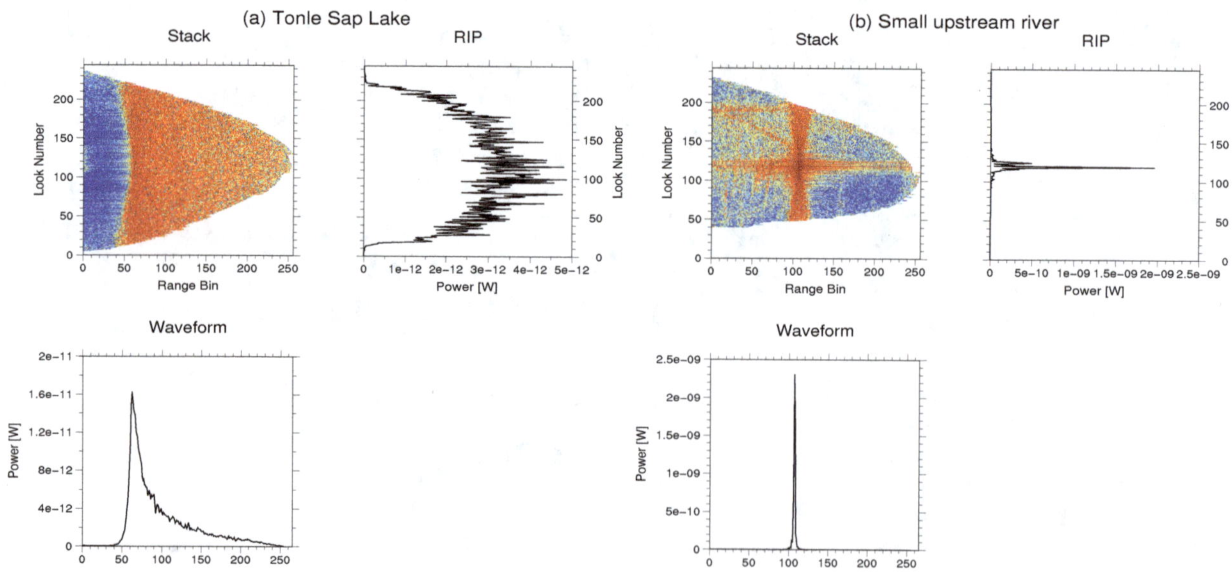

Figure 3. Two exemplary stack matrices with their RIP and waveform. The color of the stack plot indicates the power of the signal. The example on the left-hand side is measured over the Tonle Sap lake, the one on the right-hand side over a medium upstream river.

3.2. Envisat Data

Envisat water level observations are used for the validation in this study. The water levels are taken from the DAHITI database [4]. DAHITI uses a throughout outlier detection in combination with a Kalman filter approach to derive reliable water level time series for lakes and rivers. More information on the processing strategy is available in [4]. The Envisat water level time series in the Mekong Basin has a temporal resolution of 35 days, even though data gaps may occur for some cycles. For small tributaries, a correction for the Hooking effect is applied additionally in [5].

3.3. *In Situ Data*

For validation, in situ gauge data is used which is provided by the Mekong River Commission (http://ffw.mrcmekong.org/). The gauge data has a daily temporal resolution but the time series ends with 2012. No absolute height of the gauge is provided.

Additionally, we use a river polygon which is also provided by the Mekong River Commission (http://portal.mrcmekong.org/map_service). The polygon was derived from aerial images and topographic maps. The accuracy of the river mask is ~30 m, but no information about the seasonality of the polygon is given.

4. Methods

4.1. *Classification Approach*

For the medium and small rivers in our study area of the Mekong basin no reliable land–water mask is available. Thus, a classification by means of the *k-means* algorithm is performed to extract the water measurements.

The *k-means* algorithm [20] is an unsupervised method to cluster the data on the basis of different features. For the land–water classification a set of features derived from the CryoSat-2 stack data over the intermediate step of the waveform and the RIP is used. The features are summarized in Table 1. The features derived from the waveform are the *maximum power*, the *peakiness*, and the *position of the leading edge*. It is well known that waveforms of water reflections have a higher power than those of land reflections. Medium, and to a greater degree small, water bodies have smooth mirror-like surfaces which can only be measured by signals emitted close to nadir. This leads to a very peaky waveform and RIP with a high power. Following [25] the peakiness p_{wf} is calculated with

$$p_{\mathrm{wf}} = \frac{\max(\mathrm{wf}_i)}{\sum\limits_i \mathrm{wf}_i}, \tag{1}$$

where wf is the waveform and wf_i the power of the ith bin.

Table 1. Features used for the classification.

RIP Features	Waveform Features
Peakiness: p_{RIP}	p_{wf}
Standard deviation: std_{RIP}	Maximum power: max_{wf}
Width: w	Relative position of leading edge
Off-center: off	
Symmetry: s	

To estimate the relative *position of the leading edge* in the waveform, the waveform is retracked using an improved threshold retracker with a threshold of 50% on the best sub-waveform [26]. The on-board tracking system always tries to hold the leading edge of the main reflection at the nominal tracking point. This is not always possible and leads to a deviation of the leading edge from the nominal tracking point. Over wider rivers the tracking system can manage to keep the leading edge close to the tracking point. In Figure 4 (left panel), one exemplary waveform with its features *maximum power* and *position of the leading edge* is shown (the *peakiness* cannot be displayed).

Features based on the RIP are the *peakiness* p_{RIP}, the *standard deviation* std_{RIP}, the *width*, the *off-center*, and the *symmetry*. Water reflections over large water bodies result in a overall smoother RIP than water reflections over small water bodies that in turn have a smoother RIP than land reflections(see Figure 3). This smoothness is measured with a standard deviation feature std_{RIP}. The std_{RIP} is then

$$std_{\text{RIP}} = \sqrt{\frac{1}{N}\sum_i (\text{RIP}_i - mean(\text{RIP}))^2}, \tag{2}$$

where RIP_i is the ith entry of the RIP and N the number of looks in the RIP, usually 246.

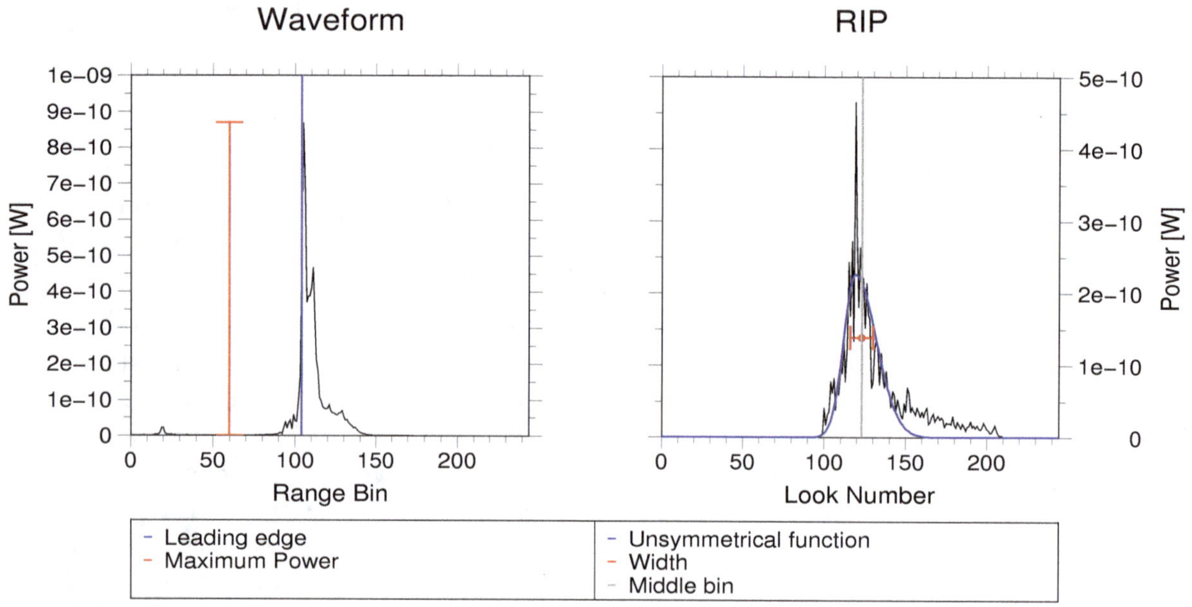

Figure 4. One example of a waveform and the corresponding RIP with some of their derived features.

As mentioned before, small and medium inland waters with a smooth surface only reflect the signal back to the satellite at near nadir. Therefore, the RIP is both very peaky and narrow. The width w is derived with:

$$w = \frac{\left(\sum_i \text{RIP}_i^2\right)^2}{\sum_i \text{RIP}_i^4}. \tag{3}$$

The off-center feature off describes the deviation of the main reflection from the nadir point. It should be close to zero for measurements of water, whereas land measurements are more disturbed and often show the maximum return in the side lobes. We measure the off-center feature off as the difference between the middle look of the RIP and the mean point of the RIP which is calculated with:

$$off = \frac{246}{2} - \frac{\sum_i i\text{RIP}_i^2}{\sum_i \text{RIP}_i^2}. \tag{4}$$

A positive off value indicates that the majority of the returning power was detected before the satellite passed the nadir position, a negative value vice versa.

The last feature is a measure of the symmetry of the RIP s. For an ideal smooth water reflection, like a small lake, the RIP should be perfectly symmetrical. However, for a sloped target, such as a river, the reflection depends on the relative orientation between the satellite and the water surface. The reflection is stronger when the satellite looks on a water surface that is sloped towards it. A positive s indicates a water surface sloped towards the approaching satellite. This effect leads to an unsymmetrical RIP. To quantify this, an unsymmetrical exponential function $\overline{\text{RIP}}$ is fitted to the RIP with

$$\overline{\text{RIP}}_i = \begin{cases} a \exp\left(\frac{(i-b)^2}{2c_1^2}\right), & \text{if } i < b \\ a \exp\left(-\frac{(i-b)^2}{2c_2^2}\right), & \text{if } i \geq b. \end{cases} \quad (5)$$

Here, a is the amplitude of the exponential function, b the look where the function reaches its maximum, and c_1 and c_2 are the two decay parameters. The *symmetry* feature is then

$$s = c_1 - c_2. \quad (6)$$

Figure 4 (right-hand side), displays an RIP with the feature w marked. The off-center feature off is too small to be visible in this example, but the symmetry, or the lack thereof, is clearly showing.

Additional to these eight features, both the whole waveform and the whole RIP are used as features. Each bin is then considered as a single feature. The waveform needs to be shifted so that the leading edge is positioned on the nominal tracking point. We found that including waveform and RIP as features only improves the results for the middle region and does not change them in the upstream regions. Therefore, the upstream region is classified without waveform and RIP as features. Since the features span different orders of magnitude, it is necessary to normalize the feature set. All of these features were chosen according to their sensitivity for the posed problem of water classification in the Mekong River Basin and independently from each other. More features were tested but discarded because they were either not sensitive for the classification or highly correlated to one of the used features.

The *k-means* algorithm is used to cluster the data on the basis of the above features in 20 classes. An unsupervised clustering algorithm is applied because no reliable training data is available. The *k-means* algorithm assumes normally distributed features with equal variance, which we ensured and tested by the normalization of the features.

The number of classes depends on the application and variation in the input features. An estimate for the number of classes can be done with knowledge of the classified data. In our study case, a look at the spatial distribution of the features tells us that only two classes, land and water, are not sufficient as altimeter measurements of land can be very diverse (this holds also for water measurements, but they are less diverse than land). The diversity of the returning waveform and RIP can be explained by the reflective properties of e.g., land, water, vegetation. With this it can be concluded that at least 10 classes are needed. We tested the classification and validated resulting water levels for several numbers of classes (10, 15, 20, 30) and found similar results for all with the results of 20 classes slightly superior.

Each of the clusters is defined by their centroid which are the mean features of all points in this cluster. New data is classified by grouping it to the closest centroid. Here, the clustering is done on one randomly drawn third of the data. The residual two-thirds of the data are then classified into the cluster classes. The clustering is not done on the whole data set due to computational efficiency. The repeatability of the clustering and classification will be validated in Section 5. After the classification, it is determined which classes represent water and land returns. This was done by visual inspection of the mean waveform and RIP for each class and the locations of the observations in each class related to the approximate location of the river known from the land–water mask (see Section 3).

As described in Section 2 the Mekong Basin is divided into different regions: upstream, middle, and downstream. We classify each of the regions separately as they are too diverse in the reflectivity properties of the water bodies to be classified together. Additionally, the classification is done only on altimeter data not further away than 20 km from the river polygon due to computational efficiency (the polygon can be seen in Figure 2).

4.2. *Water Level Estimation Approach*

4.2.1. Altimetric Water Levels

The classification results in a set of measurements considered as water returns. From these measurements the water level at each crossing of the satellite track with a river in the Mekong River Basin is determined. A river polygon is employed to locate all crossings in the river basin (see Section 3). We used all measurements less than 5 km away from the river crossing that were classified as water and retracked the SAR waveforms with an improved threshold retracker with 50% threshold [26]. The retracked range is corrected for dry and wet tropospheres, ionospheres, solid Earth and pole tides, and geoid undulation. To ensure consistency for both the Envisat and CryoSat-2 data, the same corrections are used. Instead of using a median or mean over all classified measurements, we search for a horizontal line in the heights, which is assumed to represent the water surface. It is still possible that some of the water-classified measurements do not represent the river surface and need to be excluded from the water level computation (across-track of nadir effects or water bodies surrounding the river). These outliers do not necessarily have to be at the margin of the river but can also be located in the middle due to islands or sandbanks in the river. This can restrict the use of an along-track standard deviation of the heights for outlier detection.

To find the line of equal water height, a histogram of the water levels with Doane bins [27] is used. Doane bins are more suitable for small (less than 30) non-normally distributed data sets than the classical Sturge bins [28]. The Doane bin method not only uses the number of data points to estimate the number of bins and thus bin size, but the shape of the distribution is incorporated with a skewness parameter as well. The range of the data in the data set is indirectly considered in the Doane bins. If a horizontal line is present in the heights, one of the bins is distinctively larger, e.g., contains more observations, than the others and collects the heights of nearly equal water level. The median of the heights in this bin is then taken as water level. If less than five height points are classified as water, the median of the heights is taken as the water level. The advantage of this approach is that it is better suited for rivers wider than 1 km with islands and sandbanks that cause outliers in the heights. However, in many cases this histogram approach or only taking the median of all observations delivers similar results.

4.2.2. Outlier Detection

In spite of careful data selection through the classification and in the height retrieval, some of the water levels have to be considered as outliers. The height retrieval with a horizontal line only prevents outliers if more than five observations are labeled as water observations. In particular, along the small and medium-sized rivers many crossings have less than five observations. A considerable number of crossings have only one observation. The long repeat time of the CryoSat-2 satellite of 369 days complicates the detection of outliers. Unlike altimetric water levels of short repeat time missions, such as Envisat and the Jasons, we do not have a water level time series at a location which can be investigated for outliers by the estimation of and the comparison to an annual signal.

To find the CryoSat-2 outliers we make use of the said repeat time of 369 days. With the knowledge of the very stable annual signal of the Mekong River one can assume that two measurements of the same CryoSat-2 track 369 days apart should measure a similar height [21]. Based on this, a water level is considered as an outlier if the mean difference to all other heights of the same pass is larger than 7 m; the choice of threshold will be discussed below. This is only applicable if other water level measurements of the same track exist. Due to the changing mode mask (see Section 3) some regions are only measured in the last two years. To overcome this, a second outlier detection is applied which compares the water level with water levels of other tracks that are close in space and season (day of the year). To this end, we used all measurements that were less than 10 km away along the river and less than 30 days of the year apart. If the water level was different by more than 10 m from the distance weighted mean water level of all these points, it was considered as an outlier.

The thresholds for the outlier detection were chosen as a conservative upper bound. It is expected that there is an average water level difference of 40 to 60 cm in the five days during the rising water season, but it could be as high as 4 or 5 m [29]. Additionally, some inter-annual changes in the flood season can be expected, and the slope of the river has to be considered which is has a median value of 30 cm/km for the Mekong River. Of the three thresholds used for the outlier detection the difference of 7 m w.r.t. the year is the most sensitive for the later result. The time and distance weighted mean in the second part of the outlier detection limits the sensitivity of the other threshold.

4.2.3. Merging of the Overlap Regions

From the classification we derive a set of heights for each of the different geographical regions which have a certain overlap (see Figure 2 and Section 2). In this overlap, for the same crossing two water levels were computed. Therefore, it has to be decided which height shall be used. To resolve this, we use the distance-weighted mean water level as in the outlier detection (see Section 4.2.2). The water level that is closest to this mean water level is applied.

5. Results and Validation

We applied the described methodology for the classification and water level determination on CryoSat-2 SAR data in the Mekong River Basin. In this section, both the results of the classification and the water level determination are presented and validated.

5.1. Results of the Classification

After the clustering and classification of the CryoSat-2 measurements in the Mekong Basin we select the classes of water returns. In the upstream region we identify three and in the middle region six out of 20 classes as water classes. In the downstream region the rivers are surrounded by seasonal wetland, which makes it difficult to distinguish water return related to the river from the surroundings. Therefore, the downstream region will not be investigated further.

In Figure 5 the mean waveform and mean RIP of some classes are shown (note the different power axes). The classes displayed are selected to best represent all 20 classes for the upstream and middle region. As can be seen, the shape of the mean waveform and mean RIP of water classes in the upstream region reappear in the middle region, but not as water classes. In the middle region small lakes or paddy fields have the same signature as the river upstream. This observation leads to the decision to classify the two regions separately. Some classes have very similar mean waveforms and RIPs (e.g., second water class and first land class in the middle region) but are either classified as water or as land. In this case the maximum power is the important discriminant between the classes. In the upstream region not all stacks over land are ‘full’, i.e., not every single-look recorded the returning power. This leads to strongly distorted RIPs as shown for the third land class. All mean waveforms and RIPs are displayed in Appendix A for the interested reader.

In Figure 6, a section of the river network in the upstream region with the results of the classification is shown. The course of the river is well depicted, however, not every crossing of the satellite track with the river water is identified. At some crossings no water reflection of the river is measured since the river is too narrow. On the other hand, some points classified as water are not close to the given polygon (blue line). However, the topography model (HydroSHEDs, as presented in Lehner et al. [30]) shown in the background indicates river valleys in the three circled areas. Therefore, one can assume that the classification is able to find rivers that are so small (down to 20 m width) that they are not present in the high-resolution river polygon provided by the MRC. Figure 7 shows the classification for one exemplary track in the upstream region. The measurements classified as water (red dots) line up to a nearly constant water level at all crossings of the satellite track with the river.

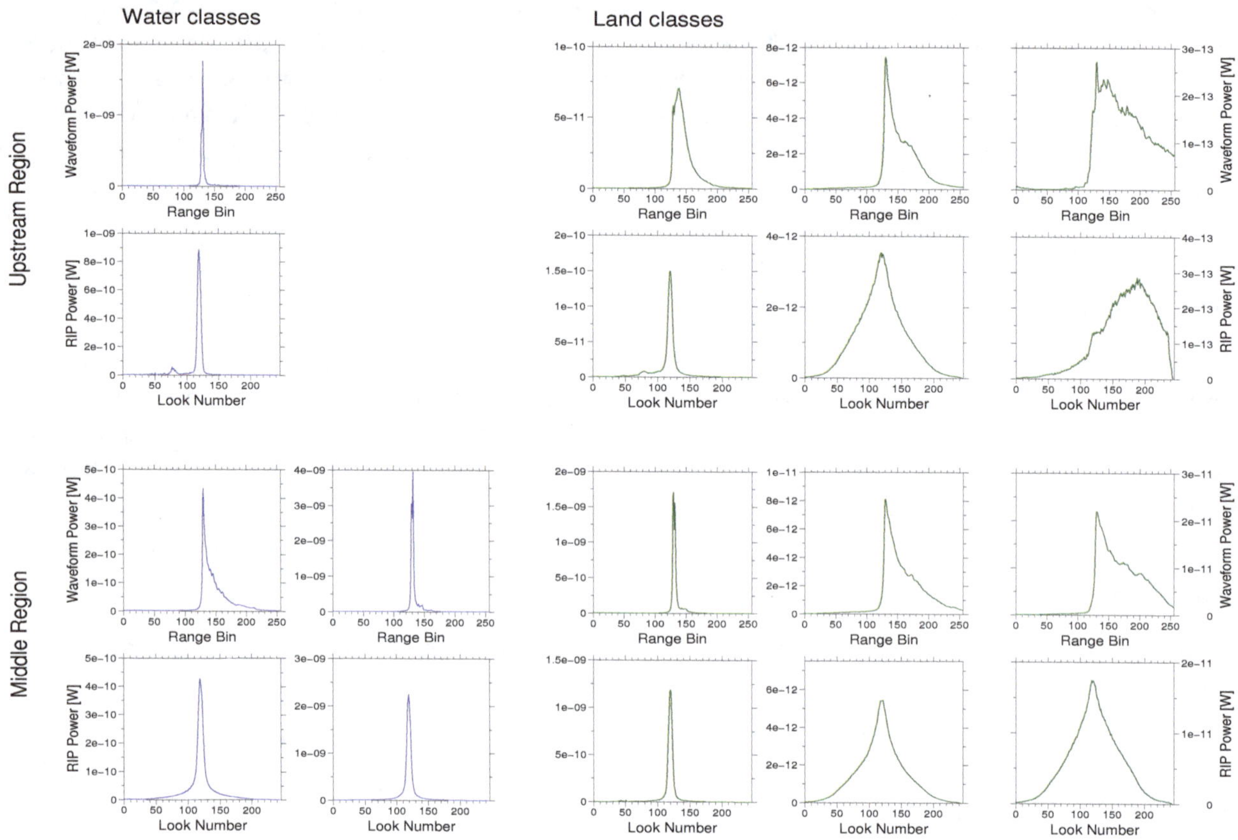

Figure 5. The mean waveforms and RIP of some selected classes; the waveforms and RIP above each other belong to the same class.

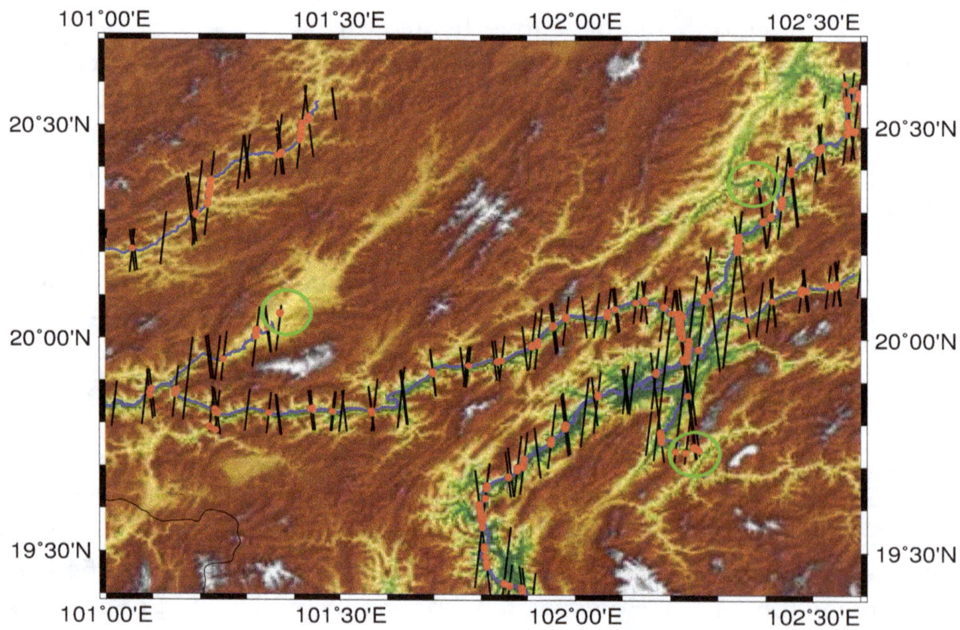

Figure 6. An example of the classification. The red dots are classified as water measurements, and black lines are the land-classified measurements. The three circles indicate areas where water was detected in river valleys which are not included in the river polygon. The background shows the HydroSHEDs topography model.

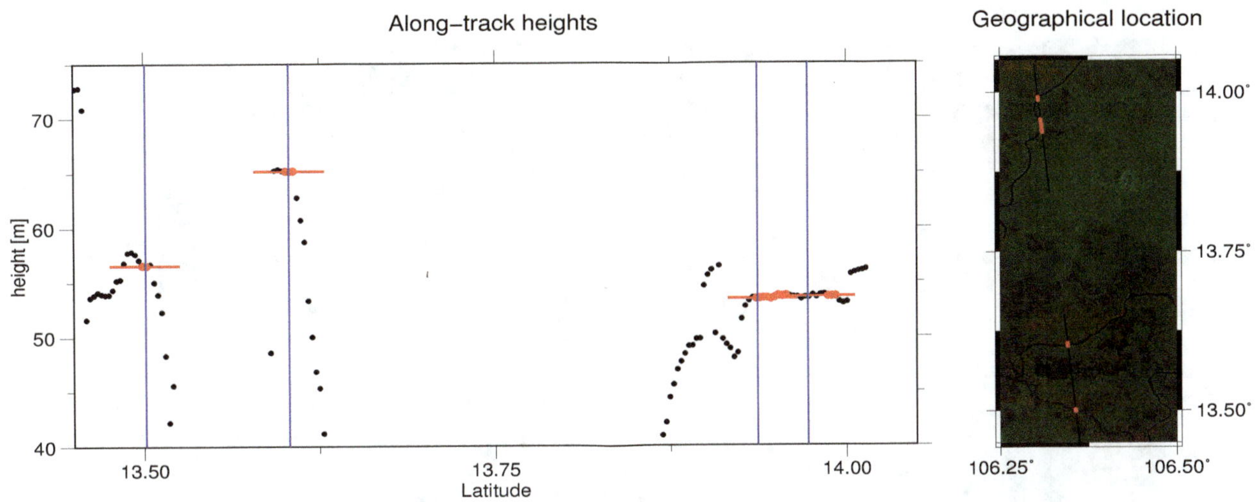

Figure 7. Heights along one track which crosses a river at four locations. The map on the right side shows the geographical surroundings with the background as a Landsat-8 scene where the rivers are well visible. The black dots are all retracked heights with the red dots indicating which measurements were classified as water. The blue vertical lines show the location of the crossing of the track with the river polygon and the horizontal lines are the estimated water level at each crossing.

5.2. Resulting Water Level

In the entire Mekong Basin we estimate water levels at more than 2000 crossings, which means approximately one measurement every 4 km along the main river (compared to 50 km for Envisat). The water level was not measured at every crossing. As mentioned before, at some crossings the river was so small that a reliable measurement could not be made in every pass, and some other water levels were discarded during outlier detection. Furthermore, at some crossings the classification failed to identify the water. However, we are still able to retrieve at least some measurements from rivers as narrow as 20 m. In Figure 8 all measured heights at all dates are presented in a map, which shows the overall topography of the river network well but cannot show details like seasonal variations.

For one track the heights and the classification are displayed in Figure 7 with a Landsat map of the river network and the surrounding area. In this track four water crossings are found where the two most northern ones are very close together with a difference of the water level of 20 cm. There the river meanders under the track which causes two crossings close together. The two southern crossings belong to two different rivers which explains the large height difference. It is visible that only few measurements are used to estimate the water level at each crossing. Approximately 180 water levels (or 8%) are estimated by just one measurement, with the majority in the upstream region. Hence, the outlier detection is more important in this area. In the upstream region, the outlier detection removes 113 of 1740 water level observations; all of them are detected by the comparison with surrounding water level measurements. The comparison with water levels of the same pass reveals only 37 of the 113 outliers. In the middle region only 13 of 529 observations are rejected as outliers: 10 by the comparison with surrounding observations, 9 by the comparison with the same pass, and 6 by both outlier detections.

For crossings with more than one water measurement we can calculate the standard deviation of the measurements used for water level estimation. More than 85% of the water levels have a standard deviation of less than 0.5 m.

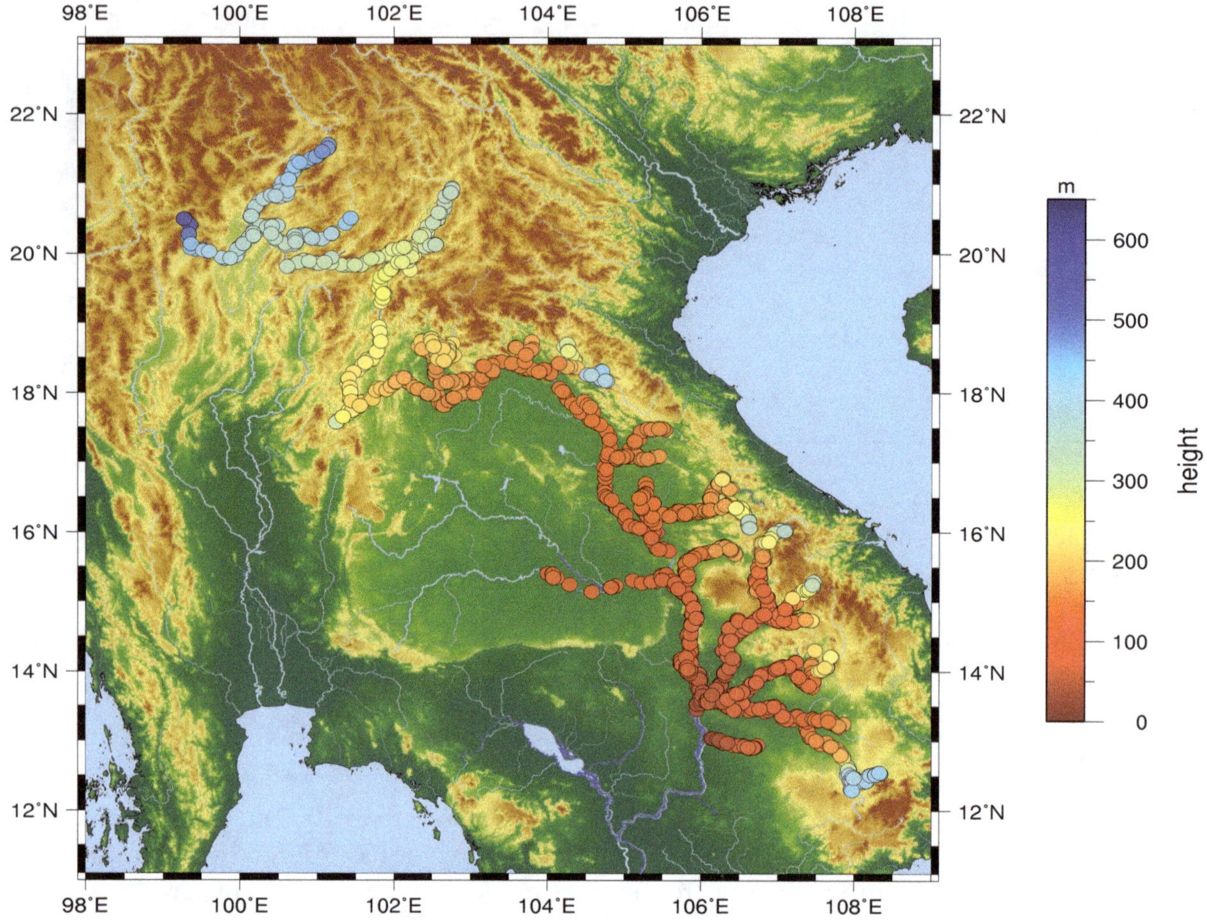

Figure 8. Resulting water levels in the Mekong River Basin.

5.3. Validation

In this section both the classification and the water levels are validated. We test and validate the repeatability of the classification with cross-validation. The altimetric water levels cannot be validated against in situ gauge data without an absolute height reference as is common for short repeat time missions because no time series can be built from CryoSat-2 data without reducing the unknown topography of the river. Therefore, we validate the resulting water levels with respect to the stable seasonal signal and compare these results with the performance of Envisat water levels and CryoSat-2 data extracted with a land–water mask in the same validation.

5.3.1. Validation of the Classification

The classification is validated with a cross-validation. To this end, we cluster the data in a first step. The resulting classes are considered as the 'true' classes. In the second step only parts of the data are clustered with the other parts classified in this second clustering. The cross-validation compares the resulting classifications of the two steps.

Table 2 summarizes the results of the cross-validation; water and non-water classes are distinguished. The overall accuracy is 97.9%. This cross-validation shows that the classification is stable and does not change with the data subset used for the clustering.

Table 2. Results of the cross-validation. The first classification is a clustering on the whole data set, and the second classification is based on clustering on a subset of data with classification of the residual data set.

		Second Classification	
		Water	**No Water**
First Classification	**Water**	7321	205
	No water	423	22,660

5.3.2. Validation of Water Levels

Unlike water level time series measured by short repeat time missions, CryoSat-2 measurements cannot be validated against the time series of in situ gauges without reducing the topography as done by [12]. The Mekong River and its tributaries have topography that is too complex to allow for reliable reduction. Besides this, the temporal overlap between the CryoSat-2 data and the gauge data is only about 1.5 years or even less (April 2011 until December 2012).

To validate the water levels we use the nearly one-year repeat time of CryoSat-2. We investigate the differences between two subsequent tracks at the same river crossing. A histogram of the differences is shown in Figure 9a. Table 3 displays the median, mean, and standard deviation of these differences for the merged results as well as for the two regions (upstream and middle) separately. The results of the validation are compared to a validation with in situ gauge data, Envisat data, and CryoSat-2 data with a land–water mask.

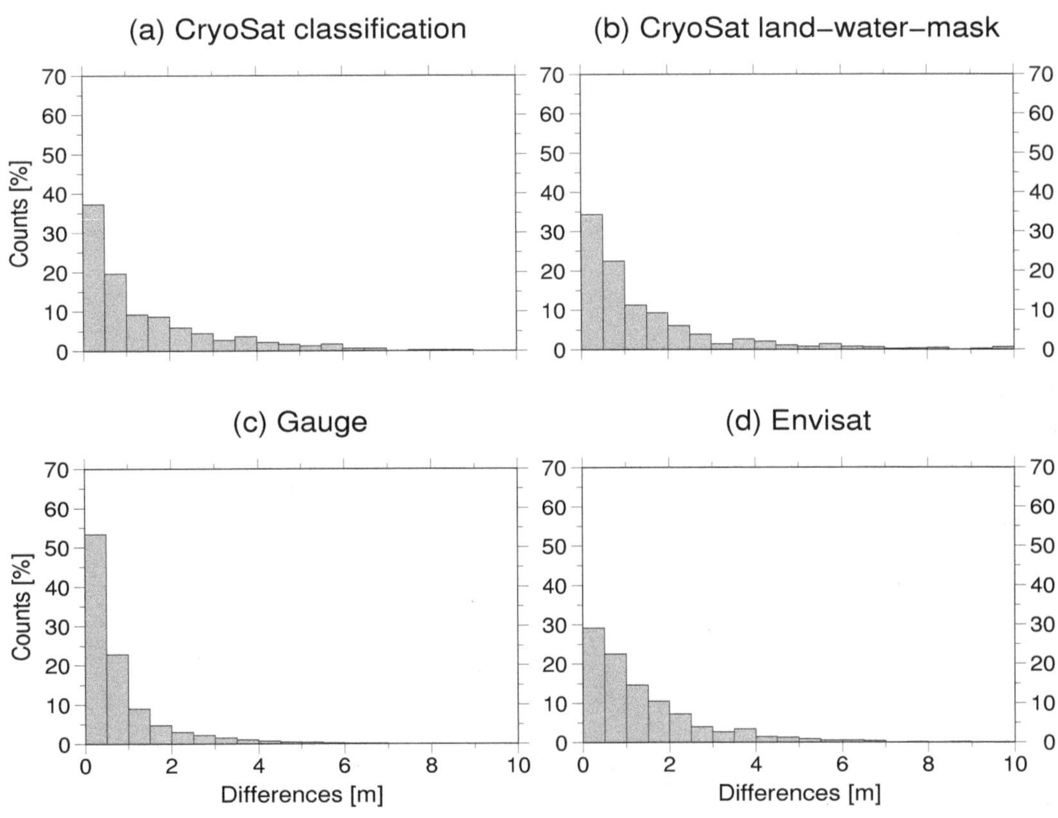

Figure 9. Histogram of the differences of height measurements 369 days apart for CryoSat-2 water levels with classification, CryoSat-2 water levels inside the land–water mask, gauge water level, and Envisat water level.

For this validation, we take the differences between gauge measurements that are 369 days apart and Envisat measurements where the day of the year has a less than 5-day difference. The gauges give a measure of how stable the annual signal is in the Mekong Basin, i.e., a benchmark on how small the differences between the water levels could be. The Envisat observations are the most commonly used data for inland waters with a pulse-limited altimeter. The comparison to the water levels which are derived from CryoSat-2 by simply averaging measurements inside the land–water mask allows for evaluation of the benefits of the classification approach. The land–water mask water levels underwent the same outlier detection as used on the results of the CryoSat-2 classification for better comparability. The median of the differences in Table 3 for the CryoSat-2 classification results are in both the upstream and the middle stream region better than the ones from the Envisat results. In the upstream region the improvements of the CryoSat-2 classification results compared to the Envisat results are more apparent than in the middle region where the validation results are more similar. However, the CryoSat-2 classification results have a higher standard deviation in the upstream regions, and thus larger differences than the Envisat results. The land–water mask approach for CryoSat-2 in the upstream region has even larger differences, which can also be seen in Figure 9. In the middle region no large differences between the validation of the two CryoSat-2 approaches are to be seen. The land–water mask approach yields to more valid water level observations in the middle region but at the same time also a higher number of outliers in both regions (see Table 4).

Table 3. Analysis of the differences of height measurements 369 days apart for the whole study area, only the upstream region, and only the middle stream region.

	Median (m)	**Mean (m)**	**Standard Deviation (m)**
	Whole study area		
CryoSat-2 classification	0.76	1.43	1.59
CryoSat-2 land–water mask	0.83	1.86	4.55
Gauge	0.45	0.82	1.09
Envisat	0.96	1.42	1.44
	Upstream region		
CryoSat-2 classification	0.79	1.54	1.70
CryoSat-2 land–water mask	0.85	2.00	5.44
Gauge	0.42	0.72	1.05
Envisat	1.01	1.46	1.49
	Middle region		
CryoSat-2 classification	0.76	1.15	1.10
CryoSat-2 land–water mask	0.84	1.55	1.87
Gauge	0.54	1.00	1.14
Envisat	0.81	1.26	1.26

Table 4. Comparison of the two approaches for CryoSat-2 water levels in terms of absolute number of water levels and number of outliers.

	Upstream Region		**Middle Region**	
	No. of Water Levels	**% of Outliers**	**No. of Water Levels**	**% of Outliers**
Classification approach	1703	2	520	2
Land–water mask approach	1534	7	1364	4

6. Discussion

In the results and validation section above we demonstrated the advantages and limitations of the classification approach for CryoSat-2 SAR data over the Mekong Basin. Compared to CryoSat-2 water levels extracted with a fixed land–water mask, the classification approach yields water levels with a higher quality according to the used measures with fewer outliers. The difference between the two approaches is most pronounced in the upstream region with rivers as small as 20 m in width. However, the land–water mask approach yields more water levels in the middle region. This reveals the opportunity that SAR altimetry provides for rivers in the Mekong Basin which are too small to be reliably identified in optical (e.g., Landsat, resolution: 30 m) or SAR (e.g., Sentinel-1, resolution: 10 m) images. As shown in Section 5.1 and Figure 6 the classification of SAR altimetry is able to identify rivers which are not visible in the land–water mask derived from satellite images. On the other hand, the classification fails to identify all crossings with a river, which leads to data gaps in the water level measurements. Some of the undetected rivers are very small and shadowed by surrounding river gorges. At other crossings the returning signal is too noisy and disturbed to be identified correctly even by visual inspection.

The better performance of the classification approach in the upstream part of the river is probably caused by the feature selection which is more focused on the reflective properties of small water bodies. The main drawback of the classification approach is its failure in the downstream area. The classification is not able to discern between river water and water in seasonal wetland regions. This problem can only be solved by using a high-resolution seasonal land–water mask that could be derived from Sentinel-1 or 2 data for example.

The existence of outliers after the classification process indicates a certain amount of misclassification. Some of these might be related to reflections from river banks, concrete areas, or other flat regions causing river-like waveforms. Moreover, water surfaces close to the river (such as paddy fields) can provoke outliers.

The inevitable outlier detection is challenging due to the long-repeat orbit of CryoSat-2 preventing the application of the usual virtual station concept. In this study, the outlier detection approach exploits the stable annual water level cycle within the Mekong Basin. Hence, this method is not transferable to other river systems with less stable annual signals. The second outlier detection approach of this study relies on a certain density of water level observations along the river. Without any observation close by a water level cannot be checked to be an outlier.

The validation inherits the problems of the outlier detection as no direct comparison with in situ gauge data is possible. As the gauge data has no absolute reference, not even the absolute height of CryoSat-2 water levels in the vicinity of a gauging station can be validated.

Using CryoSat-2 over rivers improves the knowledge for the river topography, as with this mission the spatial distribution of water level observations is much denser than with any prior satellite altimetry mission (as can be seen in Figure 8). Even small tributaries are observed successfully with CryoSat-2 SAR, which allows monitoring of the entire river network. However, this high spatial resolution entails a sparse repeat time and thus, precludes the application of the classical virtual station concept for data screening, interpretation, and validation.

Nevertheless, using the derived CryoSat-2 data set—in combination with short repeat altimetry mission results—for building a mathematical water level model (as done in [31]) or for assimilation in hydrological models [32] will certainly improve our knowledge of the water regime in the Mekong region.

7. Conclusions

We demonstrate in this study the advantages of CryoSat-2 SAR altimetry data for measuring rivers in the Mekong Basin identified by a classification which is independent of an accurate land–water mask. The classification uses features derived not only from the waveform but also from the RIP. The RIP contains more information about the reflecting surface than the waveform on its own can provide. This improves the classification and allows us to identify even very small rivers with a width as small as 20 m. In fact, the classification works better on medium and small rivers than on large rivers. The cross-validation of the classification shows that it is stable and repeatable. However, we were not able to use this classification to isolate the river in the downstream region where the Mekong River is surrounded by seasonal wetlands.

The classification in water and land measurements is used to derive water levels at the crossings of the CryoSat-2 track with a river in the whole Mekong Basin. Overall, more than 2000 water levels are measured, after outlier detection. However, it is not possible to derive a water level at every crossing. The altimeter is not able to measure a water return at every possible river crossing due to too-small rivers or overly-disturbed returns. Additionally, some measured water levels are discarded in the outlier detection.

The water levels are validated using the 369-day return time of CryoSat-2 and the very stable annual signal in the basin. The same validation is performed on Envisat water levels, gauge measurements, and by using a precise land–water mask on CryoSat-2 data. In particular, for small rivers in the upstream region the classification improves the water level determination compared to the use of a land–water mask: The median differences between consecutive passes in the upstream region are 0.79 m compared to 0.85 m for the land–water mask approach. The reduction of the standard deviation by a factor of four is even more significant. Moreover, the number of outliers is significantly smaller for the classification approach than for the land–water mask approach, with only 2% for the classification but 5.6% for the CryoSat-2 land–water mask.

Compared to Envisat water levels the CryoSat-2 water levels are of higher quality in the whole river basin due to the smaller footprint of the SAR compared to pulse-limited altimeter on Envisat. The median differences for Envisat are as high as 1.01 m in the upstream region, and 0.96 m for the whole basin (CryoSat-2 classification 0.76 m).

The classification of CryoSat-2 data allows to exploit the full along-track resolution of the data set. Therefore, it is possible to observe the water levels of rivers as small as 20 m that are usually not correctly mapped in land–water masks. Moreover, due to the dense spatial resolution of CryoSat-2 most of the smaller tributaries are observed, which is not possible with the short repeat missions Envisat, SARAL, or Jason-2 and Jason-3. Information on the water level variations for all rivers within the basin helps to further understand the hydrology of the river network.

The resulting water levels of this study will be used in combination with other altimetric water levels following the ideas of [31] to build basin-wide multi-mission water level time series. With CryoSat-2 data we will be able to significantly improve the spatial resolution of the water level observations and better close the data gap between the end of the Envisat mission and the launch of the SARAL mission. Since the launch of Sentinel-3A in February 2016, SAR altimetry data with a short repeat time of 27 days has been available. This enables testing of the transferability of the application and validation of the proposed classification approach based on a new data set.

Acknowledgments: We want to thank the ESA GPOD team who provided us with the full data set through the SARvatore database and who were very helpful with questions regarding the data. This work was supported by the German Research Foundation (DFG) through the TUM International Graduate School of Science and Engineering (IGSSE) and the Technical University of Munich (TUM) in the framework of the Open Access Publishing Program.

Author Contributions: Eva Boergens developed the method, conducted the data analysis, and wrote the majority of the paper. Karina Nielsen helped with the development of the method and validation. Ole B. Andersen and Denise Dettmering contributed to the discussion of the method and results and the writing of the manuscript. Florian Seitz supervised the research and contributed to manuscript writing and organization.

Appendix A. Mean Waveforms and RIPs

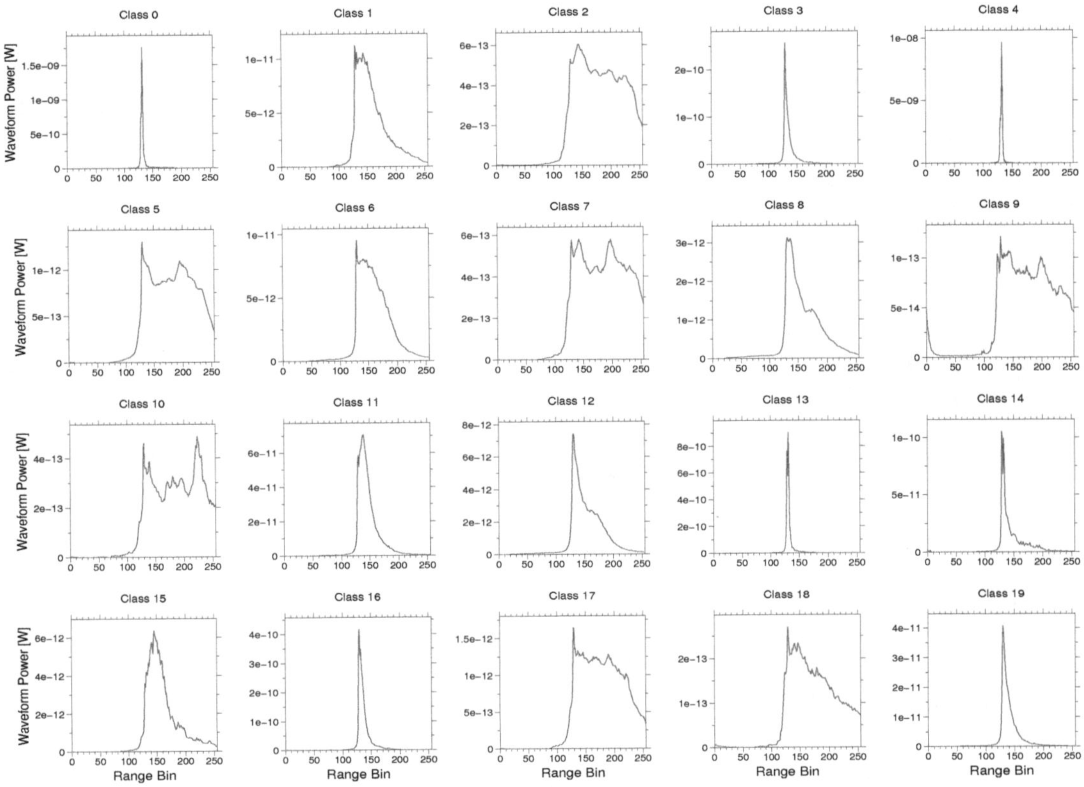

Figure A1. Upstream region mean waveforms. Water classes: 0, 4, 13.

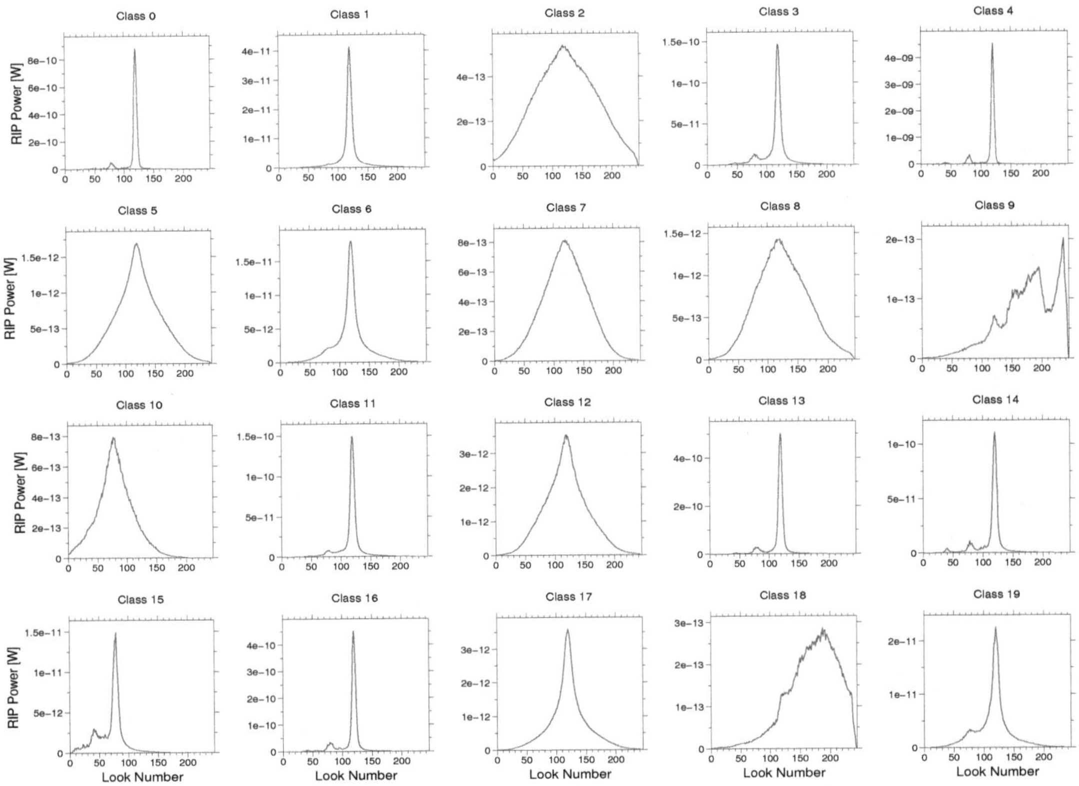

Figure A2. Upstream region mean RIPs. Water classes: 0, 4, 13.

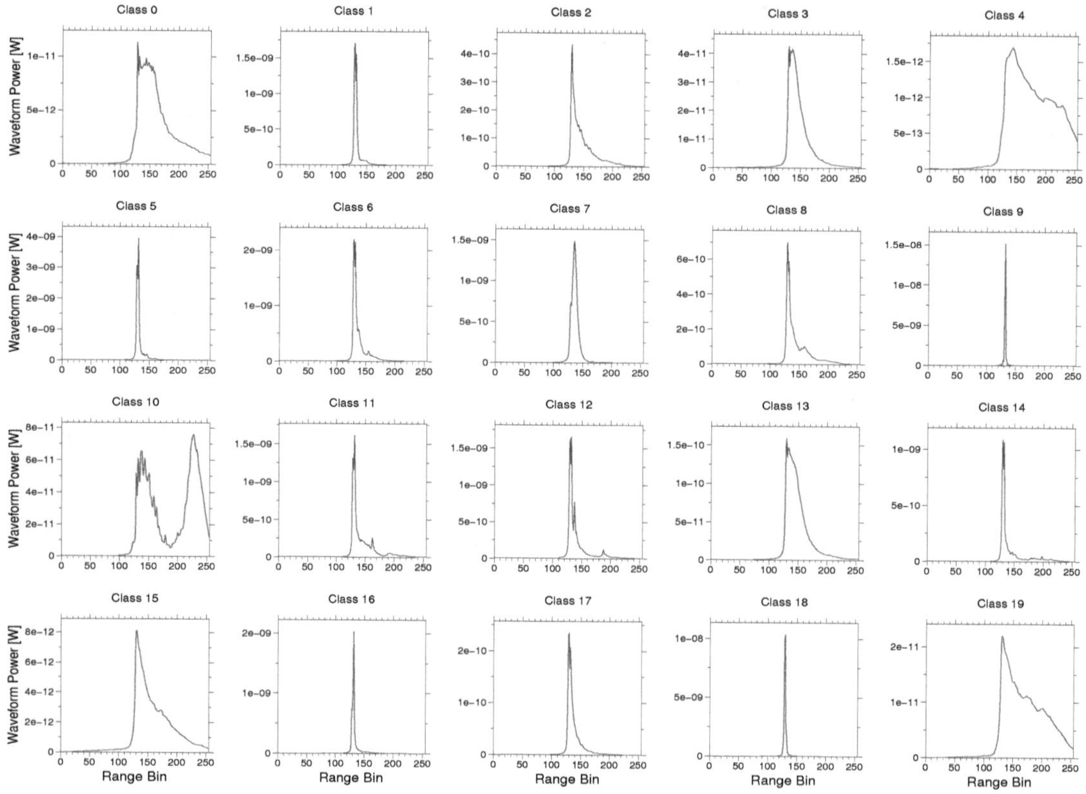

Figure A3. Middle region mean waveforms. Water classes: 2, 5, 6, 8, 11, 12, 14.

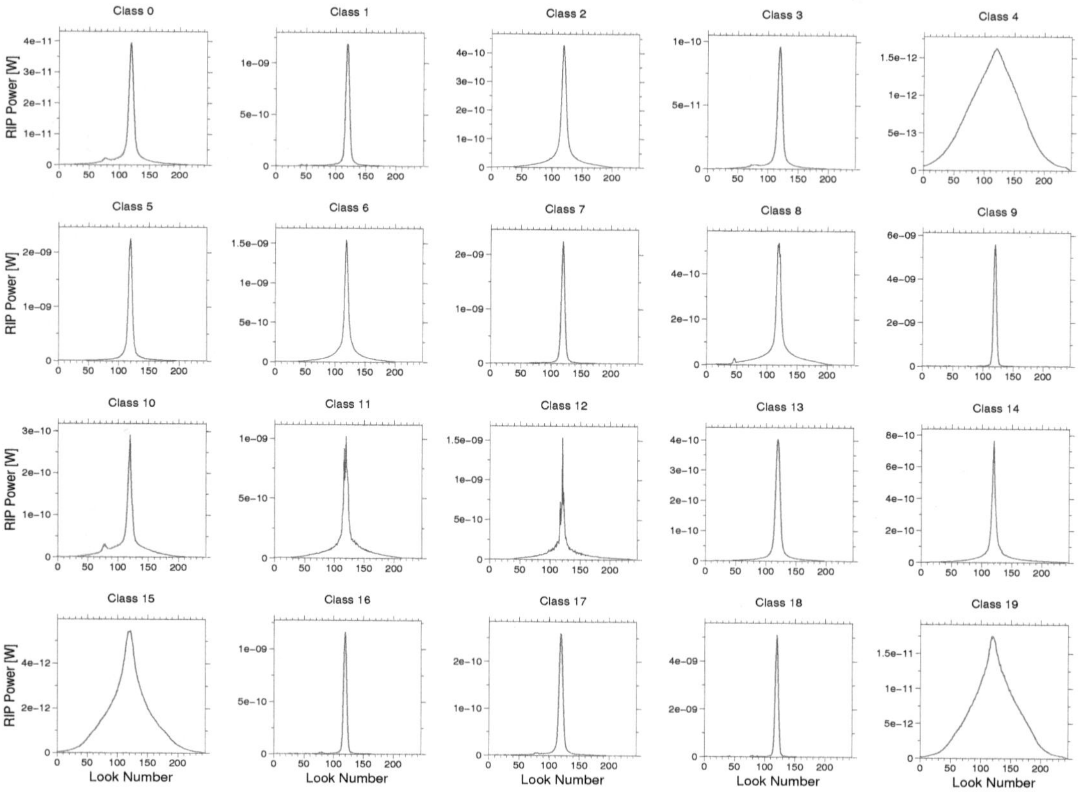

Figure A4. Middle region mean RIPs. Water classes: 2, 5, 6, 8, 11, 12, 14.

References

1. Global Runoff Data Center. *Long-Term Mean Monthly Discharges and Annual Characteristics of GRDC Stations*; Technical Report; Federal Institute of Hydrology: Koblenz, Germany, 2013.
2. Birkett, C.M. Contribution of the TOPEX NASA radar altimeter to the global monitoring of large rivers and wetlands. *Water Resour. Res.* **1998**, *34*, 1223–1239.
3. Santos da Silva, J.; Calmant, S.; Seyler, F.; Rotunno Filho, O.C.; Cochonneau, G.; Mansur, W.J. Water levels in the Amazon basin derived from the ERS-2 and ENVISAT radar altimetry missions. *Remote Sens. Environ.* **2010**, *114*, 2160–2181.
4. Schwatke, C.; Dettmering, D.; Bosch, W.; Seitz, F. DAHITI—An innovative approach for estimating water level time series over inland waters using multi-mission satellite altimetry. *Hydrol. Earth Syst. Sci.* **2015**, *19*, 4345–4364.
5. Boergens, E.; Dettmering, D.; Schwatke, C.; Seitz, F. Treating the Hooking Effect in satellite altimetry data: A case study along the Mekong River and its tributaries. *Remote Sens.* **2016**, *8*, 91.
6. Frappart, F.; Do Minh, K.; L'Hermitte, J.; Cazenave, A.; Ramillien, G.; Le Toan, T.; Mognard-Campbell, N. Water volume change in the lower Mekong from satellite altimetry and imagery data. *Geophys. J. Int.* **2006**, *167*, 570–584.
7. Biancamaria, S.; Frappart, F.; Leleu, A.S.; Marieu, V.; Blumstein, D.; Desjonquères, J.D.; Boy, F.; Sottolichio, A.; Valle-Levinson, A. Satellite radar altimetry water elevations performance over a 200 m wide river: Evaluation over the Garonne River. *Adv. Space Res.* **2016**, *59*, 1–19.
8. Raney, R.K. The delay/doppler radar altimeter. *IEEE Trans. Geosci. Remote Sens.* **1998**, *36*, 1578–1588.
9. Nielsen, K.; Stenseng, L.; Andersen, O.B.; Villadsen, H. Validation of CryoSat-2 SAR mode based lake levels. *Remote Sens. Environ.* **2015**, *171*, 162–170.
10. Kleinherenbrink, M.; Lindenbergh, R.C.; Ditmar, P.G. Monitoring of lake level changes on the Tibetan Plateau and Tian Shan by retracking Cryosat SARIn waveforms. *J. Hydrol.* **2015**, *521*, 119–131.
11. Göttl, F.; Dettmering, D.; Müller, F.L.; Schwatke, C. Lake level estimation based on CryoSat-2 SAR altimetry and multi-looked waveform classification. *Remote Sens.* **2016**, *8*, 1–16.
12. Villadsen, H.; Andersen, O.B.; Stenseng, L.; Nielsen, K.; Knudsen, P. CryoSat-2 altimetry for river level monitoring—Evaluation in the Ganges-Brahmaputra River basin. *Remote Sens. Environ.* **2015**, *168*, 80–89.
13. Bercher, N.; Dinardo, S.; Lucas, B.M.; Fleury, S.; Picot, N. A review of cryosat-2/siral applications for the monitoring of river water levels. *ESA Living Planet* **2013**, *2013*, 1–7.
14. Berry, P.A.M.; Garlick, J.D.; Freeman, J.A.; Mathers, E.L. Global inland water monitoring from multi-mission altimetry. *Geophys. Res. Lett.* **2005**, *32*, 1–4.
15. Desai, S.; Chander, S.; Ganguly, D.; Chauhan, P.; Lele, P.; James, M. Waveform Classification and Water-Land Transition over the Brahmaputra River using SARAL/AltiKa & Jason-2 Altimeter. *Indian Soc. Remote Sens.* **2015**, *43*, 475–485.
16. Dettmering, D.; Schwatke, C.; Boergens, E.; Seitz, F. Potential of ENVISAT Radar Altimetry for Water Level Monitoring in the Pantanal Wetland. *Remote Sens.* **2016**, *8*, 596.
17. Villadsen, H.; Deng, X.; Andersen, O.B.; Stenseng, L.; Nielsen, K.; Knudsen, P. Improved inland water levels from SAR altimetry using novel empirical and physical retrackers. *J. Hydrol.* **2016**, *537*, 234–247.
18. Armitage, T.W.K.; Davidson, M.W.J. Using the interferometric capabilities of the ESA CryoSat-2 mission to improve the accuracy of sea ice freeboard retrievals. *IEEE Trans. Geosci. Remote Sens.* **2014**, *52*, 529–536.
19. Wingham, D.J.; Francis, C.R.; Baker, S.; Bouzinac, C.; Brockley, D.; Cullen, R.; de Chateau-Thierry, P.; Laxon, S.W.; Mallow, U.; Mavrocordatos, C.; et al. CryoSat: A mission to determine the fluctuations in Earth's land and marine ice fields. *Adv. Space Res.* **2006**, *37*, 841–871.
20. MacQueen, J. Some methods for classification and analysis of multivariate observations. *Proc. Fifth Berkeley Symp. Math. Stat. Probab.* **1967**, *1*, 281–297.
21. Mekong River Commission. (Ed.) *Overview of the Hydrology of the Mekong Basin*; Mekong River Commission: Phnom Penh, Cambodia, 2005.
22. ESA. *CryoSat-2 Geographical Mode Mask*; ESA: Paris, France, 2016.
23. Scagliola, M. *CryoSat Footprints*. 2013. Available online: https://earth.esa.int/documents/10174/125271/CryoSat_Footprints_TN_v1.1.pdf/2a5d996b-8b77-4d1c-ae7b-fbf93848c35d;jsessionid=B1FF8C50A1B0F2A0879F6FA028844644.eodisp-prod4040?version=1.0 (accessed on 23 November 2017).

24. Cullen, R.A.; Wingham, D.J. CryoSat Level 1b Processing Algorithms and Simulation Results. In Proceedings of the 2002 IEEE International Geoscience and Remote Sensing Symposium, Toronto, ON, Canada, 24–28 June 2002; pp. 1762–1764.
25. Laxon, S.W. Sea-Ice Altimeter Processing Scheme at the EODC. *Int. J. Remote Sens.* **1994**, *15*, 915–924.
26. Gommenginger, C.; Thibaut, P.; Fenoglio-Marc, L.; Quartly, G.; Deng, X.; Gómez-Enri, J.; Challenor, P.; Gao, Y. Retracking altimeter waveforms near the coasts. In *Coastal Altimetry*; Benveniste, J., Cipollini, P., Kostianoy, A.G., Vignudelli, S., Eds.; Springer: Berlin, Germany, 2011; pp. 61–101.
27. Doane, D.P. Aesthetic frequency classifications. *Am. Stat.* **1976**, *30*, 181–183.
28. Sturges, H. The choice of a class interval. *J. Am. Stat. Assoc.* **1926**, *21*, 65–66.
29. Mekong River Commision. *Annual Mekong Flood Report 2008*; Mekong River Commission: Phnom Penh, Cambodia, 2009.
30. Lehner, B.; Verdin, K.; Jarvis, A. New global hydrography derived from spaceborne elevation data. *Eos* **2008**, *89*, 93–94.
31. Boergens, E.; Buhl, S.; Dettmering, D.; Klüppelberg, C.; Seitz, F. Combination of multi-mission altimetry data along the Mekong River with spatio-temporal kriging. *J. Geodesy* **2017**, *91*, 519–534.
32. Schneider, R.; Nygaard Godiksen, P.; Villadsen, H.; Madsen, H.; Bauer-Gottwein, P. Application of CryoSat-2 altimetry data for river analysis and modelling. *Hydrol. Earth Syst. Sci.* **2017**, *21*, 751–764.

11

Lead Detection in Polar Oceans—A Comparison of Different Classification Methods for Cryosat-2 SAR Data

Denise Dettmering *, Alan Wynne, Felix L. Müller, Marcello Passaro and Florian Seitz

Deutsches Geodätisches Forschungsinstitut, Technische Universität München, Arcisstraße 21, 80333 Munich, Germany; alan.wynne@tum.de (A.W.); felix-lucian.mueller@tum.de (F.L.M.); marcello.passaro@tum.de (M.P.); florian.seitz@tum.de (F.S.)

* Correspondence: denise.dettmering@tum.de

Abstract: In polar regions, sea-ice hinders the precise observation of Sea Surface Heights (SSH) by satellite altimetry. In order to derive reliable heights for the openings within the ice, two steps have to be fulfilled: (1) the correct identification of water (e.g., in leads or polynias), a process known as lead classification; and (2) dedicated retracking algorithms to extract the ranges from the radar echoes. This study focuses on the first point and aims at identifying the best available lead classification method for Cryosat-2 SAR data. Four different altimeter lead classification methods are compared and assessed with respect to very high resolution airborne imagery. These methods are the maximum power classifier; multi-parameter classification method primarily based on pulse peakiness; multi-observation analysis of stack peakiness; and an unsupervised classification method. The unsupervised classification method with 25 clusters consistently performs best with an overall accuracy of 97%. Furthermore, this method does not require any knowledge of specific ice characteristics within the study area and is therefore the recommended lead detection algorithm for Cryosat-2 SAR in polar oceans.

Keywords: leads; satellite altimetry; CryoSat-2; classification; peakiness; polar ocean

---

1. Introduction

Satellite altimeter missions have been providing precise measurements of the Sea Surface Height (SSH) on a nearly global scale for more than 25 years. This data set has significantly improved our understanding of the sea level dynamics. However, in the polar regions, the altimetry measurements are contaminated by the sea-ice resulting in a degradation in the precision or even usefulness of the results. Radar altimeter observations in the vicinity of sea-ice typically result in an over-estimation of SSH [1]. Knowledge of the locations of openings within the sea-ice, known as leads, can help improve the SSH estimation and the understanding of the global climate. For instance, the SSH derived through leads is assumed to be the instantaneous ocean elevation which can be used to estimate the sea-ice freeboard [2,3]. Therefore, it is clear to see that a reliable SSH is essential for the estimation of the freeboard thickness. Freeboard estimates can be negatively biased when an unreliable lead detection is imposed and the SSH precision is degraded by ice samples detected as leads [1]. Accordingly, the reliable identification of leads is an essential criteria for precisely monitoring the Arctic ice volume decrease, as well as for Arctic sea level change studies.

Indications to the usefulness of satellite altimetry for finding open water within sea-ice were provided as early as 1980 by Dwyer and Godin [4]. They were able to determine the sea-ice boundary using GEOS-3 satellite altimetry data with the knowledge that within sea-ice regions, the altimetry backscattering properties are in stark contrast to that of open water. The altimeter returned signal, known as a waveform, provides this backscattering information. Leads are known to produce high power and single peak specular returns in altimeter waveforms. This is because leads typically have a smooth (relative to the radar wavelength) surface as they generally are absent of any wave field, meaning a large percentage of the radiated energy is reflected back towards the antenna of the satellite [4,5].

To date, many methods to find leads based on altimeter observations have been implemented. Laxon [5] developed a parameter based on the maximum power of ERS satellite altimeter waveforms, pulse peakiness, which could be used for charting the open water areas within the Arctic regions. Connor et al. [6] also used the pulse peakiness parameter with Envisat data to find leads in order to make estimates of the sea-ice freeboard. Röhrs and Kaleschke [7] used a maximum power threshold on Cryosat-2 (CS-2) SAR altimeter waveforms to identify leads. Zygmuntowska et al. [8] developed a supervised classification method based on Bayesian classification [9] applied to ASIRAS (airborne version of the SIRAL instrument on-board CS-2) waveform shape parameters. Passaro et al. [10] developed a parameter based on the stack waveforms of CS-2 known as the stack peakiness. This parameter also considers multiple consecutive observations in an attempt to isolate a lead at the nadir position. Müller et al. [11] developed an unsupervised classification method based on K-medoids partitional clustering [12] of model waveform features and classification of observation data relative to the model using K-nearest neighbor (KNN) classification [13]. Shen et al. [14] aimed to find an optimal classifier–feature assembly for sea-ice classification. They tested a number of waveform feature statistical methods with different combinations of waveform features. Just recently, Lee et al. [15] proposed a promising new method for detecting leads using CS-2 waveforms. Their method is based on a waveform mixture analysis, which assumes that each footprint of the waveform is a linear combination of leads and various types of sea-ice.

Most of the above mentioned studies include a validation of the presented results. However, since they are based on different input data, different ground truth data sets and different accuracy metrics, they are not directly comparable. To overcome this deficiency and to recommend the best classification method for further applications is the motivation of this investigation.

Within this study, four lead classification methods are tested for CS-2 SAR data and evaluated using high resolution imagery acquired during CS-2 dedicated underflights. The aim is to find the best existing CS-2 SAR altimeter lead classification algorithm that provides low false classification. The four classification methods under investigation are (1) the waveform maximum power as introduced, e.g., by Röhrs and Kaleschke [7]; (2) the multi-parameter classification method primarily based on pulse peakiness developed by Ricker et al. [16]; (3) a method based on the stack peakiness presented by Passaro et al. [10]; and (4) the unsupervised classification method published by Müller et al. [11].

This is the first study that compares different classification approaches using exactly the same input data and the same very high resolution (<1 m) ground truth. Previous studies have used space-based evaluation techniques that have suffered from a relatively low resolution, meaning that small scale leads are not observed by the ground truth and, therefore, they bias the classification results. Furthermore, observations with time differences of multiple hours have been utilized, also distorting the results. In contrast, here, almost time coincident ground truth data is used.

This paper is structured as follows. Section 2 describes the CS-2 altimeter SAR data sets and some of their processing parameters as well as the ground truth image data sets. A brief description of each employed lead classification method together with an overview of the threshold optimization technique is given in Section 3. A quantitative comparison of all lead classification methods relative to the ground truth combined with a discussion of the results is provided in Section 4. Finally, the conclusion summarizes the results and provides the reader with a recommendation for the "best" lead classification method based on the available data.

2. Data Sets and Study Areas

This section will discuss the CS-2 data sets, the ground truth used for evaluating the classification methods, and the corresponding study areas.

2.1. CryoSat-2 Data

The CS-2 mission is the first satellite altimeter utilizing the so-called Delay-Doppler or SAR concept in some geographical regions, among them large parts of the Polar Oceans (depending on a changing mode mask [17]). The basic idea behind the SAR principle is that the antenna size is synthetically increased (in the along-track direction) allowing to create a coherence of the emitted pulses [18]. SAR altimeters are described as being beam-limited in the along-track direction while being pulse-limited in the across-track direction. Along-track processing improves the respective resolution and furthermore, the multi-looked processing greatly improves the signal-to-noise ratio [19]. While the footprint of CS-2 yields about 300 m in the along-track direction, in the cross-track direction up to 15 km are covered by the signal [20].

For this study, in addition to the averaged multi-looked waveforms, full stack data is needed. ESA grants access to an online service called Grid Processing On-Demand (GPOD), which provides the full stack (L1B-S) for the CS-2 mission. The CS-2 data sets used within this study (mainly multi-looked waveforms and derived parameters) are all taken from the SAR Versatile Altimetric Toolkit for Ocean Research & Exploitation (SARvatore) GPOD service (see acknowledgments). The GPOD service allows for free and easy access to two main data products, namely ESA Baseline C L1B-S and L2 data [21]. The online service allows for tailored processing of each product. Dinardo [22] has described in detail the processing steps for L1B data. For the present study, a Hamming window is applied before the along-track Fast Fourier Transform in an attempt to stop signals coming from leads that are outside the synthetic along-track 3 dB beam-width (main-lobe), as recommended by Passaro et al. [10].

2.2. Ground Truth Data

For evaluating lead classification results, the contemporaneity between altimetry measurements and ground truth observations is of great importance in order to exclude any differences due to sea-ice drifts. The NASA Operation IceBridge (OIB) mission [23] includes several CS-2 underflights, following the CS-2 groundtracks in Arctic and Antarctica with very short time discrepancy in overpassing [24]. Moreover, airborne sensors acquire cloud-free images with a very high resolution.

The ground truth validation data used within this study is the Operation IceBridge Digital Mapping System (DMS) L1B geolocated and orthorectified images [25]. These images are panchromatic and provide a very high resolution, making them an ideal choice for lead detection. The pixel resolution for a flight height of 450 m is about 10 cm $\times$ 10 cm, resulting in an image size of approximately 775 m $\times$ 690 m. Within this study, the images are resampled into 1 m resolution in order to reduce the computational costs. This resolution is deemed sufficient for detecting even small leads, especially since the CS-2 resolution is significant lower (see Section 2.1).

2.3. Flight Lines

In order to ensure a reliable validation, CS-2 observations are selected on tracks that are simultaneous and almost time-coincident with the airborne underflights from the NASA Operation IceBridge mission. In total, 12 data sets were found defining the validation data set from polar regions within this study (eight on the northern hemisphere, four on the southern hemisphere). Figure 1 illustrates the locations of these data sets for the Arctic and Antarctic regions, respectively. The majority of the data sets are taken from the spring season.

Airborne images that were acquired later than 90 min after the satellite overpass and earlier than 90 min before the satellite overpass are discarded from the analysis. This threshold is chosen as an optimum balance between having sufficient data to perform this study and preventing lead positional errors due to the sea-ice motion which may bias the statistical results. Additionally, adapting the methodology from Ricker et al. [16] to coarsely discriminate between open ocean and sea-ice, only those data observed with a sea-ice concentration greater than 70% are used, since the theoretical basis for lead detection may no longer hold in low sea-ice concentration areas. The sea-ice concentration values used in this study are the interpolated sea-ice concentration values available from the Ocean and Sea-Ice Satellite Application Facility (OSI SAF) (see acknowledgments).

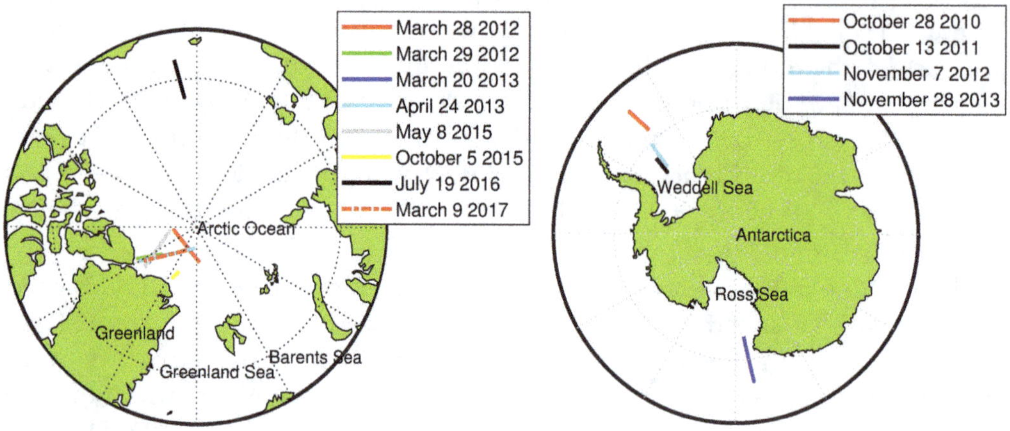

Figure 1. Illustration of Operational IceBridge (OIB) flight lines taken for validation in the Arctic Region (**left**) and the Antarctic Region (**right**).

3. Methodology

This section gives an overview of the employed four lead classification methods, whose performances will be compared within this paper. Moreover, the methods used for ground truth image classification and for optimal threshold definition will be described.

3.1. Maximum Power Classifier (MAX)

The maximum power classifier [7] has the simplest implementation of all the classification methods used within this study. It relies on the application of a threshold to the maximum power of multi-looked altimetry waveforms. If the maximum power is above the threshold, then the corresponding observation is classified as a lead. In a similar previous study, Wernecke and Kaleschke [1] found the maximum power classifier to be the best lead classification method with their ground truth validation data—MODIS satellite imagery. Although the maximum power classifier has an easy implementation, it does require knowledge of the study area in order to correctly define the threshold, which is usually not assured. Moreover, the maximum waveform power—and thus the correct threshold—is known to vary substantially across different regions and seasons [10].

3.2. Multi-Parameter Classification Method (MULTI)

The second lead classification method used in this study is the one proposed by Ricker et al. [16], which uses six parameters in order to discriminate a lead observation from one over sea-ice. The parameters are:

- Pulse Peakiness (PP): Defined as the ratio of the waveform maximum power to the accumulated power and also in this case scaled by the number of range bins. Larger values are indicative of the presence of a lead within the altimeter footprint.
- Stack Kurtosis (K): An additional measure of the peakiness of the range integrated stack power [26]. A high value suggests that the distribution is prone to an outlier, e.g., caused by the presence of a lead within the altimeter footprint.
- Stack Standard Deviation (SSD): This value provides information to the variation of surface backscattering power with incidence angle [26]. Therefore, small values of SSD are used as an indicator for the presence of leads.

- Modified PP (two parameters PP_l and PP_r to consider the bins "left" and "right" of the maximum power bin): This is meant to disregard lead observations which are not at nadir. The assumption here is that off-nadir lead reflections do no show as specular a reflection as an observation at nadir.
- Sea-ice concentration: This is used as a coarse discrimination between ocean and ice areas. Only observations in areas with significant sea-ice concentration (>70%) are allowed as leads in case all other conditions are met.

The MULTI method also requires the definition of thresholds for all parameters, which are used to discriminate lead observations from sea-ice observations. Furthermore, the correct setup of the thresholds requires knowledge of the sea-ice characteristics and are usually not transferable between different regions.

3.3. *Stack Peakiness Classifier (STACK)*

The stack kurtosis and SSD can be used to classify a waveform that might contain a lead, but they are independently insufficient in isolating a nadir return from a stack of waveforms which has observed a lead backscatter. This was the motivation of Passaro et al. [10] to develop the Stack Peakiness (SP) parameter that is defined using the Range Integrated Power (RIP) normalized against its maximum value. For CS-2 Baseline C, the SP parameter must be computed using the stack data available from the GPOD website, as it is not available in the official Baseline C L2 data product. However, it is included in the Baseline D L2 data.

The assumption behind the algorithm is that larger local maxima in SP are likely to be evoked by a lead at nadir. Therefore, a multi-observation analysis occurs before a lead classification is made, which considers the current, forward and aft observations. The forward and aft observations must be above the median SP (SP_{median}) plus Median Absolute Deviation MAD (SP_{MAD}) of all CS-2 tracks within this study for the current observation to be considered a lead. This is done in an attempt to reduce false lead observations as a lead within the altimeter footprint influences the SP value even when not at nadir. The current observation must also be above a predefined minimum threshold ($SP_{minlead}$) in order for it to be classified as a lead.

The SP strategy developed in Passaro et al. [10] had the objective of isolating one single return in each lead, without distinguishing other reflecting surfaces. In contrast, within this study all observations above a lead are considered as part of the statistical analysis. This therefore increases the percentage of points considered as erroneous ice detections in this study (see Section 4).

The performance of this classification method depends on the correct threshold definition as much as the two previous methods. Section 3.5 will handle the threshold optimization.

3.4. *Unsupervised Classifier (UNSU)*

The unsupervised classification method developed by Müller et al. [11] is believed to be the first attempt using this method applied to satellite altimetry over sea-ice areas. There are many different approaches to unsupervised learning including clustering (e.g., K-means, K-medoids [27]), latent variable models (e.g., Hidden Markov Models [28]) and others. Müller et al. [11] employed the clustering K-medoids algorithm for their unsupervised classification on pulse-limited altimetry waveforms (Envisat and SARAL). In this study, the unsupervised classification method is tested for the first time with SAR altimeter data.

The first step in the unsupervised classification process is to cluster or group model data applying the K-medoids algorithm. The six waveform features defined by Müller et al. [11] are also used for the clustering in this study, with minor changes in some computation parameters:

- Waveform maximum (Wm): As described in Section 3.1, the maximum power can be used to characterize the surface below the satellite altimeter.

- **Trailing edge decline (Ted):** The Ted is a characterization of the trailing edge of the waveform, i.e., from the maximum power range bin to the last range bin, by means of a fitting to a power series model. A rapid decay (low value) would be associated to the typical waveform shape of a lead.
- Waveform noise (Wn): In the context presented here, the Wn represents the MAD of the residuals to the fitting of the power series model. Very small values for specular lead type returns are expected.
- Waveform width (Ww): The amount of range bins with their power greater than 1% of the waveform maximum is used to determine the waveform width. A small waveform width is expected in the presence of a lead.
- **Leading edge slope (Les):** The first waveform bin containing more than 12.5% of the waveform maximum power subtracted from the maximum power bin provides relative information regarding the Les. Again, low values are indicative of a lead surface.
- Trailing edge slope (Tes): Conversely, the Tes is obtained by a subtraction of the maximum waveform power bin from the last bin position containing more than 12.5% of the waveform maximum power. The characteristics of the Tes is similar to the Les for single peak waveforms, but it can also be used to identify strong multiple peaks.

The features produce varying orders of magnitudes, which causes biased weighting within the K-medoids algorithm. Therefore, before used for clustering, all features are reduced using the standard score (sometimes referred to as z-score) to give more reliable clustering results [29].

In order to have a model that is representative of as many waveform types as possible, a large collection of waveforms is required from the CS-2 mission. This has to include waveforms from sea-ice, leads and ocean surfaces. The maximal extent of the Arctic sea-ice is in March [30] and therefore, the beginning of the melt-season (April/May) is assumed as the most likely period to cover as many sea-ice types as possible. Figure 2 shows the location of the model data taken for this study from the Greenland Sea. The data used corresponds to when CS-2 was operating in SAR mode during the time period of April through to May 2015, corresponding to 361,977 model waveforms.

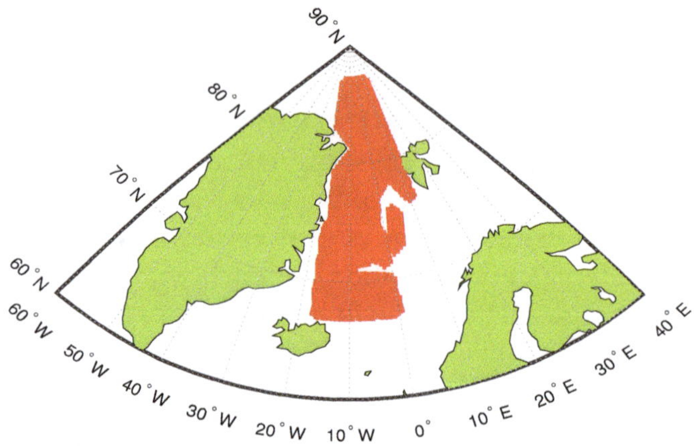

Figure 2. Map of the locations (red areas) of CS-2 waveforms used for the unsupervised classification model taken within the time period of April–May 2015.

To initialize the K-medoids algorithm, the number of clusters k needs to be decided. The value for k should be larger than the final amount of surface classifications, i.e., $k > 3$ (lead, sea-ice and ocean). Müller et al. [11] used $k = 30$ which was empirically derived by means of visual analysis of the resulting clusters and sums of distances. Within this study, a test of seven cluster numbers is performed ranging from $k = 5$ to $k = 35$ in steps of 5.

After the K-medoids algorithm has finished, the clusters have to be classified manually to one of the surface types. The final step is then to classify observation data relative to the classified cluster model. This is done using K-Nearest Neighbor (KNN) classification. The KNN size must also be chosen and this is done using a 10-fold cross-validation method as described in detail in Müller et al. [11].

3.5. Threshold Optimization

Three of the four classification methods under investigation (namely MAX, MULTI, and STACK) depend on one or more predefined thresholds that can be determined using a reliable ground truth. Wernecke and Kaleschke [1] developed an optimization technique, which is also applied here. It minimizes a cost function to derive the threshold value Θ

$$\text{cost}(\Theta) = w \times \text{False Ice}(\Theta) + \text{False Lead}(\Theta) \quad (1)$$

where w is a weighting defining how the false classifications are minimized. "False Ice" are observations classified as ice but are actually lead samples and "False Leads" are observations classified as leads but are actually ice samples. For small values of w (i.e., <1) False Lead observations are primarily reduced and for $w = 1$, the total false classifications are minimized. The ground truth is used to randomly separate the data into a training (50%) and testing (50%) subset with an equal number of sea-ice and lead samples within each subset. Θ is derived from the training subset by testing all potential threshold values as initial guesses to the Nelder and Mead [31] minimization algorithm. The random sub-sampling is repeated 200 times in order to find the global minimum of the cost function and to have an estimation of the spread of performance on each testing subset. This is done for different values of w.

For MULTI, only the PP and modified PP parameters thresholds are derived using this optimization. This is done in an attempt to reduce computation time. The sea-ice concentration value does not depend on the satellite altimeter so the threshold of 70% remains. The thresholds of 40 and 4 from Ricker et al. [16] for K and SSD respectively also remain as it was found that the vast majority of lead observations fell inside these thresholds.

For STACK, the SP_{median} and SP_{MAD} found through the ground tracks used in this study resulted in $SP_{median} = 9.76$ and $SP_{MAD} = 6.33$, meaning a slightly lower sum compared to Passaro et al. [10]. For this reason, it was decided to use the values from the literature. When developing the $SP_{minlead}$, due to the random sampling of the data, the multi-observation analysis was not conducted for the threshold optimization.

The optimized threshold values and their corresponding weight value are given in Table 1. All of them are used to generate different classification results that are shown and discussed in Section 4.

Table 1. Optimized thresholds and their corresponding weight.

w	**Max. Power**	PP	PP_l	PP_r	**SP**
0.5	4.356×10^{-11} W	120.75	175.08	30.84	345.02
1	2.714×10^{-11} W	116.18	152.58	158.11	300.43
2	1.892×10^{-11} W	99.93	63.25	145.29	156.61
3	1.051×10^{-11} W	99.67	63.38	124.28	102.86
4	4.513×10^{-12} W	96.71	66.05	69.28	91.28
5	4.958×10^{-12} W	97.01	63.92	51.59	78.61
6	3.609×10^{-12} W	97.03	64.15	50.89	73.47
7	2.219×10^{-12} W	97.00	62.25	56.19	71.14
8	1.654×10^{-12} W	97.15	53.80	51.43	70.41
9	1.399×10^{-12} W	90.18	63.20	53.98	70.41
10	1.359×10^{-12} W	80.47	57.32	30.40	54.03

3.6. Ground Truth Image Classification

In order to be able to compare and validate the different altimeter lead classification methods, a reliable ground truth image classification must be used. In optical images, such as the ones from IceBridge used here, leads are known to have typically dark optical features that translates to low image pixel intensities. Lead attributes can be dynamic due to different types of thin ice within the lead itself, formed by rapid ice growth due to the cold polar temperatures. Additionally, leads with thin ice may even be covered by snow and frost flowers [32].

The ground truth data for this study is set up applying the automatic Sea-Ice Lead Detection Algorithm using Minimum Signal (SILDAMS) developed by Onana et al. [33], which has been shown to give favorable results using NASA Operation IceBridge image data. The SILDAMS algorithm is based on an affine time-frequency distribution analysis which uses the minimum signal transform in order to perform a localization around low frequencies. The minimum signal is given by

$$S^{z}_{min}(f) = f^{2\pi z} e^{-2\pi z f} \tag{2}$$

where f represents the frequency or image pixel intensities and z is a positive parameter for adjusting the spectral width (bandwidth). With a careful selection of the z parameter, it is possible to highlight features of interest from the image, such as leads. The bandwidth parameter is defined as $z = 0.001$, since this value has been shown by Onana et al. [33] to provide good results for images acquired in clear conditions with daylight. As one of the OIB data sets (taken on 28 October 2010) was acquired in ever decreasing lighting conditions. From a visual analysis of the images, it was decided to change z according to the lighting conditions from $z = 0.003$ for the first 450 images, to $z = 0.005$ (from 451 to 1035), $z = 0.006$ (from 1036 to 1103), and $z = 0.015$ (for the remainder of the images).

The last step in the image classification is to apply a threshold to the minimum signal transformation of the image. Within this study, only a binary classification is used between sea-ice and lead/open water. In accordance with Onana et al. [33], a threshold of 0.3 is used for water detection (smaller values = sea ice; larger values = lead). This results in a binary map for each image with a one indicating a lead pixel and a zero indicating a sea-ice pixel. Nearest-neighbor interpolation is then used to associate the binary classification to the CS-2 surface sample ground track.

All images are assumed to be cloud-free since a manual inspection of completeness of laser measurements from the laser-based Airborne Topographic Mapper (ATM) revealed almost no missing values.

Onana et al. [33] specify the lead detection capability of SILDAMS to be 99%. The manual inspection of the ground truth data set established in the present study shows that all open water areas are correctly flagged without any false detections. However, the distinction between open water and thin ice was only correct at about 90%. For this reason, the open water class also contains thin ice—which is uncritical for altimetry waveform classification which can also not differentiate very thin ice from leads [6].

4. Results and Discussion

In order to be able to assess each altimeter classification method, they must be compared with the same ground truth, which is derived from the image classification described earlier. This section provides and discusses the results of the ground truth evaluation and comparison between each altimeter classification method. In total, there are 14,231 satellite observations from all the data sets, and of these, 365 observations (i.e., about 2.5%) are defined as leads (including leads with thin ice) according to the ground truth.

4.1. Quantitative Comparison between Altimeter Classification and Ground Truth

A Receiver Operating Characteristics (ROC) graph [34] is used to visualize the performance of the different classifiers (see Figure 3). It shows the change in performance of the different altimeter classification methods relative to the ground truth depending on different thresholds. Two criteria are computed for the evaluation: the True Lead Rate (TLR), which is the percentage of leads classified by the altimeter classification and confirmed by the ground truth classification (see Equation (3)) and the False Lead Rate (FLR), which is the percentage of altimeter lead classifications that are classified as sea-ice by the ground truth to the total sea-ice ground truth samples (see Equation (4)).

$$\mathrm{TLR} = \frac{\text{True Lead}}{\text{True Lead} + \text{False Ice}} \tag{3}$$

$$\mathrm{FLR} = \frac{\text{False Lead}}{\text{False Lead} + \text{True Ice}} \tag{4}$$

True Lead is the number of observations classified as lead confirmed by the ground truth. False Ice is classified as ice observations but are lead samples according to the ground truth. False Lead observations are classified as a lead but are ice samples according to the ground truth. True Ice are observations classified as ice samples and are confirmed by the ground truth.

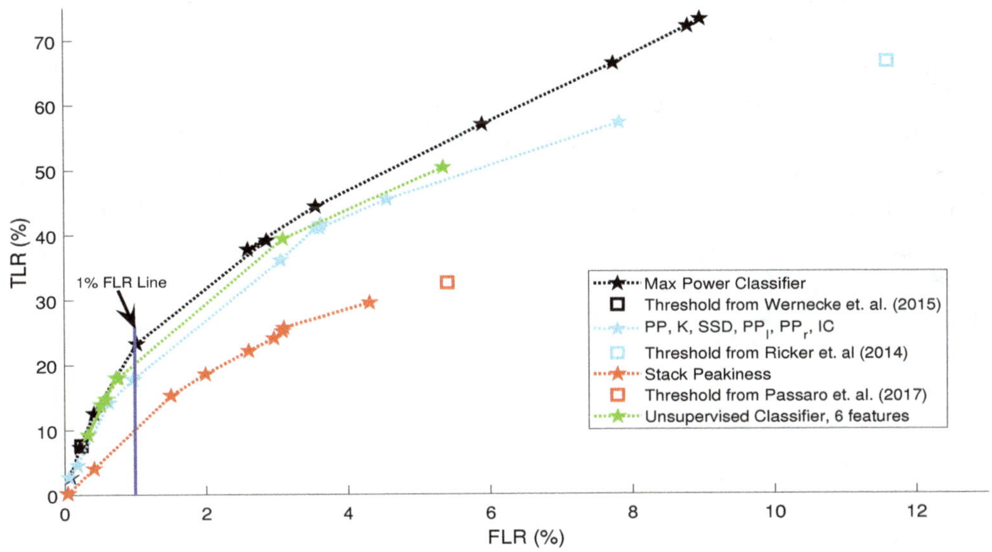

Figure 3. ROC graph depicting the change in performance for different thresholds and different cluster numbers. Each line and marker represents a different altimeter classification method and the different weight for optimized threshold respectively. The square markers represent thresholds that were determined in previous studies. The 1% False Lead Rate (FLR) line is also shown.

The aim of the ROC graphs in Figure 3 is threefold: It helps to find the best threshold for each classification method, it can be used for an inter-comparison of the methods, and it provides a quantitative performance measure for all classifiers. In general, with an increasing weight w (as defined in Equation (1)) there is an increase in the TLR and also the FLR for all classifiers with a higher rate of increase in the TLR. However, this behavior is not linear, and there seems to be a maximum TLR reachable for each method. When further increasing the weight (i.e., lowering the threshold) this only results in a larger FLR. This effect was already shown by Wernecke and Kaleschke [1].

Simply from looking at Figure 3, it is not immediately clear what is the best threshold/number of clusters to use. This strongly depends on the application, i.e., on the acceptable FLR. In case it is important to reach a very low FLR, a conservative threshold (small weight) should be used. This comes at the cost of identified leads. However, if the application requires as many leads as possible, one should use a higher w. All three classifiers reach TLR between 30% and 55% when FLR of 5% are acceptable. From Figure 3, it seems that the MAX classification method performs best for the majority of weights compared to all other classification methods, followed closely by UNSU and MULTI. The STACK classification method appears to be the poorest performing classifier, showing a narrow ROC curve with a small TLR/FLR ratio that is about half of that from MAX. The highest overall TLR of about 75% is achieved by MAX (with FLR of about 10%).

In order to quantify and compare the performance of the different classifiers, a fixed FLR is defined and the corresponding TLR are compared. A FLR of 1% is chosen as an optimum balance between a relatively high TLR and low number of false open water observations. For the MAX classifier, this results in a TLR at 23.05% (closest threshold $w = 3$; 1.05×10^{-11} W). The MULTI method results in a 18.09% TLR (closest threshold $w = 3$; $PP = 99.67$, $PP_l = 63.38$, $PP_r = 124.28$). The STACK gives a performance of 10.38% TLR (closest threshold $w = 3$; $SP_{minlead} = 102.86$). The above TLR values were derived by a polynomial curve fitting to the points of each threshold classifier. From all threshold based classifiers, this appears to show that the maximum power classifier is performing the best relative to the ground truth.

The UNSU classification method is independent from any threshold definition. Instead, the classification performance clearly depends on the number of clusters used within the process (see Section 3.4). It can be noted that for smaller k, there is typically a larger TLR and FLR. However, in contrast to the threshold-based methods, a continuous change in performance is not observed for a change in cluster number. This is illustrated in Figure 4; the figure shows the ROC graph for UNSU in dependence of k. The optimal cluster number depends on the size and variability of the input waveform data set. If only a small number of clusters is allowed, the number of observations per cluster (i.e., the cluster size) increases and the probability for correct classification decreases. In other words: the clustering is too coarse [35]. This is the case for $k \leq 10$. If k is large enough, the results only change marginally, depending on the manual preprocessing step assigning each cluster to one surface type (lead, ice, or ocean). The cluster number with FLR closest to 1% will be chosen as the optimum for this study. This corresponds to $k = 25$ and results in a 18.08% TLR and 0.73% FLR. The interpolated value at 1% FLR is 22.37% TLR.

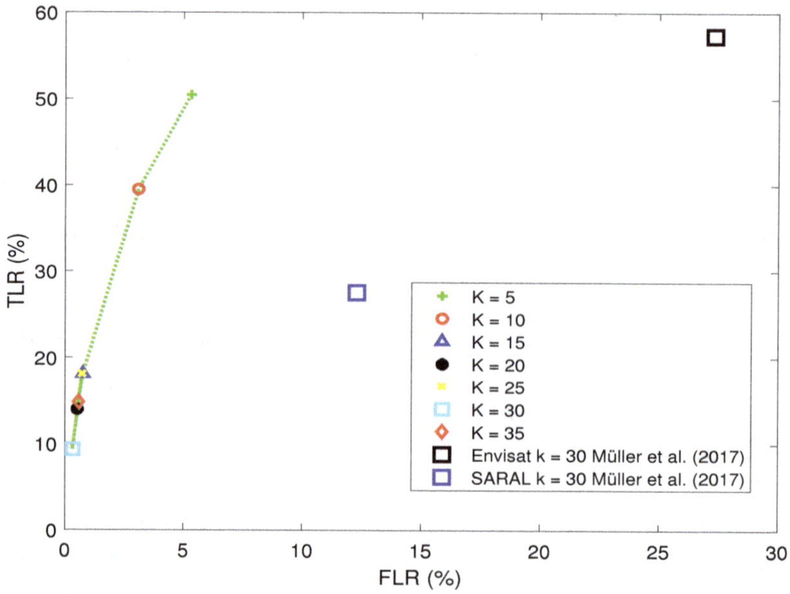

Figure 4. ROC graph depicting the change in performance for different cluster sizes. The performances calculated from the study from Müller et al. [11] are also shown.

ROC graphs are useful for analyzing the performance of classifiers relative to a ground truth, and to identify the best parameters (i.e., thresholds or number of clusters) for the respective application. However, in order to understand the overall performance of the different altimeter classifiers, different accuracy metrics must be analyzed. Producer and user accuracies are used for this purpose for both sea-ice and lead samples. While producer accuracies give an indication of the probability that a ground truth sample will be correctly classified, the user accuracy is an indication to the reliability of the classification and is defined as the ratio of the sum of correct classifications (leads and ice) to the total number of classifications [36].

Table 2 lists the overall lead and sea-ice accuracies for each altimeter classification with respect to the ground truth. The largest overall accuracy is achieved by the unsupervised classification method at 97.2%. This method also yields the best lead user accuracy and the largest ice producer accuracy.

However, the lead producer accuracy scores only second behind the maximum power classifier. Even if the numbers in some of the metrics only slightly differ, it is obvious that the two methods utilizing the maximum power for the classification (UNSU and MAX) score better than the other two (MULTI and STACK). While interpreting the numbers, it should be kept in mind,that all values have been optimized to score a FLR of about 1%. Since only integer weighting factors w are used, slightly different FLR are achieved and the numbers are not fully comparable. The good performance of UNSU is even emphasized, taking the lower FLR into account.

Table 2. Accuracies for each classifier in (%). Values provided for FLR of about 1% ($w = 3$, $k = 25$). The best performance for each of the accuracy metrics is marked in bold.

	MAX	MULTI	STACK	UNSU
FLR	**1.02**	**0.79**	**1.51**	**0.73**
Overall Accuracy	97.03	96.94	96.36	**97.18**
Lead User Accuracy	37.44	32.50	21.13	**39.29**
Lead Producer Accuracy (TLR)	**23.29**	17.81	15.34	18.08
Ice User Accuracy	**98.00**	97.86	97.79	97.87
Ice Producer Accuracy	98.98	99.03	98.49	**99.26**

4.2. Comparison to Other Studies

The statistics presented above are not directly comparable to results from previous studies. This is mainly due to different input data sets (e.g., different missions [11] or different Cryosat-2 Baselines [1]), different thresholds used within the studies [1,10,16], and a different ground truth used for validation. Most of the older studies used ground truth data sets whose resolutions were limited to many meters (40 m to 250 m; instead of 1 m used here) and therefore, potentially, many leads have been missed. Furthermore, the time between image acquisition and satellite overpass has been limited here to 90 min, meaning there is less of a possibility of induced sea-ice drift related biases in the ground truth results.

In comparison to Passaro et al. [10], the validation approach differs significantly. Their study was focused on the detection of leads with different sizes, whereas here, the detection of altimetry returns over leads is the aim. Thus, instead of counting each lead observation separately, they considered consecutive lead observations to be the same lead. If a single observation from a lead with multiple observations was correctly classified, the overall lead was assumed to be correctly classified. This directly influences the corresponding statistics and for that reason, the results presented in this study appear poorer. However, for lead prospection, they found that the MULTI classification performed on par with the STACK when compared using the ratio of correctly classified leads to false detections.

Müller et al. [11] used exactly the same metrics to validate their results, however, a different ground truth was applied. From Figure 4 it seems that in this study a better performance can be achieved. This is presumably related to the fact that CS-2 has an improved along-track resolution with respect to Envisat and SARAL (i.e., SAR stack data compared to classical pulse-limited altimetry performs better). Though, in order to prove this, the validation of Müller et al. [11] should be repeated based on a high-resolution ground truth.

4.3. Discussion of Altimeter Classification Methods

Threshold Definition: From the previous sections it became clear that those classification methods based on the maximum waveform power perform better than the rest. However, the success of MAX strongly depends on the defined threshold. Many factors affect the backscattering power received at the altimeter antenna. This is clearly evident in Figure 5, which shows the mean and standard deviation of the maximum power at lead surfaces (at nadir) according to the ground truth for each data set used within this study. The mean power varies greatly across each data set which can be a completely different region or season. The backscattering power is not just a function of the lead within the altimeter footprint but also the sea-ice, lead width, lead orientation, refrozen areas of the lead, melt-ponds and more within the illuminated area. These all contribute to varying statistics across different study areas and also make it very difficult to establish an absolute threshold that can be used for all regions and seasons. Figure 5 also displays the threshold developed for the MAX classifier. Since it is optimized based on all data, it matches relatively well. However, some leads are completely missed. This can be even worse for other regions, seasons, or conditions not included in this study.

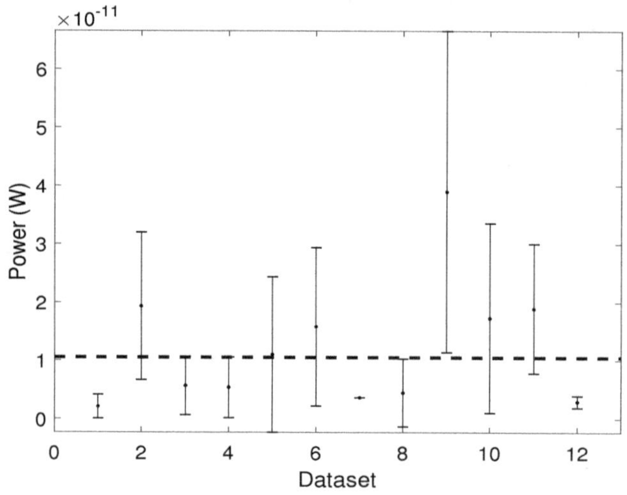

Figure 5. Mean and standard deviation of maximum power for leads across all data sets with the optimized threshold (1.05×10^{-11} W) for FLR = 1% developed here shown as a dotted line.

Nevertheless, it is clear that the maximum power is a good discriminator between leads and sea-ice compared to other index-based classification methods. The definition of a relative power threshold as proposed by Passaro et al. [10] could account for the problem of finding an absolute value and still exploit the returned power as classifier. In that study, the relative power threshold was calculated as the ratio between the maximum power of CS-2 returns classified as leads to the median of the maximum power observations for a given region/segment. A fixed power ratio threshold of 10 was derived a posteriori based on a comparison with Sentinel-1 SAR images across different regions and time of the year. The subsequent classification based on this threshold gave the best ratio of correctly classified leads to false detections among the methods trialled in their study.

Lead Location: Another important effect influencing the performance of all classifiers are off-nadir leads. When located within the satellites footprint but not directly in nadir, they strongly impact the returned radar echo even if not detected by the ground truth. This effect becomes more evident when high-resolution ground truth is used. MULTI and STACK both define dedicated parameters to disregard off-nadir leads. These methods rely on the assumption that the multi-looked waveform (in case of MULTI) or the RIP (in case of STACK) will be less peaky for a lead that is not at the nadir position. However, this is not always true. From Figure 6, it can clearly be seen that even for sea-ice observations (waveform #1159), the multi-looked waveform still exhibits a very peaky shape and therefore, the PP_l and PP_r will unlikely disregard these waveforms. This is confirmed by the classification result of MULTI, which identifies all measurements except the first one as lead observation (waveforms #1156 to #1159). Once a lead falls inside the altimeter footprint, the backscatter return is highly influenced by the presence of this lead and therefore, doubt is cast in the ability of the modified PP in disregarding off-nadir returns as the multi-looked waveform can still be influenced by leads in off-nadir positions.

The STACK method also claims to handle off-nadir leads. However, it relies on the assumption that a lead within the vicinity of the altimeter will be crossed by the satellite track (e.g., will be the local maximum), which is not always the case. Cross-track off-nadir leads cannot be detected by this method.

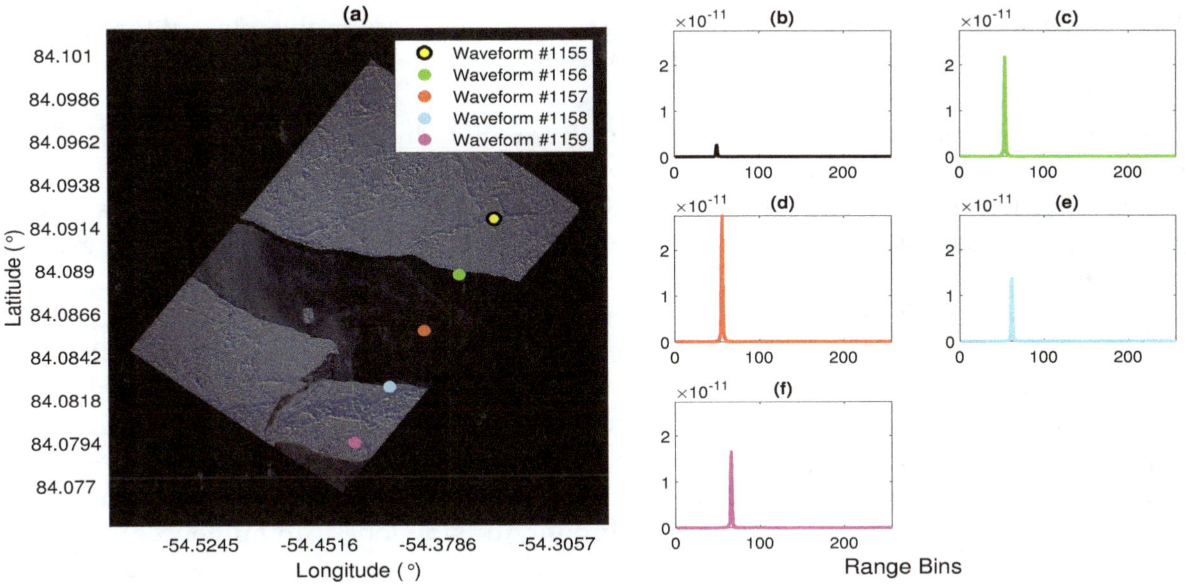

Figure 6. CS-2 descending track #15632 (taken on 20 March 2013) showing five waveform surface sample locations (**a**). Image is acquired at 17:03:15 GPS time, approximately 30 min before satellite overpass. Waveforms associated with the locations are also shown. (**b**) is waveform #1155; (**c**) #1156; (**d**) #1157; (**e**) #1158 and (**f**) #1159.

This is evident in Figure 7: local maxima in SP (i.e., measurements identified as leads; blue triangles) do not always correspond to a lead at the nadir position (red squares). An example of when the SP works well is visible between 84.05° and 84.1° latitude. Here, the two SP peaks correspond to lead surfaces at nadir according to the ground truth. These particular lead observations are orientated along-track relative to the satellite and belong to the same lead.

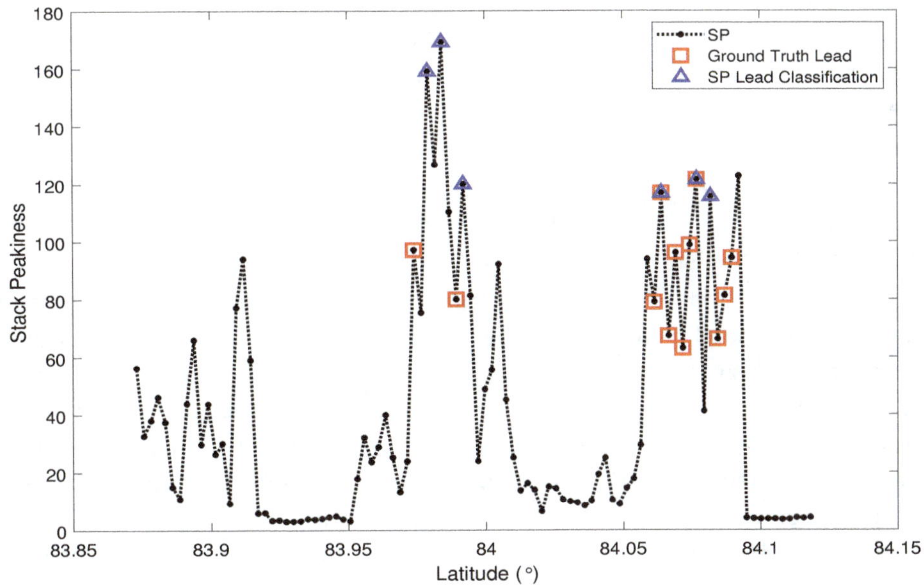

Figure 7. Evolution of CS-2 stack peakiness in data set taken on 20 March 2013. Observations classified as leads by the STACK classifier are marked in blue. Actual nadir leads identified by the ground truth are highlighted in red.

All results presented here are also influenced by a potential CS-2 mispointing angle in along-track direction (pitch angle) related to biases of the on-board star trackers [37]. Such an effect may result in the location of the CS-2 surface sample also being biased. This is especially relevant for narrow leads where, for example, the SP would peak in the observation before the lead (assuming satellite pitched up) resulting in a false classification.

In order to unambiguously isolate nadir leads from off-nadir leads, the usage of the CS-2 InSAR mode will be beneficial. The exploitation of signals from both antennas will allow for an exact location of the reflection point of the signal (in cross track direction). However, depending on the CS-2 geographically mode mask, currently, these data sets are only available in a few areas, mostly not in sea-ice ocean regions.

Stack Information: From the four methods under investigation, STACK and MULTI use the full CS-2 stack information, with respectively derived parameters (SSD and SP). Theoretically, one may expect that this additional information helps to improve the lead classification. However, it seems that using only stack data without taking the maximum power into account is not an optimal method. This might also be related to residual side-lobe effects in the RIP even after the application of the Hamming window function. Passaro et al. [10] have shown that without the use of the Hamming window, leads that have influenced the signal through the side-lobes can influence the statistics of the stack data. The SP value is being influenced by high power returns in the non-nadir look angles. This has the effect of making the RIP return for these observations wider than what might be expected for a lead at nadir and therefore, it contributes to a lower SP value. An analysis of the RIP for the locations shown in Figure 6 found that the sea-ice (#1155) SP was much larger than that of the central lead location (#1157) and this was attributed to residual side-lobe effects as much larger RIP was observed in the non-nadir look angles. In spite of this, it will be useful to combine stack information with maximum power information to develop an improved classification approach.

Unsupervised Classification: From all the lead classification methods presented in this study, the UNSU method has performed the best when aiming at a low FLR of better than 1%. In general, the UNSU method has a similar performance as the MAX classifier, and the difference between the two classification methods is minimal. The most prominent feature among the six features used in the unsupervised classification method is the maximum power for distinguishing lead surfaces. This may explain the similar performances when using the MAX approach.

The most distinct advantage of the unsupervised classification method is that no training data is necessary and that the approach does not rely on the development of a threshold. Therefore, no prior knowledge of the study area is required. The only requirement is that the general characteristics of altimeter waveforms are understood and meaningful features are defined. The clustering can be done in one region (e.g., the Greenland Sea) and transferred to other study areas (such as the Arctic Ocean or the Weddell Sea) in case all relevant waveform types are included in the clustering process. This will not affect the classification results.

The used cluster number k has only marginal influence on the classification results in case the order of magnitude is correct and matches to the number of waveform features and variability of waveforms used for the clustering process. Moreover, the approach is applicable to different altimetry missions (e.g., Envisat and SARAL [11]) and different measurement techniques (i.e., traditional pulse-limited and SAR altimetry) with only slight modifications to the waveform features used for the clustering. In contrast, all threshold-based methods must be tuned for each mission separately. Last but not least, the UNSU method can be adopted for the marginal sea-ice zones to include ocean returns in the classification whereas the other methods are not easily adaptable to ocean retrievals.

5. Conclusions and Outlook

The aim of this paper is to find a CS-2 altimeter lead classification algorithm that provides low false classification. Four altimeter classification methods are trialled and compared to high resolution automatically classified airborne images. Twelve dedicated airborne underflight image data sets acquired within 90 min before and after satellite overpass are used for the validation of all altimeter classification methods.

The unsupervised classification method (UNSU) is found to be the best classification method within this study with an overall accuracy at 97%. The maximum power classification method (MAX) closely followed with basically the same overall accuracy. The achieved lead user accuracies are 39% and 37%, respectively. Both methods perform very similar and in the case that larger false detection rates (larger than about 1%) are acceptable, the maximum power classifier is superior. The performance is expected to be further improved when a relative power threshold is applied instead a fixed one for all data sets. It is clear from this study that the maximum power index remains the most favorable index for reducing false lead observations. The two methods that do not use this parameter perform worse within this study.

In contrast to all other methods under investigation, the unsupervised classification method does not rely on any knowledge of the study area, especially on ground truth for threshold optimization. Moreover, it is applicable to all altimetry missions independent of the used measurement technique (pulse-limited or SAR). For these reasons, the unsupervised classification method (with 20 to 30 clusters) is the recommended lead classification method for Cryosat-2 SAR data.

The tested altimeter classifications within this paper have high misclassification compared to previous studies. This can be primarily associated with a much more robust and high resolution ground truth validation, which is much higher than the CS-2 along-track resolution.

The CS-2 waveforms are found to be very sensitive to the presence of leads within the altimeter footprint even when the lead does not cross the nadir point. This influences each waveform feature in the vicinity of a lead and thus, can result in many false classifications. The modified pulse peakiness (MULTI) and the stack peakiness (STACK) were designed to disregard off-nadir leads. However, these parameters are still strongly influenced by the presence of leads within the altimeter footprint, especially if these are located cross-track (in parallel to the satellite's track).

The utilization of CS-2 stack data in modified pulse peakiness and the stack peakiness methods cannot outperform the approaches using the maximum power parameter (alone or in connection with other parameters). Thus, a combination of stack data with a relative maximum power is expected to help to improve the existing classifiers.

The CS-2 mission also provides the opportunity to operate in SAR-interferometric (SARIn) mode which may help to reliably separate all types of off-nadir leads from nadir leads. Armitage and Davidson [38] have shown that the localization of nadir leads can be improved compared to the use of SAR mode. However, currently, the availability of SARIn mode is limited to certain regions.

Author Contributions: A.W. implemented the classification and validation methods, conducted the data analysis, and wrote the first version of the paper. D.D. supervised the study, wrote the final manuscript and made major contributions to the discussion and interpretation of the classification methods and results. F.L.M. and M.P. both helped with the discussion of the classification methods and results and also contributed to the manuscript writing. F.S. supervised the research and also contributed to the discussion of the applied methods and results.

Acknowledgments: This work was supported by the German Research Foundation (DFG) and the Technical University of Munich (TUM) in the framework of the Open Access Publishing Program. The authors would like to acknowledge the European Space Agency for allowing access to the CryoSat-2 data from their G-POD website (https://gpod.eo.esa.int/). Also, thanks go to NASA and the NSIDC for the distribution of the airborne image data key to this study (https://nsidc.org/icebridge/portal/map). Furthermore, thanks are afforded to the OSI SAF for the sea-ice concentration products (http://osisaf.met.no/p/ice/index.html).

References

1. Wernecke, A.; Kaleschke, L. Lead Detection in Arctic Sea Ice from Cryosat-2: Quality Assessment, Lead Area Fraction and Width Distribution. *Cryosphere* **2015**, *9*, 2167–2200. [CrossRef]
2. Laxon, S.; Peacock, N.; Smith, D. High Interannual Variability of Sea Ice Thickness in the Arctic Region. *Lett. Nat.* **2003**, *425*, 947–950. [CrossRef] [PubMed]
3. Farrell, S.L.; Laxon, S.W.; McAdoo, D.C.; Yi, D.; Zwally, H.J. Five Years of Arctic Sea Ice Freeboard Measurements from the Ice, Cloud and land Elevation Satellite. *J. Geophys. Res.* **2009**, *114*. [CrossRef]

4. Dwyer, R.E.; Godin, R.H. *Determining Sea-Ice Boundaries and Ice Roughness using GEOS-3 Altimeter Data*; Technical Report; NASA Wallops Flight Center: Wallops Island, VA, USA, 1980.
5. Laxon, S. Sea Ice Altimeter Processing Scheme at the EODC. *Int. J. Remote Sens.* **1994**, *15*, 915–924, doi:10.1080/01431169408954124. [CrossRef]
6. Connor, L.N.; Laxon, S.W.; Ridout, A.L.; Krabill, W.B.; McAdoo, D.C. Comparison of Envisat Radar and Airborne Laser Altimeter Measurements Over Arctic Sea Ice. *Remote Sens. Environ.* **2009**, *113*, 563–570, doi:10.1016/j.rse.2008.10.015. [CrossRef]
7. Röhrs, J.; Kaleschke, L. An Algorithm to Detect Sea Ice Leads by Using AMSR-E Passive Microwave Imagery. *Cryosphere* **2012**, *6*, 343–352, doi:10.5194/tc-6-343-2012. [CrossRef]
8. Zygmuntowska, M.; Khvorostovsky, K.; Helm, V.; Sandven, S. Waveform Classification of Airborne Synthetic Aperture Radar Altimeter over Arctic Sea Ice. *Cryosphere* **2013**, *7*, 1315–1324, doi:10.5194/tc-7-1315-2013. [CrossRef]
9. Friedman, N.; Kohavi, R. Bayesian Classification. In *Handbook of Data Mining and Knowledge Discovery*; Oxford University Press: New York, USA, 2002; pp. 282–288
10. Passaro, M.; Müller, F.L.; Dettmering, D. Lead Detection using CryoSat-2 Delay Doppler Processing and Sentinel-1 SAR images. *Adv. Space Res.* **2017**, doi:10.1016/j.asr.2017.07.011. [CrossRef]
11. Müller, F.L.; Dettmering, D.; Bosch, W.; Seitz, F. Monitoring the Arctic Seas: How Satellite Altimetry can be used to Detect Open Water in Sea-Ice Regions. *Remote Sens.* **2017**, *9*, 551. [CrossRef]
12. Park, H.S.; Jun, C.H. A Simple and Fast Algorithm for K-medoids Clustering. *Expert Syst. Appl.* **2009**, *36*, 3336–3341, doi:10.1016/j.eswa.2008.01.039. [CrossRef]
13. Cover, T.M.; Hart, P.E. Nearest Neighbor Pattern Classification. *IEEE Trans. Inf. Theory* **1967**, *13*. [CrossRef]
14. Shen, X.; Zhang, J.; Zhang, X.; Meng, J.; Ke, C. Sea Ice Classification Using Cryosat-2 Altimeter Data by Optimal Classifier–Feature Assembly. *IEEE Geosci. Remote Sens. Lett.* **2017**, *14*, 1948–1952. [CrossRef]
15. Lee, S.; Kim, H.C.; Im, J. Arctic Lead Detection using a Waveform Mixture Algorithm from CryoSat-2 Data. *Cryosphere* **2018**, *12*, 1665–1679, doi:10.5194/tc-12-1665-2018. [CrossRef]
16. Ricker, R.; Hendricks, S.; Helm, V.; Skourup, H.; Davidson, M. Sensitivity of CryoSat-2 Arctic Sea-ice Freeboard and Thickness on Radar-Waveform Interpretation. *Cryosphere* **2014**, *8*, 1607–1622, doi:10.5194/tc-8-1607-2014. [CrossRef]
17. Jiang, L.; Schneider, R.; Andersen, O.B.; Bauer-Gottwein, P. CryoSat-2 Altimetry Applications over Rivers and Lakes. *Water* **2017**, *9*, 211. [CrossRef]
18. Raney, R.K. The Delay/Doppler Radar Altimeter. *IEEE Trans. Geosci. Remote Sens.* **1998**, *36*, 1578–1588. [CrossRef]
19. Martin-Puig, C.; Ruffini, G. SAR Altimeter Retracker Performance Bound over Water Surfaces. In Proceedings of the 2009 IEEE International Geoscience and Remote Sensing Symposium, Cape Town, South Africa, 12–17 July 2009; pp. 449–452.
20. European Space Agency. Cryosat Product Handbook. 2012. Available online: http://emits.sso.esa.int/emits-doc/ESRIN/7158/CryoSat-PHB-17apr2012.pdf (accessed on 21 September 2017).
21. Scagliola, M.; Fornari, M. *Main Evolutions and Expected Quality Improvements in Baseline C Level 1b Products*; Aresys Technical Note, C2-TN-ARS-GS-5154, Issue 1.3; 2015. Available online: https://earth.esa.int/documents/10174/1773005/C2-BaselineC_L1b_improvements_1.3 (accessed on 27 July 2018).
22. Dinardo, S. *Guidelines for the SAR (Delay-Doppler) L1b Processing*; Technical Note, XCRY-GSEG-EOPS-TN-14-0042 Version 2; European Space Agency, ESRIN: Frascati, Italy, 2013. Available online: https://wiki.services.eoportal.org/tiki-download_wiki_attachment.php?attId=2540 (accessed on 30 September 2017).
23. Studinger, M.; Koenig, L.; Martin, S.; Sonntag, J. Operation icebridge: Using instrumented aircraft to bridge the observational gap between icesat and icesat-2. In Proceedings of the 2010 IEEE International Geoscience and Remote Sensing Symposium, Honolulu, HI, USA, 25–30 July 2010; pp. 1918–1919.
24. Xia, W.; Xie, H. Assessing three waveform retrackers on sea ice freeboard retrieval from Cryosat-2 using Operation IceBridge Airborne altimetry datasets. *Remote Sens. Environ.* **2018**, *204*, 456–471. [CrossRef]
25. Dominguez, R. *IceBridge DMS L1B Geolocated and Orthorectified Images, Version 1*; Updated 2017; NASA National Snow and Ice Data Center Distributed Active Archive Center: Boulder, CO, USA, 2010; doi:10.5067/OZ6VNOPMPRJ0.

26. Wingham, D.J.; Francis, C.R.; Baker, S.; Bouzinac, C.; Brockley, D.; Cullen, R.; de Chateau-Thierry, P.; Laxon, S.; Mallow, U.; Mavrocordatos, C.; et al. CryoSat: A Mission to Determine the Fluctuations in Earth's Land and Marine Ice Fields. *Adv. Space Res.* **2006**, *37*, 841–871. [CrossRef]
27. Hastie, T.; Tibshirani, R.; Friedman, J. *The Elements of Statistical Learning*; Springer-Verlag: New York, NY, USA, 2009; Volume 2.
28. Durbin, R.; Eddy, S.R.; Krogh, A.; Mitchison, G. *Biological Sequence Analysis*; Cambridge University Press: Cambridge, UK, 1998.
29. Banks, D.; House, L.; McMorris, F.; Arabie, P.; Gaul, W. (Eds.) *Classification, Clustering, and Data Mining Applications: Proceedings of the Meeting of the International Federation of Classification Societies (IFCS), Illinois Institute of Technology, Chicago, 15–18 July 2004*; Studies in Classification, Data Analysis, and Knowledge Organization; Springer: Berlin/Heidelberg, Germany, 2004.
30. Kvingedal, B. On Sea Ice Variability in the Nordic Seas. Ph.D. Thesis, University of Bergen, Bergen, Norway, 2005. Available online: http://web.gfi.uib.no/publikasjoner/pdf/Kvingedal.pdf (accessed on 3 November 2017).
31. Nelder, J.A.; Mead, R. A Simplex Method for Function Minimization. *Comput. J.* **1965**, *7*, 308–313, doi:10.1093/comjnl/7.4.308. [CrossRef]
32. Perovich, D.K.; Richter-Menge, J.A. Surface Characteristics of Lead Ice. *J. Geophyis. Res.* **1994**, *99*, 16341–16350. [CrossRef]
33. Onana, V.D.P.; Kurtz, N.T.; Farrell, S.L.; Koenig, L.S.; Studinger, M.; Harbeck, J.P. A Sea-Ice Lead Detection Algorithm for Use With High-Resolution Airborne Visible Imagery. *IEEE Trans. Geosci. Remote Sens.* **2013**, *51*, doi:10.1109/TGRS.2012.2202666. [CrossRef]
34. Fawcett, T. An Introduction to ROC analysis. *Pattern Recogn. Lett.* **2006**, *27*, 861–874, doi:10.1016/j.patrec.2005.10.010. [CrossRef]
35. Xue, R.; Wunsch, D.C.; IEEE Computational Intelligence Society. *Clustering*; IEEE Press Series on Computational Intelligence; IEEE Press: Hoboken, NJ, USA; Wiley: Piscataway, NJ, USA, 2009.
36. Story, M.; Congalton, R.G. Accuracy Assessment: A User's Perspective. *Am. Soc. Photogramm. Remote Sens. Remote Sens. Brief* **1986**, *52*, 397–399.
37. Galin, N.; Wingham, D.J.; Cullen, R.; Francis, R.; Lawrence, I. Measuring the Pitch of CryoSat-2 Using the SAR Mode of the SIRAL Altimeter. *IEEE Geosci. Remote Sens. Lett.* **2014**, *11*, 1399–1403, doi:10.1109/LGRS.2013.2293960. [CrossRef]
38. Armitage, T.W.K.; Davidson, M.W.J. Using the Interferometric Capabilities of the ESA CryoSat-2 Mission to Improve the Accuracy of Sea Ice Freeboard Retrievals. *IEEE Trans. Geosci. Remote Sens.* **2014**, *52*, doi:10.1109/TGRS.2013.2242082. [CrossRef]

Permissions

All chapters in this book were first published in MDPI; hereby published with permission under the Creative Commons Attribution License or equivalent. Every chapter published in this book has been scrutinized by our experts. Their significance has been extensively debated. The topics covered herein carry significant findings which will fuel the growth of the discipline. They may even be implemented as practical applications or may be referred to as a beginning point for another development.

The contributors of this book come from diverse backgrounds, making this book a truly international effort. This book will bring forth new frontiers with its revolutionizing research information and detailed analysis of the nascent developments around the world.

We would like to thank all the contributing authors for lending their expertise to make the book truly unique. They have played a crucial role in the development of this book. Without their invaluable contributions this book wouldn't have been possible. They have made vital efforts to compile up to date information on the varied aspects of this subject to make this book a valuable addition to the collection of many professionals and students.

This book was conceptualized with the vision of imparting up-to-date information and advanced data in this field. To ensure the same, a matchless editorial board was set up. Every individual on the board went through rigorous rounds of assessment to prove their worth. After which they invested a large part of their time researching and compiling the most relevant data for our readers.

The editorial board has been involved in producing this book since its inception. They have spent rigorous hours researching and exploring the diverse topics which have resulted in the successful publishing of this book. They have passed on their knowledge of decades through this book. To expedite this challenging task, the publisher supported the team at every step. A small team of assistant editors was also appointed to further simplify the editing procedure and attain best results for the readers.

Apart from the editorial board, the designing team has also invested a significant amount of their time in understanding the subject and creating the most relevant covers. They scrutinized every image to scout for the most suitable representation of the subject and create an appropriate cover for the book.

The publishing team has been an ardent support to the editorial, designing and production team. Their endless efforts to recruit the best for this project, has resulted in the accomplishment of this book. They are a veteran in the field of academics and their pool of knowledge is as vast as their experience in printing. Their expertise and guidance has proved useful at every step. Their uncompromising quality standards have made this book an exceptional effort. Their encouragement from time to time has been an inspiration for everyone.

The publisher and the editorial board hope that this book will prove to be a valuable piece of knowledge for researchers, students, practitioners and scholars across the globe.

List of Contributors

Jingbin Liu
State Key Laboratory of Information Engineering in Surveying, Mapping and Remote Sensing, Wuhan University, Wuhan 430079, China

Baojun Zhang
State Key Laboratory of Information Engineering in Surveying, Mapping and Remote Sensing, Wuhan University, Wuhan 430079, China
Chinese Antarctic Center of Surveying and Mapping, Wuhan University, 129 Luoyu Road, Wuhan 430079, China

Zemin Wang, Jiachun An, Fei Li and Tingting Liu
Chinese Antarctic Center of Surveying and Mapping, Wuhan University, 129 Luoyu Road, Wuhan 430079, China

Quanming Yang
Chinese Antarctic Center of Surveying and Mapping, Wuhan University, 129 Luoyu Road, Wuhan 430079, China
Power China Zhongnan Engineering Corporation Limited, Changsha 410014, China

Hong Geng
School of Resource and Environment Science, Wuhan University, Wuhan 430079, China

Ya-Lun S. Tsai, Igor Klein and Andreas Dietz
German Remote Sensing Data Center (DFD), German Aerospace Center (DLR), Muenchener Strasse 20, 82234 Wessling, Germany

Natascha Oppelt
Department of Geography, Earth Observation and Modelling, Kiel University, Ludewig-Meyn-Str. 14, 24118 Kiel, Germany

Shengjun Zhang and Defu Che
School of Resources and Civil Engineering, Northeastern University, Shenyang 110819, China

Jiancheng Li and Taoyong Jin
MOE Key Laboratory of Geospace Environment and Geodesy, School of Geodesy and Geomatics, Wuhan University, Wuhan 430072, China
Collaborative Innovation Center of Geospatial Technology, Wuhan University, Wuhan 430072, China

Pascal Bonnefond
SYRTE, Observatoire de Paris, PSL Research University, CNRS, Sorbonne Universités, UPMC University Paris 06, LNE, 75014 Paris, France

Olivier Laurain and Pierre Exertier
Géoazur, Observatoire de la Côte d'Azur, 06905 Sophia-Antipolis CEDEX, France

François Boy, Thierry Guinle and Nicolas Picot
Centre National d'Etudes Spatiales (CNES), 31401 Toulouse CEDEX, France

Sylvie Labroue and Matthias Raynal
Collecte Localisation Satellites (CLS), 31520 Ramonville Saint-Agne, France

Craig Donlon
European Space Agency/European Space Research and Technology Centre (ESA/ESTEC), 2201 AZ Noordwijk, The Netherlands

Pierre Féménias and Tommaso Parrinello
European Space Agency/European Space Research Institute (ESA/ESRIN), 00044 Frascatti, Italy

Salvatore Dinardo
HeSpace, 64293 Darmstadt, Germany

Jacques Verron
Institut des Géosciences de l'Environnement (IGE), CNRS, 38041 Grenoble, France

Lofti Aouf
Météo-France, 31057 Toulouse, France

Florence Birol, Stéphane Calmant, Jean-François Crétaux, Kevin Guerreiro, Fatma Jebri, Rosemary Morrow and Frédérique Rémy
Laboratoire d'Etudes en Géophysique et Océanographie Spatiales (LEGOS), 31400 Toulouse, France

Suchandra A. Bhowmick, A. K. Dubey, P. K. Gupta and Raj Kumar
Space Applications Centre (ISRO), Ahmedabad 380015, India

Taina Conchy
Universidade do Estado de Amazonas, Manaus 69020, Brazil

Yannice Faugère
Collecte Localisation Satellites (CLS), 31520 Ramonville Saint-Agne, France

Oscar Vergara
Laboratoire d'Etudes en Géophysique et Océanographie Spatiales (LEGOS), 31400 Toulouse, France
Collecte Localisation Satellites (CLS), 31520 Ramonville Saint-Agne, France

Mathieu Hamon and Elisabeth Rémy
Mercator Océan, 31520 Ramonville Saint-Agne, France

Ananda Pascual
Institut Mediterrani d'Estudis Avançats (IMEDEA) (CSIC-UIB), 07190 Esporles, Illes Balears, Spain

Walter H. F. Smith
Laboratory for Satellite Altimetry, NOAA, College Park, MD 20740-3818, USA

Jean Tournadre
Laboratoire d'Océanographie Physique et Spatiale (LOPS), 29280 Plouzané, France

Edward Salameh
Laboratoire d'Etudes en Géophysique et Océanographie Spatiales (LEGOS), Université de Toulouse, IRD, CNES, CNRS, UPS, 31400 Toulouse, France
Normandie University, UNIROUEN, UNICAEN, CNRS, M2C, Morphodynamique Continentale et Côtière, 76000 Rouen, France

Imen Turki and Benoit Laignel
Normandie University, UNIROUEN, UNICAEN, CNRS, M2C, Morphodynamique Continentale et Côtière, 76000 Rouen, France

Frédéric Frappart
Laboratoire d'Etudes en Géophysique et Océanographie Spatiales (LEGOS), Université de Toulouse, IRD, CNES, CNRS, UPS, 31400 Toulouse, France
Géosciences Environnement Toulouse (GET), Université de Toulouse, IRD, CNES, CNRS, UPS, 31400 Toulouse, France

Vincent Marieu, Jean-Paul Parisot and Vincent Hanquiez
Environnements et Paléoenvironnements Océaniques et Continentaux (EPOC), UMR 5805, allée Geoffroy St Hilaire, 33615 Pessac CEDEX, France

Alexandra Spodar
Environnements et Paléoenvironnements Océaniques et Continentaux (EPOC), UMR 5805, allée Geoffroy St Hilaire, 33615 Pessac CEDEX, France
Laboratoire d'Océanologie et de Géosciences (LOG), UMR 8187, 59140 Dunkerque, France

Alka Singh, Joshua B. Fisher and John T. Reager
Jet Propulsion Laboratory, California Institute of Technology, Pasadena, CA 91109, USA

Ali Behrangi
Jet Propulsion Laboratory, California Institute of Technology, Pasadena, CA 91109, USA
Department of hydrology and atmospheric sciences, The University of Arizona, Tucson, AZ 85721, USA

Wolfgang Bosch
Deutsches Geodätisches Forschungsinstitut, Technische Universität München, Arcisstraße 21, 80333 Munich, Germany

Gérald Dibarboure and Alain Lamy
Centre National d'Etudes Spatiales, 31400 Toulouse, France

Marie-Isabelle Pujol and Ghita Jettou
Collecte Localisation Satellites, 31520 Ramonville Saint-Agne, France

Eva Boergens
Deutsches Geodätisches Forschungsinstitut der Technischen Universität München (DGFI-TUM), Arcisstraße 21, 80333 Munich, Germany

Karina Nielsen and Ole Baltazar Andersen
Division of Geodesy, DTU Space, National Space Institute, DK-2800 Kongens Lyngby, Denmark

Denise Dettmering, Alan Wynne, Felix L. Müller, Marcello Passaro and Florian Seitz
Deutsches Geodätisches Forschungsinstitut, Technische Universität München, Arcisstraße 21, 80333 Munich, Germany

Index